Springer Series in Computational Mathematics

26

Springer
Berlin
Heidelberg
New York
Barcelona
Hong Kong
London
Milan
Paris
Singapore
Tokyo

Mark Ainsworth
Jeremy Levesley
Marco Marletta (Eds.)

The Graduate Student's Guide to Numerical Analysis '98

Lecture Notes from the VIII EPSRC
Summer School in Numerical Analysis

 Springer

Editors:
Mark Ainsworth
Jeremy Levesley
Marco Marletta (e-mail: mm7@mcs.le.ac.uk)

Department of Mathematics
and Computer Science
University of Leicester
University Road
LE1 7RH Leicester
United Kingdom

Cataloging-in-Publication Data applied for

Die Deutsche Bibliothek – CIP-Einheitsaufnahme

The graduate student's guide to numerical analysis '98: lecture notes from the VIII EPSRC Summer School in Numerical Analysis / Mark Ainsworth... (ed.). – Berlin; Heidelberg; New York; Barcelona; Hong Kong; London; Milan; Paris; Singapore; Tokyo: Springer, 1999
(Springer series in computational mathematics; 26)
ISBN 3-540-65752-5

Mathematics Subject Classification (1991): 15-XX, 34-XX, 35-XX, 65-XX, 70-XX, 76-XX

ISSN 0179-3632
ISBN 3-540-65752-5 Springer-Verlag Berlin Heidelberg New York

Typeset by the authors using a Springer T$_E$X macro package
Cover production: *design & production* GmbH, Heidelberg
Computer to film: Saladruck, Berlin
Binding: Buchbinderei Lüderitz & Bauer, Berlin

SPIN 10689296   46/3143 – 5 4 3 2 1 0 – Printed on acid-free paper

Preface

The Eighth EPSRC Numerical Analysis Summer School was held at the University of Leicester from the 5th to the 17th of July, 1998. This was the third Numerical Analysis Summer School to be held in Leicester. The previous meetings, in 1992 and 1994, had been carefully structured to ensure that each week had a coherent 'theme'. For the 1998 meeting, in order to widen the audience, we decided to relax this constraint. Speakers were chosen to cover what may appear, at first sight, to be quite diverse areas of numerical analysis. However, we were pleased with the extent to which the ideas cohered, and particularly enjoyed the discussions which arose from differing interpretations of those ideas. We would like to thank all six of our main speakers for the care which they took in the preparation and delivery of their lectures. In this volume we present their lecture notes in alphabetical rather than chronological order. Nick Higham, Alastair Spence and Nick Trefethen were the speakers in week 1, while Bernardo Cockburn, Stig Larsson and Bob Skeel were the speakers in week 2.

Another new feature of this meeting compared to its predecessors was that we had 'invited seminars'. A numer of established academics based in the UK were asked to participate in the afternoon seminar program. These speakers were Phil Knight (Strathclyde University), Karl Meerbergen (Rutherford Appleton Laboratory) and Françoise Tisseur (Manchester University) during the first week; John Barrett (Imperial College), Charlie Elliott (Sussex University) and Petr Plechac (Rutherford Appleton Laboratory) in the second week. We would like to thank them for supporting the Summer School by presenting their research.

The other afternoon seminars were given by Ke Chen, Stuart Hawkins, Carsten Keller, Mark Küther, Elizabeth Larsson and Milan Mihailović in week 1, and Mark Arnold, Carsten Maple and Robert Scheichl in week 2. When we announced the Summer School we also announced that in each week we would give prizes for the best presentations by PhD students. These were won by Marc Küther and Elizabeth Larsson (week 1) and Robert Scheichl (week 2), whom we warmly congratulate. Not all the people listed above were eligible for the prize competition, but we shall not embarrass the 'oldies' by listing them separately.

One tradition which we did not abandon was that of choosing two distinguished UK academics to act as 'Local Experts'. We would like to thank Jennifer Scott from Rutherford Appleton Laboratories and John Barrett from Imperial College, Local Experts in weeks 1 and 2 respectively. They ensured that the academic program ran smoothly and gave much time and thought to the organization of the seminars and selection of invited seminar speakers. They also found themselves in the difficult position of assisting us in judging the student talks for the prizes. Gerald Moore from Imperial College was also co-opted onto one of the prize adjudicating committees. We would like to thank our Local Experts and Gerald Moore for successfully carrying out the near impossible task of fair adjudication.

Finally, we thank the Engineering and Physical Sciences Research Council for their financial support, which covered all the costs of the six main speakers plus the accommodation costs of the UK participants.

Mark Ainsworth, Jeremy Levesley and Marco Marletta
Leicester, November 1998.

Table of Contents

List of Contributors

Bernardo Cockburn
School of Mathematics,
University of Minnesota,
Minneapolis,
Minnesota 55455, USA

Nicholas J. Higham
Department of Mathematics,
University of Manchester,
Manchester M13 9PL,
England

Stig Larsson
Department of Mathematics,
Chalmers University of Technology
and
Göteborg University,
SE-412 96 Göteborg, Sweden

Robert D. Skeel
Department of Computer Science and
Beckman Institute,
University of Illinois,
Urbana,
Illinois 61801, USA

Alastair Spence and Ivan G. Graham
Department of Mathematical Sciences,
University of Bath,
Claverton Down,
Bath BA2 7AY,
England

Lloyd N. Trefethen
Oxford University Computing Laboratory,
Wolfson Building,
Parks Road,
Oxford OX1 3QD,
England

A Simple Introduction to Error Estimation for Nonlinear Hyperbolic Conservation Laws

Some Ideas, Techniques, and Promising Results

Bernardo Cockburn

School of Mathematics, University of Minnesota, Minneapolis, Minnesota 55455, USA

Preface

In these notes, we present a simple introduction to the topic of a posteriori error estimation for nonlinear hyperbolic conservation laws. This is a topic of great practical interest which has been receiving increasing attention from many researchers in recent years. On the other hand, the highly complex character of its mathematics often obscures the main ideas behind the technical manipulations. Aware of this unfortunate situation, we have written these notes in an attempt to emphasize the *ideas* and simplify, as much as possible, the presentation of the *techniques*.

The reader has to be warned, however, that there is still much to be researched in this area. As a consequence, these are notes about *ongoing* research– not about fully solved problems. Thus, the purpose of these notes is not to tell the story of a battle fought long ago, but to entice the reader to participate in one that is currently taking place!

We have many people to thank. Let us start with the three organizers. Thanks to Mark Ainsworth for the invitation to give a series of lectures at the EPSRC VIII-th Summer School in Numerical Analysis, University of Leicester, UK, July 13–17, 1998, the material of which is contained in these notes; to Marco Marletta for his excellent editorial work; and, last but not least, to Jeremy Levesley for making sure everybody had a good time. Next, we must thank John Barrett, Charlie Elliot, David Silvester, and my good friend Endre Süli for their questions and remarks. Finally, we must thank all the participants to the VIII-th EPSRC Numerical Analysis Summer School for being a patient and engaging audience.

1. Introduction

In these notes, we present a simple introduction to the topic of a posteriori error estimation for nonlinear hyperbolic conservation laws. The interest in this topic stems from two facts:

- The first is that many relevant physical phenomena are modeled by conservation laws that when *formally* modified become nonlinear hyperbolic conservation laws. This modification takes place when the terms associated with some physical phenomenon that is *not dominant* are dropped from the equations. An instance of this situation occurs in secondary oil recovery when the capillarity effects are neglected; another occurs in compressible fluid flow when the viscosity and heat transfer effects are dropped.
- The second is that a good error estimate is essential for devising mathematically sound adaptive algorithms. A posteriori error estimates are important because they tell us how good a *given* approximation u is without having to know the exact solution v. The a posteriori error estimates we are interested in are of the form

$$\| v - u \| \leq \Psi(u).$$

With such an estimate, we can ensure that the error is smaller than a given tolerance τ, by simply finding an approximation u such that

$$\Psi(u) \leq \tau.$$

The problem is now how to achieve this goal with minimal computational effort. The devising of algorithms to solve this problem is probably the most important topic in modern computational mathematics for partial differential equations.

The organization and content of these notes strongly reflect the fact that this is a rapidly expanding field in which there are more questions than answers. Thus, the purpose of these notes is to bring the reader as fast as possible face to face with what we think are some of the most important current issues in this field. The reader will find (i) some *ideas* about how to approach the problem of a posteriori error estimation for nonlinear hyperbolic conservation laws, (ii) some *techniques* that allow us to implement those ideas, and (iii) some *promising results* which reflect the current state of development of the approach we consider.

- The *ideas* that we develop are:
  1. that nonlinear hyperbolic conservation laws are (in many important cases) approximations to equations of different mathematical type,
  2. that a posteriori error estimates are nothing but continuous dependence results,
  3. that in order to obtain a posteriori error estimates for nonlinear hyperbolic conservation laws, we must first obtain them for the original equations of which the hyperbolic conservation laws are an approximation.
- The *techniques* we study are *duality* techniques for a posteriori error estimates. There are many adaptive algorithms for nonlinear hyperbolic conservation laws and most, if not all, are based in heuristic arguments. We will

not concern ourselves with those heuristic arguments; instead, we consider techniques for obtaining rigorously proven a posteriori error estimates. The techniques we consider allow us:

1. not to have to actually solve an adjoint equation,
2. not to have to deal explicitly with the nonlinear convective terms,
3. to obtain continuous dependence results that hold when the singular limit process that gives rise to the hyperbolic conservation law takes place.

• The *promising results* we consider are partial results in which we numerically test the a posteriori error estimates obtained in this approach.

To develop the program just sketched, we adopt as our model problem the Cauchy problem for the nonlinear scalar convection-diffusion equation

$$v_t + \nabla \cdot \mathbf{f}(\mathbf{v}) - \nu \Delta \mathbf{v} = \mathbf{0},$$

and take the following nonlinear hyperbolic conservation law

$$v_t + \nabla \cdot \mathbf{f}(\mathbf{v}) = \mathbf{0},$$

to be its *formal* modification when the diffusion effects modeled by the term $\nu \Delta v$ become 'negligible.' Thus, in Section 3, we construct continuous dependence results for the convection-diffusion equation that will also hold for the nonlinear hyperbolic conservation law; this section constitutes the core of these notes. In Section 4, we show how these continuous dependence results single out in a natural way a unique solution of the hyperbolic conservation law, the so-called *entropy* solution. We also show how the hyperbolic conservation law *inherits* effortlessly all the continuous dependence results constructed for the original model. Then, we apply these results to obtain an a posteriori error estimate for the Engquist-Osher scheme. In Section 5, we treat the case of a *continuous* approximate solution and in Section 6, the case of a *discontinuous* approximate solution.

We have chosen the above model problem, not only because it is a simple and relevant case but also because it is the *only* case for which the approach to error estimation we present in these notes has been developed. This means that the extension of this approach to systems is an exciting challenge and a rich, wide-open field of research. In Section 7, we end these notes with some concluding remarks.

2. Some Convection-Diffusion Problems

In this section, we consider two conservation laws modeling physical phenomena of practical interest, traffic flow and phase propagation in solids, and illustrate on them the typical modification of the equations that gives rise to *ill posed* problems for nonlinear hyperbolic conservation laws.

When some of the physical phenomena are considered to be *non-dominant*, it is a wide spread practice to *formally* drop the terms in the equations modeling these phenomena. Thus, in traffic flow, when the driver's awareness of the conditions ahead is negligible, second-order terms of the equations are dropped. In phase propagation in solids, when transport is dominant, the high-order terms modeling the phase transition are dropped.

We will see that the main problem with this *formal* modification is that the resulting problem becomes *ill-posed*. We argue that this happens because, although the neglected physical phenomena can be correctly considered to be not important in most parts of the domain, *they are still crucial in small, key parts of the domain*. For example, in traffic flow, the driver's awareness of the conditions ahead is essential near a strong variation of the density of cars; as we all know, this can only take place in a small set of the space domain. Also, in phase propagation in solids, the viscosity and capillarity effects, which contain all the information of the physics of the phase transition, are crucial only near the phase transition; the phase transition occurs in small subsets of the domain. The *formal* modification of the equations is thus equivalent to the removal of essential physical information and this, not surprisingly, induces a loss of *well-posedness* of the resulting problem.

In this section, we show how this happens for traffic flow and phase transitions which are nonlinear conservation laws. We note that in the linear case, this phenomenon does not occur. We end this section by some concluding remarks.

2.1. Traffic Flow

2.1.1. The Model. If ρ represents the density of cars in a highway and v represents the flow velocity, the fact that cars do not appear out of the blue or vanish spontaneously in the middle of the highway can be written mathematically as follows:

$$\rho_t + (\rho v)_x = 0.$$

In a first approximation, the flow velocity v can be thought to be a function of the density of cars ρ only, say $V(\rho)$. In this case, the above conservation law becomes a nonlinear hyperbolic conservation law,

$$\rho_t + (f(\rho))_x = 0,$$

where $f(\rho) = \rho V(\rho)$ is the so-called density flow. It is reasonable to assume that $\rho \mapsto V(\rho)$ is a decreasing mapping and that for a given density, say $\rho^\star$, the velocity V is equal to zero; this corresponds to the situation in which the cars are bumper to bumper. The simplest case is the following:

$$V(\rho) = v_{max}\left(1 - \frac{\rho}{\rho^\star}\right),$$

where v_{max} represents the maximum velocity, and it corresponds to a quadratic density flow,

$$f(\rho) = (\rho^\star v_{max})\left(\frac{\rho}{\rho^\star}\right)\left(1 - \frac{\rho}{\rho^\star}\right).$$

A better model for the flow velocity v takes into account our tendency of avoid those crazy nuts that get too close to our car and our tendency not to get too close to a high concentration of cars. A simple way to model this tendency is to take

$$\rho v = f(\rho) - \nu \rho_x.$$

Witham [20] claims that the term our $\nu \rho_x$ models our 'awareness of conditions ahead,' since when we perceive a high density of cars ahead, we try to suitably decrease our speed to avoid a potentially dangerous situation; of course, for this to happen, the coefficient ν must be positive. With this choice of flow velocity, our original conservation law becomes

$$\rho_t + (f(\rho))_x - \nu \rho_{xx} = 0,$$

and has the good taste to give rise to mathematically *well posed* initial value problems.

2.1.2. Traveling Waves and the Diffusion Coefficient ν. It is reasonable to expect that when the convection is dominant, that is, when the number

$$\frac{\rho^\star v_{max}}{\nu},$$

is big, the effects of the term $\nu \rho_{xx}$ are negligible. This can happen in those parts of the domain in which the quantity ρ_{xx} remains not too big, but cannot happen where ρ_x changes rapidly in a small part of the space domain.

It is well known that if the density flow is linear in ρ the size of those parts of the domain are of order $\sqrt{\nu}$. In the nonlinear case we have in hand, it turns out that the size of those parts is even smaller: it is of order ν only. The simplest way to illustrate this fact is to look for solutions of our conservation law of the form

$$\rho(x,t) = \phi\left(\frac{x - ct}{\epsilon}\right).$$

These are called traveling wave solutions.

If we insert this expression for ρ in the conservation law and set

$$\epsilon = \nu,$$

we obtain a simple equation for ϕ, namely,

$$-c\phi' + (f(\phi))' - \phi'' = 0.$$

Next, if we assume that

$$\lim_{z \to \infty^\pm} \phi(z) = \rho^\pm,$$

and that

$$\lim_{z \to \infty^{\pm}} \phi'(z) = 0,$$

we can integrate once to get that ϕ must satisfy the following simple first-order ordinary differential equation:

$$\phi' = f(\phi) - \{ f(\rho^+) - c\,(\rho^+ - \phi) \},$$

where the *speed of propagation* of the traveling wave is

$$c = \frac{f(\rho^+) - f(\rho^-)}{\rho^+ - \rho^-}.$$

There is a unique solution of this equation if and only if

$$\phi'\,\mathrm{sign}(\rho^+ - \rho^-) > 0,$$

that is, if

$$\big(\, f(\varphi) - \{ f(\rho^+) - c\,(\rho^+ - \phi) \}\, \big)\,\mathrm{sign}(\rho^+ - \rho^-) > 0.$$

A simple geometric interpretation of this condition can be obtained as follows. First, we note that the function

$$g(\rho) := f(\rho^+) + c\,(\rho - \rho^+),$$

is nothing but the linear function that coincides with f at $\rho = \rho^{\pm}$. Thus, we have the following result.

Theorem 2.1 (Existence of traveling waves). *A traveling wave solution exists if and only if the graph of f on the interval (ρ^-, ρ^+) (resp., (ρ^+, ρ^-)) lies above (resp., below) the straight line joining the points $(\rho^{\pm}, f(\rho^{\pm}))$.*

In the case we are considering, f is a concave function and so the condition for the existence of a traveling wave is satisfied if

$$\rho^- < \rho^+.$$

If there is a traveling wave solution of the form

$$\phi(\frac{x - c\,t}{\nu}),$$

it is clear that strong variations of its derivative must be contained in a set of measure of order ν.

Next, we consider what happens when we let the diffusion coefficient ν go to zero.

2.1.3. Loss of Well-Posedness when ν Goes to Zero. It is very easy to see that when we let the diffusion coefficient tend to zero in the traveling wave solution, we obtain the following limit

$$\rho(x,t) = \lim_{\nu\downarrow 0} \phi\left(\frac{x-ct}{\nu}\right) = \begin{cases} \rho^+ & \text{if } x-ct > 0, \\ \rho^- & \text{if } x-ct < 0. \end{cases}$$

Since this limit can be proven to be a *weak* solution of the following Cauchy problem, also called a Riemann problem,

$$\rho_t + (f(\rho))_x = 0,$$
$$\rho(x,0) = \begin{cases} \rho^+ & \text{if } x > 0, \\ \rho^- & \text{if } x < 0; \end{cases}$$

this fact could be thought to be an indication that to *formally* drop the second-order term

$$\nu\,\rho_{xx},$$

from the equation could be mathematically justified. Indeed, it is a well known fact that piecewise-smooth *weak* solutions of the equation

$$\rho_t + (f(\rho))_x = 0,$$

are strong solutions in the parts of the domain where they are smooth, and satisfy the so-called jump-condition at the discontinuity curves $(x(t),t)$:

$$\frac{d}{dt}x(t) = \frac{f(\rho^+) - f(\rho^-)}{\rho^+ - \rho^-}.$$

Thus, it is clear that the limit of traveling wave solutions that we computed above is a *weak* solution of the Cauchy problem under consideration. However, it is easy to construct infinitely many *weak* solutions for the same Cauchy problem.

To do this, let us fix ideas and set

$$f(\rho) = \rho(1-\rho),$$

and

$$\rho^- = 1/4, \qquad \rho^+ = 3/4.$$

Note that this gives $c = 0$. In this case, the limit of the traveling wave solutions is

$$\rho(x,t) = \lim_{\nu\downarrow 0} \phi\left(\frac{x-ct}{\nu}\right) = \begin{cases} 3/4 & \text{if } x > 0, \\ 1/4 & \text{if } x < 0. \end{cases}$$

However, the following functions are also *weak* solutions of the same Cauchy problem, for *all* positive values of the parameter δ:

$$\rho(x,t) = \begin{cases} 3/4 & \text{if } c_1 < x/t, \\ 1/4 - \delta & \text{if } c_2 < x/t < c_1, \\ 3/4 + \delta & \text{if } c_3 < x/t < c_2, \\ 1/4 & \text{if } x/t < c_3, \end{cases}$$

where

$$c_1 = \delta, \qquad c_2 = 0, \qquad c3 = -\delta.$$

Note that the discontinuities $x/t = c_1$ and $x/t = c_3$ do satisfy the condition for the existence of traveling waves of the corresponding parabolic regularization. However, this is *not* true for the discontinuity $x/t = c_2$; in other words, this discontinuity does not 'remember' anything about the physics contained in the modeling of the 'awareness of the conditions lying ahead.' Because of the loss of this crucial information, to *formally* drop the second-order term

$$\nu \, \rho_{xx},$$

from the equations results in the *loss of the well-posedness* of the problem. Moreover, this unfortunate situation is present only if the density flow f is a nonlinear function; indeed, if the density flow f is linear, there is a unique *weak* solution.

2.2. Propagation of Phase Transitions

2.2.1. The Model. The so-called viscosity-capillarity model for phase transitions in van der Waal fluids was proposed independently by Truskinovsky [18] in 1982 and Slemrod [15] in 1984. In the framework of phase transitions in solids, a similar model exists which can be written as follows:

$$\gamma_t = v_x,$$
$$v_t = (\sigma(\gamma))_x + \nu \, v_{xx} - \lambda \, \gamma_{xxx},$$

where v is the velocity, γ the strain, and σ the stress; the parameter ν is the viscosity and the parameter λ is the capillarity. A very well studied case of strain-stress relations is displayed in Figure 2.1.

The material under consideration has three phases, or crystals, associated with the strain-stress curve. Phase 1 is associated with the strain γ lying in the interval $(-1, \gamma_{1,2})$, phase 2 with the interval $(\gamma_{1,2}, \gamma_{2,3})$, and phase 3 with the interval $(\gamma_{2,3}, \infty)$.

2.2.2. The Size of the Sets Containing the Phase Transitions. Just as in the case of traffic flow, it can be proven that the width of the boundary layer around a phase boundary of the solution of the system under consideration is of the order of the viscosity coefficient ν; see [16], [1], and [2]. As a consequence, the measure of the sets containing strong variations of the derivatives of both the velocity and the strain must be of the order of the viscosity coefficient ν.

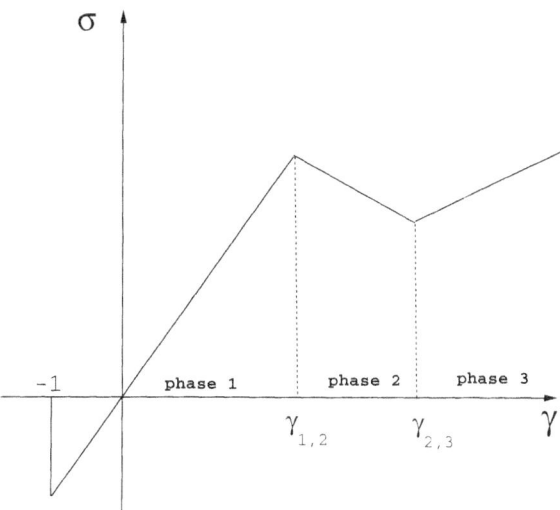

Fig. 2.1. Piecewise-linear strain-stress relation.

A simple way to see this is to construct traveling wave solutions for the system of equations under consideration. It is very simple to see that those traveling waves must depend on the parameter

$$\omega = \frac{\lambda}{\nu^2},$$

which seem to be a measure of the relative importance of viscous and capillarity effects.

2.2.3. The Loss of Well-Posedness. If we let the viscosity and capillarity coefficients ν and λ go to zero while keeping fixed the parameter ω, it is easy to see that the corresponding traveling wave solution will converge to a *weak* solution of a Riemman problem for the system

$$\gamma_t = v_x,$$
$$v_t = (\sigma(\gamma))_x,$$

which is a system of mixed type, since it is well known that when $\sigma' > 0$ the system is hyperbolic and that when $\sigma' < 0$ the system is elliptic. Since the initial condition of the Riemann problem does not depend on the parameter ω, we see immediately that for each value of this parameter, we have a *weak* solution of the Riemann problem.

Once again, we see that the uniqueness of the *weak* solution of Riemann problem has been lost. This happened because when the terms modeling the viscosity and capillarity effects are dropped from the equations, the physics of the phase transition is lost and this results, once again, in the *loss of the well-posedness* of the corresponding Cauchy problem.

Finally, let us point out that this does not occur if the problem is linear, that is, when σ is a linear function.

2.3. Concluding Remarks

We hope to have convinced the reader that to happily drop high-order terms from the equations of a well-posed problem and end up with a seemingly simpler nonlinear conservation law might be a quite dangerous sport. Indeed, even though the effects of those high-order terms can be felt on very small subsets of the domain, they contain essential information of physical processes that cannot be tossed away ... without loosing the well-posedness properties of the problem.

In what follows, we take as our model problem the following Cauchy problem for a convection-diffusion scalar conservation law:

$$v_t + \nabla \cdot \mathbf{f}(v) - \nu \Delta v = r, \qquad \text{in } \mathbb{R}^d \times (0, T),$$
$$v(t = 0) = v_0, \qquad \text{on } \mathbb{R}^d,$$

which is an extension of the Cauchy problem for the traffic flow model. In Section 3, we obtain continuous dependence results for this problem and *only then* we pass to the limit in the diffusion coefficient; this is done in Section 4.

3. Continuous Dependence for Nonlinear Convection-Diffusion

In this section, we consider the initial-value problem

$$v_t + \nabla \cdot \mathbf{f}(v) - \nu \Delta v = r, \qquad \text{in } \mathbb{R}^d \times (0, T), \qquad (3.1)$$
$$v(t = 0) = v_0, \qquad \text{on } \mathbb{R}^d, \qquad (3.2)$$

and study how to estimate the effect in the solution v induced by variations in the data. In other words, we are interested in estimating the distance between v_1 and v_2 where v_ℓ denotes the solution of the above initial-boundary value problem with $v_0 = v_{0,\ell}$, $\mathbf{f} = \mathbf{f}_\ell$, $\nu = \nu_\ell$, and $r = r_\ell$, for $\ell = 1, 2$.

This section is the main section of these notes; all the remaining results in this section are based on the estimates constructed in this section. We construct continuous dependence results that do not break down when the diffusion coefficients go to zero. As a consequence, these continuous dependence results are naturally *inherited* by the entropy solution of the corresponding nonlinear hyperbolic conservation law. To construct these estimates, we proceed in several steps; we start from the well known *duality* technique and successively modify it to fit our interest.

Thus, we start by introducing the main technique to obtain continuous dependence results: the *duality* technique. In this very well known technique, an expression of the error is obtained in terms of the data of the problem and in terms of the solution of the so-called *adjoint* problem. Since this problem is usually very difficult to study because of the *nonlinear* character of the convective term and because of the smallness of the diffusion coefficients, other techniques have been developed for which no adjoint problem has to be solved. The price to pay for this advantage, however, is that we can no longer freely choose the norms in which we measure the distance between two solutions.

The classical example is, of course, the standard energy technique for parabolic equations which gives rise to continuous dependence results in L^2-norms. However, the use of L^2-norms does not allow us to deal with the nonlinearities of the convective terms in a simple way. We show that the use of L^1-like norms does allow us to treat those terms in a simple and elegant way.

The next step is to modify the above technique to obtain continuous dependence results that (i) do not break down when the diffusion coefficients go to zero, and that (ii) can be used to compare the exact solution v with an arbitrary function u of low regularity. We show that these goals can be achieved by introducing the *doubling of the variables* technique. We show that if we *double* the space variables, we can achieve the first goals and that if we further *double* the time variable, we can achieve the second goal.

3.1. The Standard Duality Technique and the Adjoint Problem

The main technique to obtain continuous dependence results is the so-called *duality* technique. Let us introduce this technique. We start with the trivial identity for the error $e(t) = v_1(t) - v_2(t)$,

$$\int_{\mathbb{R}^d} e(T)\zeta(T)\,dx = \int_{\mathbb{R}^d} e(0)\zeta(0)\,dx + \int_0^T \int_{\mathbb{R}^d} (e_t\,\zeta + e\,\zeta_t)\,dx\,dt,$$

where the *test* function ζ is to be determined. For simplicity, let us consider the case in which $\mathbf{f}_1 = \mathbf{f}_2 = \mathbf{f}$ and $\nu_1 = \nu_2 = \nu$. In this case, we have that

$$e_t = -\nabla \cdot (\mathbf{f}(v_1) - \mathbf{f}(v_2)) + \nu(\,\Delta v_1 - \Delta v_2\,) + (r_1 - r_2),$$

and so,

$$\int_{\mathbb{R}^d} e(T)\zeta(T)\,dx = \int_{\mathbb{R}^d} e(0)\zeta(0)\,dx + \int_0^T \int_{\mathbb{R}^d} (e\,A(\zeta) + (r_1 - r_2)\zeta)\,dx\,dt,$$

where the operator A is given by

$$A(\zeta) = \zeta_t + \frac{\mathbf{f}(v_1) - \mathbf{f}(v_2)}{v_1 - v_2} \cdot \nabla\zeta + \nu\,\Delta\zeta.$$

Thus, if we take ζ as the solution of the following initial-value problem

$$A(\zeta) = 0 \qquad \text{in } \mathbb{R}^d \times (0, T), \tag{3.3}$$
$$\zeta(T) = \psi \qquad \text{on } \mathbb{R}^d, \tag{3.4}$$

we will have that

$$\int_{\mathbb{R}^d} e(T)\psi \, dx = \int_{\mathbb{R}^d} e(0)\zeta(0) \, dx + \int_0^T \int_{\mathbb{R}^d} (r_1 - r_2) \, \zeta \, dx \, dt.$$

Now, a simple application of Hölder's inequality gives us the following continuous dependence result in the L^p-norm, for $1 \le p < \infty$:

$$|e(T)\|_{L^p(\mathbb{R}^d)} \le C \left\{ \| e(0)\|_{L^p(\mathbb{R}^d)} + \int_0^T \| r_1(t) - r_2(t) \|_{L^p(\mathbb{R}^d)} \, dt \right\},$$

where

$$C = \sup_{\psi \in L^q(\mathbb{R}^d)} \frac{\| \zeta \|_{L^\infty(0,T;L^q(\mathbb{R}^d))}}{\| \psi \|_{L^q(\mathbb{R}^d)}}, \qquad 1/p + 1/q = 1.$$

In this way, the study of the continuous dependence properties of the solutions of the equation (3.1) is reduced to the study of the solutions of the adjoint problem (3.3) and (3.4). This, however, is not a simple matter, not only because of the presence of the nonlinearity $\mathbf{f}$ and the non-divergence form of the equation, but also because of the smallness of the diffusion coefficient ν.

Next, we show that it is possible to obtain continuous dependence results *without* having to solve an adjoint problem.

3.2. A Technique to Bypass the Resolution of the Adjoint Problem

The main idea is to try to obtain a continuous dependence result by assuming that the test function ζ is of the form

$$\zeta = H(e),$$

where H has to be chosen in such a way that the resolution of the adjoint problem could be bypassed. Once again, for the sake of simplicity, we present this technique for the case in which $\mathbf{f}_1 = \mathbf{f}_2 = \mathbf{f}$ and $\nu_1 = \nu_2 = \nu$. In this case, we have the following equation for the error e:

$$\int_{\mathbb{R}^d} e(T) \, H(e(T)) \, dx = \int_{\mathbb{R}^d} e(0) \, H(e(0)) \, dx + \Gamma(H)$$
$$+ \int_0^T \int_{\mathbb{R}^d} (r_1 - r_2) \, H(e) \, dx \, dt,$$

where

$$\Gamma(H) = \int_0^T \int_{\mathbb{R}^d} e\, A(H(e))\, dx\, dt.$$

Note that in the standard duality argument, the term $\Gamma(H)$ does not appear in the equation for the error since ζ is chosen to satisfy $A(\zeta) = 0$. This is not the case in this approach; in fact, we have

$$e\, A(H(e)) = e\, H'(e)\, e_t + e\, H'(e)\, (\mathbf{f}(v_1) - \mathbf{f}(v_2)) \cdot \nabla e + \nu\, e\, \Delta H(e),$$

and so

$$\Gamma(H) = \int_0^T \int_{\mathbb{R}^d} e\, H'(e)\, I\, dx\, dt - \nu \int_0^T \int_{\mathbb{R}^d} |\nabla e|^2\, H'(e)\, dx\, dt,$$

where

$$I = e_t + \frac{\mathbf{f}(v_1) - \mathbf{f}(v_2)}{v_1 - v_2} \cdot \nabla e.$$

Finally, noting that

$$e\, H'(e)\, e_t = \left(e\, H(e) - \int_0^e H(s)\, ds \right)_t,$$

and setting

$$U(e) = \int_0^e H(s)\, ds,$$

we rewrite the equation for the error as follows:

$$\int_{\mathbb{R}^d} U(e(T))\, dx = \int_{\mathbb{R}^d} U(e(0))\, dx + Z(U) + \int_0^T \int_{\mathbb{R}^d} (r_1 - r_2)\, U'(e)\, dx\, dt, \quad (3.5)$$

where

$$Z(U) = -\int_0^T \int_{\mathbb{R}^d} R(e)\, S\, dx\, dt - \nu \int_0^T \int_{\mathbb{R}^d} |\nabla e|^2\, U''(e)\, dx\, dt,$$

in which

$$R(e) = \int_0^e s\, U''(s)\, ds \quad \text{and} \quad S = \frac{\mathbf{f}(v_1) - \mathbf{f}(v_2)}{v_1 - v_2} \cdot \nabla e.$$

In the classical energy technique for parabolic equations, the *test* function ζ is taken to be

$$\zeta = e \equiv H(e).$$

This means that the classical energy technique corresponds to the technique we are presenting with the following choice of the function U:

$$U(e) = \frac{1}{2}\, e^2.$$

Indeed, in this case, we have

$$Z(U) = -\frac{1}{2} \int_0^T \int_{\mathbb{R}^d} e^2 \, \nabla \cdot \left(\frac{\mathbf{f}(v_1) - \mathbf{f}(v_2)}{v_1 - v_2} \right) dx \, dt - \nu \int_0^T \int_{\mathbb{R}^d} |\nabla e|^2 \, dx \, dt, \quad (3.6)$$

and (3.5) becomes

$$\| e \|_{2,T,\nu}^2 = \frac{1}{2} \| e(0) \|_{L^2(\mathbb{R}^d)}^2 + \int_0^T \int_{\mathbb{R}^d} (r_1 - r_2) \, e \, dx \, dt$$

$$- \frac{1}{2} \int_0^T \int_{\mathbb{R}^d} e^2 \, \nabla \cdot \left(\frac{\mathbf{f}(v_1) - \mathbf{f}(v_2)}{v_1 - v_2} \right) dx \, dt,$$

where

$$\| e \|_{2,T,\nu}^2 = \frac{1}{2} \| e(T) \|_{L^2(\mathbb{R}^d)}^2 + \nu \int_0^T \int_{\mathbb{R}^d} |\nabla e|^2 \, dx \, dt.$$

From this equality, it is possible to obtain continuous dependence results with respect to the initial data and with respect to the right-hand side by using standard manipulations. Thus, we have shown that it is possible to compare two solutions of (3.1) and (3.2) *without* having to solve an adjoint problem. The price we must pay for this is that we do not have much control on the norm in which we measure the error. In these notes, we assume that this is a reasonable price to pay for not solving the adjoint equation.

Note that the term

$$\frac{1}{2} \int_0^T \int_{\mathbb{R}^d} e^2 \, \nabla \cdot \left(\frac{\mathbf{f}(v_1) - \mathbf{f}(v_2)}{v_1 - v_2} \right) dx \, dt,$$

is equal to zero if $\mathbf{f}(v) = \mathbf{a} v$ and $\mathbf{a}$ is a constant vector. However, in general, this term is not easy to handle. This is why we would like to explore the possibility of finding another function U for which the term involving the convective nonlinearity $\mathbf{f}$ has a simpler form. We do this next.

3.3. A Very Simple Way of Handling the Convective Nonlinearity f

A simple glance at the expression (3.2) for $Z(U)$ suggests that, in order to get rid of the term that contains the nonlinearity $\mathbf{f}$, we should find a non-trivial function U such that $e \, U''(e) = 0$. Of course, such a function does *not* exist but we can still achieve our goal if we take U'' to be the limit of a a sequence of smooth functions $\{ U_\epsilon \}_{\epsilon > 0}$ such that

$$\lim_{\epsilon \to 0} e \, U_\epsilon''(e) = 0.$$

It is easy to see that U must be a continuous function, with a kink at the origin and linear elsewhere.

Thus, if we want to consider

$$U(e) = e^+ \equiv \max\{0, e\},$$

we can take, for any positive parameter ϵ,

$$U_\epsilon(e) = \int_0^{e/\epsilon} (e - \epsilon\, s)\, \mu(s)\, ds$$

where μ is a smooth nonnegative function with support in $[0, 1]$ and integral equal to one. Then, for every fixed value of e, it is easy to verify that

$$\lim_{\epsilon \to 0} e\, U_\epsilon''(e) = 0,$$
$$\lim_{\epsilon \to 0} U_\epsilon'(e) = \mathrm{sign}(e)^+,$$
$$\lim_{\epsilon \to 0} U_\epsilon''(e) \equiv \mathrm{sign}'(e)^+ \equiv \delta(e) \geq 0,$$

where δ is the Dirac delta, and that

$$Z(U) = -\nu \int_0^T \int_{\mathbb{R}^d} \nabla e \cdot \nabla \mathrm{sign}(e)^+\, dx\, dt,$$

by the Lebesgue dominated convergence theorem. We thus see that the term containing the convective nonlinearity $\mathbf{f}$ has completely disappeared, as we wanted.

The equation (3.5) becomes the following simple equality:

$$\| e \|_{1,T,\nu}^+ = \| e^+(0) \|_{L^1(\mathbb{R}^d)} + \int_0^T \int_{\mathbb{R}^d} \mathrm{sign}(v_1 - v_2)^+ (r_1 - r_2)\, dx\, dt, \qquad (3.7)$$

where

$$\| e \|_{1,T,\nu}^+ = \| e^+(T) \|_{L^1(\mathbb{R}^d)} + \nu \int_0^T \int_{\mathbb{R}^d} \nabla e \cdot \nabla \mathrm{sign}(e)^+\, dx\, dt. \qquad (3.8)$$

Similar results with $(-e)^+$ and $|e|$ instead of e^+ can be obtained if we take μ to have support in $[-1, 0]$ and if we take μ to be an even function with support in $[-1, 1]$, respectively. This shows that the use of L^1-like norms leads naturally to a simple treatment of the nonlinear convective terms.

Note that the term

$$\nu \int_0^T \int_{\mathbb{R}^d} \nabla e \cdot \nabla \mathrm{sign}(e)^+\, dx\, dt \equiv \lim_{\epsilon \to 0} \nu \int_0^T \int_{\mathbb{R}^d} \nabla e \cdot \nabla U_\epsilon'(e)\, dx\, dt,$$

is a nonnegative term. This strange term is an L^1-version of the more familiar term

$$\nu \int_0^T \int_{\mathbb{R}^d} |\nabla e|^2\, dx\, dt,$$

obtained by working with L^2-norms.

We have thus shown a way to avoid having to solve an adjoint problem and obtain a very simple continuous dependence result that can easily handle the convective nonlinearity $\mathbf{f}$. From that estimate, several important results can be obtained, as we show next.

3.4. Continuous Dependence Results in L^1-like Norms

In this section, we obtain three types of results: (i) a priori estimates on the exact solution v of (3.1) and (3.2), (ii) continuous dependence results, and (iii) a posteriori error estimates.

3.4.1. A Priori Estimates of v. From (3.7) and (3.8), several properties of the solution v of (3.1) and (3.2) can be obtained. We are only interested in the following two:

$$v(x,t) \in I(v_0, r) = [a, b], \tag{3.9}$$

where

$$a = \inf_{x \in \mathbb{R}^d} v_0(x) - \| (-r)^+ \|_{L^1(0,T;L^\infty(\mathbb{R}^d))},$$

$$b = \sup_{x \in \mathbb{R}^d} v_0(x) + \| r^+ \|_{L^1(0,T;L^\infty(\mathbb{R}^d))},$$

and

$$\| v \|_{L^\infty(0,T;TV(\mathbb{R}^d))} \leq \| v_0 \|_{L^\infty(0,T;TV(\mathbb{R}^d))} + \| r \|_{L^1(0,T;TV(\mathbb{R}^d))}, \tag{3.10}$$

in which

$$|v|_{TV(\mathbb{R}^d)} = \sup_{\| \mathbf{p} \|_{L^\infty(\mathbb{R}^d, \mathbb{R}^d)} = 1} \int_\mathbb{R} v \nabla \cdot \mathbf{p}.$$

Note how these estimates are independent of the size of the diffusion coefficient ν.

The first estimate follows from (3.7) in two steps. First, we take $v_1 = v$ and

$$v_2(x,t) = c_2(t) := \sup_{x \in \mathbb{R}^d} v_0(x) + \| r^+ \|_{L^1(0,t;L^\infty(\mathbb{R}^d))}$$

and obtain that $v(x,t) \leq c_2(t)$. Then, we take $v_2 = v$ and

$$v_1(x,t) = c_1(t) := \inf_{x \in \mathbb{R}^d} v_0(x) - \| (-r)^+ \|_{L^1(0,t;L^\infty(\mathbb{R}^d))}$$

to obtain that $v(x,t) \geq c_1(t)$.

To obtain the second estimate (3.10), we first obtain from (3.7) that

$$\| e(T) \|_{L^1(\mathbb{R}^d)} \leq \| e(0) \|_{L^1(\mathbb{R}^d)} + \int_0^T \int_{\mathbb{R}^d} |r_1 - r_2| \, dx \, dt.$$

Then, taking $v_1 = v$ and $v_2(x,t) = v(x + h e_i, t)$, where e_i is one of the canonical Euclidean basis vectors, we get that

$$\| v_{x_i}(T) \|_{L^1(\mathbb{R}^d)} \leq \| v_{x_i}(0) \|_{L^1(\mathbb{R}^d)} + \int_0^T \int_{\mathbb{R}^d} |r_{x_i}| \, dx \, dt.$$

Summing from $i = 1$ to $i = d$, we get the result. [Note the relationship between $\sum_i \|v_{x_i}\|_{L^1(\mathbb{R}^d)}$ and $|v|_{TV(\mathbb{R}^d)}$, a simple consequence of integration by parts.]

3.4.2. Continuous Dependence. To obtain a general continuous dependence result, we simply rewrite the equation for v_2 as

$$(v_2)_t + \nabla \cdot \mathbf{f}_1(v_2) - \nu_1 \Delta v_2 = r_2 + \nabla \cdot (\mathbf{f}_1(v_2) - \mathbf{f}_2(v_2)) - (\nu_1 - \nu_2)\Delta v_2,$$

and apply the estimate (3.7) to get

$$\begin{aligned}
\| e \|^+_{1,T,\nu_1} \leq{} & \| e^+(0) \|_{L^1(\mathbb{R}^d)} + \int_0^T \int_{\mathbb{R}^d} (r_1 - r_2)^+ \, dx \, dt \\
& + \int_0^T \int_{\mathbb{R}^d} (\nabla \cdot (-\mathbf{f}_1(v_2) + \mathbf{f}_2(v_2)))^+ \, dx \, dt \\
& + \int_0^T \int_{\mathbb{R}^d} ((\nu_1 - \nu_2)\Delta v_2)^+ \, dx \, dt.
\end{aligned}$$

Note that the term

$$\int_0^T \int_{\mathbb{R}^d} |\nabla \cdot (\mathbf{f}_1(v_2) - \mathbf{f}_2(v_2))| \, dx \, dt,$$

can be bounded uniformly with respect to the diffusion coefficient ν_2 by

$$\| \mathbf{f}_1'(w) - \mathbf{f}_2'(w) \|_{L^\infty(I(v_{2,0}, r_2))} \left(\| v_{2,0} \|_{L^\infty(0,T;TV(\mathbb{R}^d))} + \| r_2 \|_{L^1(0,T;TV(\mathbb{R}^d))} \right).$$

This means that the above continuous dependence result for the nonlinearities does hold when the diffusion coefficients tend to zero.

Unfortunately, this is not true for the dependence with respect to the diffusion coefficients themselves since the term

$$|\nu_1 - \nu_2| \int_0^T \int_{\mathbb{R}^d} |\Delta v_2| \, dx \, dt,$$

blows up when the diffusion coefficient ν_2 goes to zero.

3.4.3. An a Posteriori Error Estimate. To obtain an a posteriori error estimate of the difference between the exact solution v of (3.1) and (3.2), and any *smooth* approximation u, we first note that u satisfies the equation

$$u_t + \nabla \cdot \mathbf{f}(u) - \nu \Delta u = r_2,$$

where r_2 is trivially defined by the left-hand side of the above equation. Then, we set $v_1 = v$ and $v_2 = u$ in (3.7) and obtain the following very simple a posteriori error equation:

$$\begin{aligned}
\| e \|^-_{1,T,\nu} ={} & \int_0^T \int_{\mathbb{R}^d} \mathrm{sign}(v - u)^+ (r - u_t + \nabla \cdot \mathbf{f}(u) - \nu \Delta u) \, dx \, dt \\
& + \| e^+(0) \|_{L^1(\mathbb{R}^d)},
\end{aligned}$$

Note that this equality does hold if we let the diffusion coefficient ν go to zero. However, in this case, it becomes *useless* for computational purposes since the quantity

$$\| e \|_{1,T,\nu}^+ = \| e^+(T) \|_{L^1(\mathbb{R}^d)} + \nu \int_0^T \int_{\mathbb{R}^d} \nabla e \cdot \nabla \mathrm{sign}(e)^+ \, dx \, dt,$$

might converge to an order one quantity. Indeed, since the approximate solution u is independent of the diffusion coefficient ν, the expression

$$\lim_{\nu \to 0} \nu \nabla e = (\lim_{\nu \to 0} \nu \nabla v),$$

might be different from zero when the function $\mathbf{f}$ is strictly nonlinear; an example of this situation will be considered in section 5.3.

An interpretation of this unfortunate situation is the following. If the approximate solution u is to be a good approximation of the exact solution v, it must resolve the strong gradients of v well. Typically, the strong gradients of v are concentrated in very small sets whose measure is proportional to a positive power of the diffusion coefficient ν. If we let ν go to zero, it is clear that the approximate solution u will be totally unable to resolve those structures while remaining *smooth*.

In summary, we have been able to obtain (i) a priori estimates on the exact solution v of (3.1) and (3.2), (ii) general continuous dependence results, and (iii) a posteriori error estimates. However, we run into essential difficulties when the diffusion coefficients go to zero. In the next section, we show how to overcome this difficulty.

3.5. Allowing the Diffusion Coefficients to Go to Zero

To deal with this difficult case, we have to modify the technique we have been working with. This modification has two main ingredients. First, the function v_1 is evaluated at the space variable x whereas the function v_2 is evaluated at a *different* space variable x'; this trick is called *doubling* of the variables and will allow us to deal properly with the second-order terms.

Since it is not possible to compare v_1 and v_2 at the same points anymore, instead of dealing with the quantity

$$\| (v_1(t) - v_2(t))^+ \|_{L^1(\mathbb{R}^d)},$$

we will consider the quantity

$$\rho_{\epsilon_x}^+(v_1(t), v_2(t)) = \int_{\mathbb{Q}} (v_1(x,t) - v_2(x',t))^+ \varphi_{\epsilon_x}(x - x') \, dx \, dx',$$

where $\mathbb{Q} = \mathbb{R}^d \times \mathbb{R}^d$,

$$\varphi_{\epsilon_x}(z) = \Pi_{i=1}^d \frac{1}{\epsilon_x} \omega(\frac{z_i}{\epsilon_x}),$$

and ω is an even, nonnegative function with support $[-1, 1]$ and integral equal to one. The choice of the form of $\rho_{\epsilon_x}^+(v_1(t), v_2(t))$ is appropriate since it is easy to prove that

$$\left| \rho_{\epsilon_x}^+ (v_1(t), v_2(t)) - \| e^+(t) \|_{L^1(\mathbb{R}^d)} \right| \leq \epsilon_x \, | v_2(t) |_{TV(\mathbb{R}^d)}. \qquad (3.11)$$

We are now ready to obtain our new continuous dependence results; we proceed in several steps.

Step 1. We start exactly as in the standard duality argument with a trivial identity for $\rho_{\epsilon_x}^+ (v_1(t), v_2(t))$:

$$\rho_{\epsilon_x}^+ (v_1(T), v_2(T)) = \rho_{\epsilon_x}^+ (v_1(0), v_2(0)) + \int_0^T \!\! \int_Q e_t^+ \, \varphi_{\epsilon_x} \, dx \, dx' \, dt.$$

For the sake of simplicity, we start by considering the case in which $\mathbf{f}_1 = \mathbf{f}_2 = \mathbf{f}$ and $r_1 = r_2 = 0$. In this case

$$
\begin{aligned}
e_t^+ \;=\; & \operatorname{sign}(v_1 - v_2)^+ \left(-\nabla_x \cdot \mathbf{f}(v_1) + \nu_1 \Delta_x v_1 \right) \\
& + \operatorname{sign}(v_1 - v_2)^+ \left(+\nabla_{x'} \cdot \mathbf{f}(v_2) - \nu_2 \Delta_{x'} v_2 \right).
\end{aligned}
$$

Next, we rewrite the above expression by using the fact that $v_2 = v_2(x', t)$ does not depend on x, and that $v_1 = v_1(x, t)$ does not depend on x'. We obtain that

$$
\begin{aligned}
\operatorname{sign}(v_1 - v_2)^+ \nabla_x \cdot \mathbf{f}(v_1) &= \nabla_x \cdot \mathbf{f}^+(v_1, v_2), \\
-\operatorname{sign}(v_1 - v_2)^+ \nabla_{x'} \cdot \mathbf{f}(v_2) &= \nabla_{x'} \cdot \mathbf{f}^+(v_1, v_2),
\end{aligned}
$$

where

$$\mathbf{f}^+(a, b) = (\mathbf{f}(a) - \mathbf{f}(b)) \operatorname{sign}^+(a - b),$$

and that

$$
\begin{aligned}
\operatorname{sign}(e)^+ \nu_1 \Delta_x v_1 &= \nu_1 \Delta_x \, e^+ - \nu_1 \operatorname{sign}'(e)^+ \, | \nabla_x v_1 |^2, \\
-\operatorname{sign}(e)^+ \nu_2 \Delta_{x'} v_2 &= \nu_2 \Delta_{x'} \, e^+ - \nu_2 \operatorname{sign}'(e)^+ \, | \nabla_{x'} v_2 |^2.
\end{aligned}
$$

As a consequence,

$$
\begin{aligned}
e_t^+ = \; & -(\nabla_x + \nabla_{x'}) \cdot F^+(v_1, v_2) \\
& + (\nu_1 \Delta_x + \nu_2 \Delta_{x'}) \, (v_1 - v_2)^+ \\
& - \operatorname{sign}'(v_1 - v_2)^+ (\nu_1 \, | \nabla_x v_1 |^2 + \nu_2 \, | \nabla_{v'} v_2 |^2).
\end{aligned}
$$

Hence, after simple integrations by parts, we get

$$
\begin{aligned}
\rho_{\epsilon_x}^+ (v_1(T), v_2(T)) = \; & \rho_{\epsilon_x}^+ (v_1(0), v_2(0)) \\
& + \int_0^T \!\! \int_Q F^+(v_1, v_2) \cdot (\nabla_x + \nabla_{x'}) \varphi_{\epsilon_x} \, dx \, dx' \, dt \\
& + \int_0^T \!\! \int_Q (v_1 - v_2)^+ (\nu_1 \Delta_x + \nu_2 \Delta_{x'}) \varphi_{\epsilon_x} \, dx \, dx' \, dt \\
& - \int_0^T \!\! \int_Q \operatorname{sign}'(v_1 - v_2)^+ (\nu_1 \, | \nabla_x v_1 |^2 + \nu_2 \, | \nabla_{x'} v_2 |^2) \varphi_{\epsilon_x} \, dx \, dx' \, dt.
\end{aligned}
$$

Finally, using the fact that $\varphi_{\epsilon_x} = \varphi_{\epsilon_x}(x - x')$,

$$\rho_{\epsilon_x}^+(v_1(T), v_2(T)) = \rho_{\epsilon_x}^+(v_1(0), v_2(0))$$

$$+(\nu_1 + \nu_2) \int_0^T \int_Q (v_1 - v_2)^+ \Delta_{x'} \varphi_{\epsilon_x} \, dx \, dx' \, dt$$

$$- \int_0^T \int_Q \text{sign}'(v_1 - v_2)^+ (\nu_1 |\nabla_x v_1|^2 + \nu_2 |\nabla_{x'} v_2|^2) \varphi_{\epsilon_x} \, dx \, dx' \, dt.$$

Step 2. Note that when the diffusion coefficients are equal, we do not immediately recover the estimate (3.7) for the case $\nu_1 = \nu_2$. To work toward that goal, we write

$$\nu_1 + \nu_2 = (\sqrt{\nu_1} - \sqrt{\nu_2})^2 + 2\sqrt{\nu_1 \nu_2},$$

and rewrite the last equality of **Step 1** as follows:

$$\rho_{\epsilon_x}^+(v_1(T), v_2(T)) = \rho_{\epsilon_x}^+(v_1(0), v_2(0))$$

$$+(\sqrt{\nu_1} - \sqrt{\nu_2})^2 \int_0^T \int_Q (v_1 - v_2)^+ \Delta_{x'} \varphi_{\epsilon_x} \, dx \, dx' \, dt$$

$$- \int_0^T \int_Q \text{sign}'(v_1 - v_2)^+ (\nu_1 |\nabla_x v_1|^2 + \nu_2 |\nabla_{x'} v_2|^2) \varphi_{\epsilon_x} \, dx \, dx' \, dt$$

$$+2\sqrt{\nu_1 \nu_2} \int_0^T \int_Q (v_1 - v_2)^+ \Delta_{x'} \varphi_{\epsilon_x} \, dx \, dx' \, dt.$$

Next, we must work on the last term of the right-hand side which we denote by Ψ. Using once again the facts that $v_2 = v_2(x', t)$ does not depend on x, that $v_1 = v_1(x, t)$ does not depend on x', and that $\varphi_{\epsilon_x} = \varphi_{\epsilon_x}(x - x')$, we get

$$\begin{aligned}
\Psi &= 2\sqrt{\nu_1 \nu_2} \int_0^T \int_Q (v_1 - v_2)^+ \Delta_{x'} \varphi_{\epsilon_x} \, dx \, dx' \, dt \\
&= -2\sqrt{\nu_1 \nu_2} \int_0^T \int_Q (v_1 - v_2)^+ \nabla_x \cdot \nabla_{x'} \varphi_{\epsilon_x} \, dx \, dx' \, dt \\
&= 2\sqrt{\nu_1 \nu_2} \int_0^T \int_Q \text{sign}(v_1 - v_2)^+ \nabla_x v_1 \cdot \nabla_{x'} \varphi_{\epsilon_x} \, dx \, dx' \, dt \\
&= 2\sqrt{\nu_1 \nu_2} \int_0^T \int_Q \text{sign}'(v_1 - v_2)^+ \nabla_x v_1 \cdot \nabla_{x'} v_2 \, \varphi_{\epsilon_x} \, dx \, dx' \, dt.
\end{aligned}$$

Hence, we get

$$\rho_{\epsilon_x}^+(v_1(T), v_2(T)) + \mathbb{FLT}_{\epsilon_x, T}^+(v_1, v_2) = \rho_{\epsilon_x}^+(v_1(0), v_2(0)) + \Xi_{\epsilon_x, T}^+(v_1, v_2)$$

where

$$\mathbb{FLT}^+_{\epsilon_x,T}(v_1, v_2) =$$
$$\int_0^T \int_{\mathbb{Q}} \mathrm{sign}'(v_1 - v_2)^+ \, |\sqrt{\nu_1}\,\nabla_x v_1 - \sqrt{\nu_2}\,\nabla_{x'} v_2\,|^2 \, \varphi_{\epsilon_x} \, dx \, dx' \, dt,$$

and

$$\Xi^+_{\epsilon_x,T}(v_1, v_2) = (\sqrt{\nu_1} - \sqrt{\nu_2})^2 \int_0^T \int_{\mathbb{Q}} \mathrm{sign}(v_1 - v_2)^+ \, \nabla_{x'} v_2 \cdot \nabla_{x'} \varphi_{\epsilon_x} \, dx \, dx' \, dt$$

If we now set $\nu_1 = \nu_2$ and let ϵ_x go to zero, we immediately obtain the estimate (3.7).

Step 3. In the general case of different nonlinearities and different right-hand sides, it is a simple exercise to see that the above estimate holds with

$$\Xi^+_{\epsilon_x,T}(v_1, v_2) = \int_0^T \int_{\mathbb{Q}} \mathrm{sign}(v_1 - v_2)^+ \, \nabla_{x'} \cdot (-\mathbf{f}_1(v_2) + \mathbf{f}_2(v_2))\varphi_{\epsilon_x} \, dx \, dx' \, dt$$
$$+ (\sqrt{\nu_1} - \sqrt{\nu_2})^2 \int_0^T \int_{\mathbb{Q}} \mathrm{sign}(v_1 - v_2)^+ \, \nabla_{x'} v_2 \cdot \nabla_{x'} \varphi_{\epsilon_x} \, dx \, dx' \, dt$$
$$+ \int_0^T \int_{\mathbb{Q}} \mathrm{sign}(v_1 - v_2)^+ \, (r_1 - r_2) \, \varphi_{\epsilon_x} \, dx \, dx' \, dt. \qquad (3.12)$$

Step 4. Now, it is very simple to get an error estimate in $L^\infty(0, T; L^1(\mathbb{R}^d))$. Since, by (3.11),

$$\rho^+_{\epsilon_x}(v_1(T), v_2(T)) \geq \| e^+(T) \|_{L^1(\mathbb{R}^d)} - \epsilon_x \, |\, v_2(T)\,|_{TV(\mathbb{R}^d)},$$
$$\rho^+_{\epsilon_x}(v_1(0), v_2(0)) \leq \| e^+(0) \|_{L^1(\mathbb{R}^d)} + \epsilon_x \, |\, v_2(0)\,|_{TV(\mathbb{R}^d)},$$

we obtain

$$\| e \|^+_{1,\epsilon_x,\nu_1,\nu_2,T} \leq \quad \| e^+(0) \|_{L^1(\mathbb{R}^d)} + 2\,\epsilon_x \, |\, v_2\,|_{L^\infty(0,T;TV(\mathbb{R}^d))} \qquad (3.13)$$
$$+ \Xi^+_{\epsilon_x,T}(v_1, v_2),$$

where

$$\| e \|^+_{1,\epsilon_x,\nu_1,\nu_2,T} = \| e^+(0) \|_{L^1(\mathbb{R}^d)} + \mathbb{FLT}^+_{\epsilon_x,\nu_1,\nu_2,T}(v_1, v_2). \qquad (3.14)$$

This is the estimate we were looking for. In the next section, we display the new continuous dependence results that can be obtained from the above estimate and show that they do *not* break down when the diffusion coefficients go to zero.

3.6. New Continuous Dependence Results

From the estimate obtained in the previous section, we will deduce (i) a new continuous dependence result with respect to the diffusion coefficients and (ii) a new a posteriori error estimate that holds for less smooth approximate solutions.

3.6.1. Continuous Dependence Results with Respect to the Diffusion Coefficients. To obtain this result, we take $v_{0,1} = v_{0,2} = v_0$, $f_1 = f_2$ and $r_1 = r_2 = r$ in (3.13), (3.14), and (3.12), and use the following estimates, proven in the appendix,

$$\int_0^T \int_{\underset{\sim}{}} (v_1 - v_2)^+ \, \Delta_{x'} \varphi_{\epsilon_x} \, dx \, dx' \, dt \leq \frac{1}{\epsilon_x} \| v_2 \|_{L^1(0,T;TV(\mathbb{R}^d))},$$

and

$$\int_0^T \int_{\underset{\sim}{}} \text{sign}(v_1 - v_2)^+ (r_1 - r_2) \, \varphi_{\epsilon_x} \, dx \, dx' \, dt \leq \epsilon_x \| r \|_{L^1(0,T;TV(\mathbb{R}^d))},$$

to get that

$$\| e^-(T) \|_{L^1(\mathbb{R}^d)} \leq 2\epsilon_x \| v_2 \|_{L^\infty(0,T;TV(\mathbb{R}^d))} + \frac{|\sqrt{\nu_1} - \sqrt{\nu_2}|}{\epsilon_x} \| v_2 \|_{L^1(0,T;TV(\mathbb{R}^d))}.$$

Minimizing with respect to the parameter ϵ_x, we finally get that

$$\| e(T) \|_{L^1(\mathbb{R}^d)} \leq C \, |\sqrt{\nu_1} - \sqrt{\nu_2}|,$$

where

$$C^2 = 8 \| v_2 \|_{L^\infty(0,T;TV(\mathbb{R}^d))} \| v_2 \|_{L^1(0,T;TV(\mathbb{R}^d))}.$$

This constant C can be bounded in terms of the initial data v_0 and in terms of the right-hand side r, by the property (3.10).

An immediate consequence of the above estimate is that the sequence $\{v_\nu\}_{\nu>0}$ of exact solutions of the parabolic initial value problem (3.1) and (3.2), is a Cauchy sequence in $\mathcal{C}^0(0,T;L^1(\mathbb{R}^d))$. Hence, it converges to a unique limit v^* that also belongs to $\mathcal{C}^0(0,T;L^1(\mathbb{R}^d))$. This limit is a *weak* solution of the following initial value problem:

$$v_t^* + \nabla \cdot \mathbf{f}(v^*) = r, \qquad \text{on } \mathbb{R}^d \times (0,T),$$
$$v^*(t=0) = v_0, \qquad \text{on } \mathbb{R}^d,$$

and is called the entropy solution. We will discuss this point in detail in the next section.

3.6.2. A New a Posteriori Error Estimate. Here, we are going to obtain an a posteriori error estimate of the difference between the exact solution v of (3.1) and (3.2), and an approximate solution u.

We assume that it is feasible to resolve the strong gradients of v numerically when the viscosity coefficient ν is not smaller than the value $\hat{\nu}$. In other words, we define an approximate solution u that tries to drive to zero the residual

$$u_t + \nabla_{x'} \cdot \mathbf{f}(u) - \hat{\nu} \Delta_{x'} u - r,$$

instead of the standard residual

$$u_t + \nabla_{x'} \cdot \mathbf{f}(u) - \nu \Delta_{x'} u - r.$$

Thus, to obtain our a posteriori error estimate, we write

$$u_t + \nabla_{x'} \cdot \mathbf{f}(u) - \hat{\nu} \Delta_{x'} u = r_2,$$

where r_2 is trivially defined by the left-hand side of the above equation, and set $v_1 = v$ and $v_2 = u$ in (3.13) to obtain

$$
\begin{aligned}
\| e \|_{1,\epsilon_x,\nu,\hat{\nu},T}^+ \quad \leq \quad & \| e^+(0) \|_{L^1(\mathbb{R}^d)} + 2\,\epsilon_x \, | u |_{L^\infty(0,T;TV(\mathbb{R}^d))} \\
& + \frac{(\sqrt{\nu} - \sqrt{\hat{\nu}})^2}{\epsilon_x} \, \| u \|_{L^1(0,T;TV(\mathbb{R}^d))} \\
& + \mathrm{RES}_{\epsilon_x,T}^+(v,u),
\end{aligned}
$$

where the *residual form* $\mathrm{RES}_{\epsilon_x,T}^+(v,u)$ is given by

$$\int_0^T \int_Q \mathrm{sign}(v-u)^+ \left(r - \left(u_t + \nabla_{x'} \cdot \mathbf{f}(u) - \hat{\nu} \Delta_{x'} u \right) \right) \varphi_{\epsilon_x} \, dx \, dx' \, dt. \quad (3.15)$$

Noting that

$$\mathrm{RES}_{\epsilon_x,T}^+(v,u) \leq \int_0^T \int_{\mathbb{R}^d} \left(r - u_t - \nabla_{x'} \cdot \mathbf{f}(u) + \hat{\nu} \Delta_{x'} u \right)^+ dx \, dt,$$

and minimizing with respect to ϵ_x, we get

$$
\begin{aligned}
\| e \|_{1,\epsilon_x,\nu,\hat{\nu},T}^+ \quad \leq \quad & \| e^+(0) \|_{L^1(\mathbb{R}^d)} + C(u) \, | \sqrt{\hat{\nu}} - \sqrt{\nu} | \\
& + \int_0^T \int_{\mathbb{R}^d} \left(r - u_t - \nabla_{x'} \cdot \mathbf{f}(u) + \hat{\nu} \Delta_{x'} u \right)^+ dx \, dt,
\end{aligned}
$$

where

$$C^2(u) = 8 \, | u |_{L^\infty(0,T;TV(\mathbb{R}^d))} \, | u |_{L^1(0,T;TV(\mathbb{R}^d))}.$$

We see that this a posteriori error estimate has three main terms in its right-hand side. The first measures the error made in the choice of the initial condition. the second measures the error made when a different diffusion coefficient is taken, and the last measures the residual that the approximate solution is trying to drive to zero.

The above a posteriori error estimate does hold when the diffusion coefficient ν tends to zero, as we wanted. However, one disadvantage of the above a posteriori error estimate is that the approximate solution u has to be *smooth* so that the quantity

$$r - u_t - \nabla_{x'} \cdot \mathbf{f}(u) + \hat{\nu} \Delta_{x'} u,$$

be integrable. This is certainly inconvenient in practical applications.

It is possible, however, to relax the regularity conditions on u with respect to the space variables thanks to the fact that the space variables have been doubled. Indeed, since $v = v(x,t)$ is not a function of x', we can rewrite

$$\text{sign}(v - u)^+ \left(r - u_t - \nabla_{x'} \cdot \mathbf{f}(u) + \hat{\nu} \Delta_{x'} u \right)$$

as

$$\text{sign}(v - u)^+ \left(r - u_t \right) + \nabla_{x'} \cdot \mathbf{f}^+(v, u) - \hat{\nu} \Delta_{x'} (v - u)^+ + \hat{\nu} \text{sign}'(v - u)^\top |\nabla_{x'} u|^2 .$$

Hence

$$
\begin{aligned}
\text{RES}^+_{\epsilon_x, T}(v, u) \;=\; & \int_0^T \!\! \int_Q \text{sign}(v - u)^+ \left(r - u_t \right) \varphi_{\epsilon_x} \, dx \, dx' \, dt \\
& - \int_0^T \!\! \int_Q F^+(v, u) \cdot \nabla_{x'} \varphi_{\epsilon_x} \, dx \, dx' \, dt \\
& - \hat{\nu} \int_0^T \!\! \int_Q (v - u)^+ \Delta_{x'} \varphi_{\epsilon_x} \, dx \, dx' \, dt \\
& + \hat{\nu} \int_0^T \!\! \int_Q \text{sign}'(v - u)^+ |\nabla_{x'} u|^2 \varphi_{\epsilon_x} \, dx \, dx' \, dt .
\end{aligned}
$$

We can see that it is no longer necessary to impose smoothness conditions on the Laplacian of u, only on its gradient. However, we still require smoothness of the time derivative of u. In the next section, we show how to relax this requirement.

3.7. Relaxing the Smoothness in Time of the Approximate Solution u

In the previous computation, it is clear that we could not relax the smoothness in time of u because we could not integrate by parts in time. We can do that if we double the time variables.

So, we evaluate v_1 at (x,t) and v_2 at (x',t') and instead of dealing with

$$\| e^+(\tau) \|_{L^1(\mathbb{R}^d)},$$

or with

$$\rho^+_{\epsilon_x}(v_1(t), v_2(t)) = \int_Q \left(v_1(x,t) - v_2(x',t) \right)^+ \varphi_{\epsilon_x}(x - x') \, dx \, dx',$$

we consider the quantity

$$
\begin{aligned}
\rho^+_{\epsilon_t, \epsilon_x}(v_1, v_2; \tau) = \; & \\
& \int_0^T \!\! \int_Q (v_1(x,\tau) - v_2(x',t'))^+ \varphi_{\epsilon_x}(x - x') \varphi_{\epsilon_t}(\tau - t') \, dx \, dx' \, dt' \\
& + \int_0^T \!\! \int_Q (v_1(x,t) - v_2(x',\tau))^+ \varphi_{\epsilon_x}(x - x') \varphi_{\epsilon_t}(t - \tau) \, dx \, dx' \, dt,
\end{aligned}
$$

in which

$$\varphi_{\epsilon_t}(z) = \frac{1}{\epsilon_t} \, \eta\left(\frac{z}{\epsilon_t}\right),$$

where η is an even, nonnegative function with support in $[-1, 1]$ and integral equal to one.

To obtain the new estimate, we proceed in several steps.

Step 1. We start as in the standard duality argument with the following trivial identity for $\rho^+_{\epsilon_t, \epsilon_x}(v_1, v_2; \tau)$:

$$\rho^+_{\epsilon_t, \epsilon_x}(v_1, v_2; T) = \rho^+_{\epsilon_t, \epsilon_x}(v_1, v_2; 0) + \int_0^T \frac{d}{d\tau} \rho^+_{\epsilon_t, \epsilon_x}(v_1, v_2; \tau)\, d\tau.$$

If we take into account that

$$\left(\frac{d}{dt} + \frac{d}{dt'}\right)\varphi_{\epsilon_t}(t - t') = 0,$$

we obtain that

$$\rho^+_{\epsilon_t, \epsilon_x}(v_1, v_2; T) = \rho^+_{\epsilon_t, \epsilon_x}(v_1, v_2; 0)$$
$$+ \int_0^T \int_0^T \int_Q \mathrm{sign}(v_1 - v_2)^+ \left((v_1)_t - (v_2)_{t'}\right)\varphi_{\epsilon_x}\,\varphi_{\epsilon_t}\, dx\, dx'\, dt\, dt'.$$

From this point, we proceed exactly as in the case in which only the space variables were doubled and easily obtain the following result:

$$\rho^+_{\epsilon_t, \epsilon_x}(v_1, v_2; T) + \mathbb{FLT}^+_{\epsilon_t, \epsilon_x, \nu_1, \nu_2, T}(v_1, v_2))$$
$$= \rho^+_{\epsilon_t, \epsilon_x}(v_1, v_2; 0) + \Xi^+_{\epsilon_t, \epsilon_x, \nu_1, \nu_2, T}(v_1, v_2)$$

where $\mathbb{FLT}^+_{\epsilon_t, \epsilon_x, \nu_1, \nu_2, T}(v_1, v_2)$ is given by

$$\int_0^T \int_0^T \int_Q \mathrm{sign}'(v_1 - v_2)^+ \, | \sqrt{\nu_1}\, \nabla_x v_1 - \sqrt{\nu_2}\, \nabla_{x'} v_2 |^2 \, \varphi_{\epsilon_x}\, \varphi_{\epsilon_t}\, dx\, dx'\, dt\, dt'$$

and $\Xi^+_{\epsilon_t, \epsilon_x, \nu_1, \nu_2, T}(v_1, v_2)$ is given by

$$\int_0^T \int_0^T \int_Q \mathrm{sign}(v_1 - v_2)^+ \, \nabla_{x'}(\mathbf{f}_1(v_2) - \mathbf{f}_2(v_2)) \cdot \nabla_{x'}\varphi_{\epsilon_x}\,\varphi_{\epsilon_t}\, dx\, dx'\, dt\, dt'$$
$$+ (\sqrt{\nu_1} - \sqrt{\nu_2})^2 \int_0^T \int_0^T \int_Q \mathrm{sign}(v_1 - v_2)^+ \, \nabla_{x'} v_2 \cdot \nabla_{x'}\varphi_{\epsilon_x}\,\varphi_{\epsilon_t}\, dx\, dx'\, dt\, dt'$$
$$+ \int_0^T \int_0^T \int_Q \mathrm{sign}(v_1 - v_2)^+ \, (r_1 - r_2)\, \varphi_{\epsilon_x}\,\varphi_{\epsilon_t}\, dx\, dx'\, dt\, dt'.$$

It only remains to relate the term $\rho^+_{\epsilon_t, \epsilon_x}(v_1, v_2; \tau)$ to the error $\| e^+(\tau) \|_{L^1(\mathbb{R}^d)}$ for $\tau = 0$ and $\tau = T$. In what follows, we present two ways of doing that.

Step 2. If we assume that both v_1 and v_2 are smooth, it can be proven that the term

$$\left| \rho^+_{\epsilon_t,\epsilon_x}(v_1, v_2; T)/\mathbb{N}(T) - \| e^+(T) \|_{L^1(\mathbb{R}^d)} \right|$$

is bounded by

$$\frac{1}{2} \left(| v_1(T) |_{TV(\mathbb{R}^d)} + | v_2(T) |_{TV(\mathbb{R}^d)} \right) \epsilon_x$$
$$+ \frac{1}{2} \left(\| (v_1)_t \|_{L^\infty(T-\epsilon_t, T; L^1(\mathbb{R}^d))} + \| (v_2)_{t'} \|_{L^\infty(T-\epsilon_t, T; L^1(\mathbb{R}^d))} \right) \epsilon_t,$$

and that the term

$$\left| \rho^-_{\epsilon_t,\epsilon_x}(v_1, v_2; 0)/\mathbb{N}(T) - \| e^+(0) \|_{L^1(\mathbb{R}^d)} \right|$$

is bounded by

$$\frac{1}{2} \left(| v_1(0) |_{TV(\mathbb{R}^d)} + | v_2(0) |_{TV(\mathbb{R}^d)} \right) \epsilon_x$$
$$+ \frac{1}{2} \left(\| (v_1)_t \|_{L^\infty(0,\epsilon_t; L^1(\mathbb{R}^d))} + \| (v_2)_{t'} \|_{L^\infty(0,\epsilon_t; L^1(\mathbb{R}^d))} \right) \epsilon_t,$$

where

$$\mathbb{N}(T) = 2 \int_0^{T/\epsilon_t} \eta(s) \, ds.$$

Note that for $T > \epsilon_t$, we have that $\mathbb{N}(T) = 1$.

Thus, we obtain the following result:

$$\| e \|^+_{1,\epsilon_t,\epsilon_x,\nu_1,\nu_2,T} \leq \| e^+(0) \|_{L^1(\mathbb{R}^d)}$$

$$+ \left(| v_1 |_{L^\infty(0,T;TV(\mathbb{R}^d))} + | v_2 |_{L^\infty(0,T;TV(\mathbb{R}^d))} \right) \epsilon_x$$

$$+ \left(\| (v_1)_t \|_{L^\infty(0,T;L^1(\mathbb{R}^d))} + \| (v_2)_{t'} \|_{L^\infty(0,T;L^1(\mathbb{R}^d))} \right) \epsilon_t, \tag{3.16}$$

$$+ \Xi^+_{\epsilon_t,\epsilon_x,\nu_1,\nu_2,T}(v_1, v_2)/\mathbb{N}(T).$$

where

$$\| e \|^-_{1,\epsilon_t,\epsilon_x,\nu_1,\nu_2,T} = \| e^+(T) \|_{L^1(\mathbb{R}^d)} + \mathbb{FLT}^+_{\epsilon_t,\epsilon_x,\nu_1,\nu_2,T}(v_1, v_2)/\mathbb{N}(T) \tag{3.17}$$

Step 3. If we assume that only v_1 is smooth, we have to proceed in a very different way which is described in detail in the appendix. The result is the following:

$$\lim_{\eta \to \frac{1}{2}\chi_{[-1,1]}} \| e \|^+_{1,\epsilon_t,\epsilon_x,\nu_1,\nu_2,T} \leq 2\| e^+(0) \|_{L^1(\mathbb{R}^d)}$$

$$+ 4 \lim_{\eta \to \frac{1}{2}\chi_{[-1,1]}} \Xi^+_{\epsilon_t,\epsilon_x,\nu_1,\nu_2,T}(v_1, v_2)/\mathbb{N}(T) \tag{3.18}$$

$$+ 4| v_1 |_{L^\infty(0,T;TV(\mathbb{R}^d))} \epsilon_x + 4\| (v_1)_t \|_{L^\infty(0,T;L^1(\mathbb{R}^d))} \epsilon_t.$$

Let us compare this result with the one obtained in the previous step. Note how the moduli of continuity of v_2, namely $|v_2|_{L^\infty(0,T;TV(\mathbb{R}^d))}$ and $\|(v_2)_{t'}\|_{L^\infty(0,T;L^1(\mathbb{R}^d))}$, are not present anymore, and that the small price we are paying for this is the presence of the factor 2 in the right-hand side and that we must let the auxiliary function η tend to half the characteristic function of the interval $[-1, 1]$, $\frac{1}{2}\chi_{[-1,1]}$.

3.8. The a Posteriori Error Estimate for Non-Smooth u

From the estimates obtained in the previous section, we can obtain the a posteriori error estimate sought. To do that, we simply set $v_1 = v$ and $v_2 = u$ in the estimate (3.16) or in the estimate (3.18). For example, if we choose the estimate (3.18), we have

$$\lim_{\eta \to \frac{1}{2}\chi(-1,1)} \| e \|^+_{1,\epsilon_t,\epsilon_x,\nu,\hat{\nu},T} \leq 2 \| e^+(0) \|_{L^1(\mathbb{R}^d)}$$

$$+ 4 | v_1 |_{L^\infty(0,T;TV(\mathbb{R}^d))} \, \epsilon_x$$

$$+ 4 \| (v_1)_t \|_{L^\infty(0,T;L^1(\mathbb{R}^d))} \, \epsilon_t,$$

$$+ 4 \frac{(\sqrt{\nu} - \sqrt{\hat{\nu}})^2}{\epsilon_x} \| u \|_{L^1(0,T;TV(\mathbb{R}^d))}$$

$$+ 4 \lim_{\eta \to \frac{1}{2}\chi(-1,1)} \mathrm{RES}_{\epsilon_t,\epsilon_x,T}(v,u)/\mathbb{N}(T).$$

where the residual form $\mathrm{RES}_{\epsilon_t,\epsilon_x,T}(v,u)$ can be expressed in two different ways. If the residual

$$R(u) = u_{t'} + \nabla_{x'} \cdot \mathbf{f}(u) - \hat{\nu}\Delta_{x'} u - r,$$

is an integrable function, we can write

$$\mathrm{RES}_{\epsilon_t,\epsilon_x,T}(v,u) = - \int_0^T \!\! \int_{\mathbb{R}} \Sigma(x',t') \, R(u(x',t')) \, dx' \, dt',$$

where

$$\Sigma(x',t') = \int_0^T \!\! \int_{\mathbb{R}} \mathrm{sign}(v(x,t) - u(x',t')) \, \varphi(x,t,x',t') \, dx \, dt,$$

and

$$\varphi(x,t,x',t') = \varphi_{\epsilon_x}(x - x')\varphi_{\epsilon_t}(t - t').$$

If we want to relax the smoothness of the approximate solution u, we write the residual form as follows:

$$\mathrm{RES}_{\epsilon_t,\epsilon_x,T}(v,u) = \int_0^T \!\! \int_{\mathbb{R}} \Theta(v,u;x,t) \, dx \, dt$$

where

$$\Theta(c, u; x, t) \;=\; \int_0^T \int_{\mathbb{R}} \text{sign}(c - u)^+ \, r\, \varphi \, dx' \, dt'$$

$$+ \int_0^T \int_{\mathbb{R}} \text{sign}(c - u)^+ \, \varphi_{t'} \, dx' \, dt'$$

$$- \int_{\mathbb{R}} \text{sign}(c - u(t' = T))^+ \, \varphi(t' = T) \, dx'$$

$$+ \int_{\mathbb{R}} \text{sign}(c - u(t' = 0))^+ \, \varphi(t' = 0) \, dx'$$

$$- \int_0^T \int_{\mathbb{R}} F^+(c, u) \cdot \nabla_{x'} \varphi \, dx' \, dt'$$

$$- \hat{\nu} \int_0^T \int_{\mathbb{R}} (c - u)^+ \Delta_{x'} \varphi \, dx' \, dt'$$

$$+ \hat{\nu} \int_0^T \int_{\mathbb{R}} \text{sign}'(c - u)^+ \, |\nabla_{x'} u|^2 \varphi \, dx' \, dt'.$$

This is the form of the a posteriori error estimate we were looking for.

3.9. Concluding Remarks

In this section, we have shown how to obtain continuous dependence results for the solution of the following initial value problem:

$$v_t + \nabla \cdot \mathbf{f}(v) - \nu \Delta v = r, \qquad \text{in } \mathbb{R}^d \times (0, T),$$
$$v(t = 0) = v_0 \qquad\qquad\quad \text{on } \mathbb{R}^d.$$

We have seen that the standard way to do that is through a *duality* argument. This requires the study of the solution of the so-called adjoint problem which is not a trivial matter. To avoid having to deal with the adjoint problem, a technique has been developed which, however, does not allow us to freely choose the norm in which we evaluate the error.

The classical example is the well known L^2-energy technique that has been widely used to analyze parabolic problems. Unfortunately, this technique does not treat the nonlinear convective terms in a convenient way. The use of L^1-like norms, however, naturally leads to a simple treatment of the nonlinear convective terms. To obtain continuous dependence results in these L^1-like norms that do not break down when the diffusion coefficient go to zero, we introduced the technique of the *doubling* of the space variables. Finally, to be able to compare the exact solution v with a non-smooth function u, we introduced the *doubling* of the time variables.

In this way, we have obtained a theory of continuous dependence results for solutions of the initial value problem for nonlinear convection-diffusion equations. This theory allows us to obtain (i) continuous dependence results with respect to the initial data, with respect to the nonlinearities, and with respect to the right-hand side, (ii) regularity results for the exact solution, and (iii) a posteriori error estimates that are independent of the numerical scheme used to compute the approximate solution u. Since these results do not break down when the diffusion coefficients go to zero, they trivially hold for the entropy solution, as we show in the next section.

4. Continuous Dependence for Nonlinear Convection

In this section, we study continuous dependence results for the entropy solution of the following initial-value problem:

$$v_t + \nabla \cdot \mathbf{f}(v) = r, \qquad \text{in } \mathbb{R}^d \times (0, T), \tag{4.1}$$
$$v(t = 0) = v_0, \qquad \text{on } \mathbb{R}^d. \tag{4.2}$$

First, we show how the entropy solution of the above problem *inherits* the continuous dependence results of the solutions of the parabolic problem (3.1) and (3.2). Then, we deduce from them regularity properties of the *entropy* solution and the a posteriori error estimate that we sought.

4.1. Existence and Uniqueness of the Entropy Solution

It is well known that for smooth enough data, the solution v_ν of the parabolic problem

$$(v_\nu)t + \nabla \cdot \mathbf{f}(v_\nu) - \nu \, \Delta v_\nu = r, \qquad \text{in } \mathbb{R}^d \times (0, T),$$
$$v_\nu(t = 0) = v_0, \qquad \text{on } \mathbb{R}^d,$$

belongs to $\mathcal{C}^0(0, T; L^1(\mathbb{R}^d))$. In the last section, we saw that there is a constant C independent of ν such that

$$\| v_{\nu_1} - v_{\nu_2} \|_{L^\infty(0,T;L^1(\mathbb{R}^d))} \leq C \, | \sqrt{\nu_1} - \sqrt{\nu_2} | \, .$$

As we said in the previous section, this means that the sequence $\{ v_\nu \}_{\nu>0}$ is a Cauchy sequence in $\mathcal{C}^0(0, T; L^1(\mathbb{R}^d))$ and so it converges to a *unique* limit that we denote by v. This limit belongs to $\mathcal{C}^0(0, T; L^1(\mathbb{R}^d))$ and is a *weak* solution of our initial value problem (4.1) and (4.2), as can easily be proven. This weak solution is the so-called entropy solution. Moreover, we immediately get that

$$\| v_\nu - v \|_{L^\infty(0,T;L^1(\mathbb{R}^d))} \leq C \sqrt{\nu} \, .$$

It is now clear that to obtain continuous dependence results for the entropy solutions, we only have to let one of the diffusion coefficients go to zero in the continuous dependence results obtained for the parabolic case.

4.2. The Inherited Continuous Dependence Results

We start by letting the diffusion coefficient ν to go to zero in the estimate (3.7), (3.8). We obtain

$$\| e \|_{1,T,0}^{+} = \| e^{+}(0) \|_{L^1(\mathbb{R}^d)} + \int_0^T \int_{\mathbb{R}^d} \operatorname{sign}(v_1 - v_2)^{+} (r_1 - r_2) \, dx \, dt, \quad (4.3)$$

in which

$$\| e \|_{1,T,0}^{+} = \| e^{+}(T) \|_{L^1(\mathbb{R}^d)} + \lim_{\nu \to 0} \nu \int_0^T \int_{\mathbb{R}^d} \nabla e \cdot \nabla \operatorname{sign}(e)^{+} \, dx \, dt, \quad (4.4)$$

where, of course. v_ℓ is the entropy solution of the initial value problem (4.1) and (4.2) with initial data $v_{0,\ell}$ and with right-hand side r_ℓ for $\ell = 1, 2$. As in the case of the solutions of the parabolic problem, from the above estimate, we can deduce a priori estimates for the entropy solution as well as a continuous dependence with respect to the initial data, the nonlinearity $\mathbf{f}$, and the right-hand side r.

4.2.1. A Priori Estimates of the Entropy Solution. It is a trivial exercise to see that we have

$$v(x, t) \in I(v_0, r) = [a, b], \quad (4.5)$$

where

$$
\begin{aligned}
a &= \inf_{x \in \mathbb{R}^d} v_0(x) - \| (-r)^{+} \|_{L^1(0,T;L^\infty(\mathbb{R}^d))}, \\
b &= \sup_{x \in \mathbb{R}^d} v_0(x) + \| r^{+} \|_{L^1(0,T;L^\infty(\mathbb{R}^d))},
\end{aligned}
$$

and

$$\| v \|_{L^\infty(0,T;TV(\mathbb{R}^d))} \le \| v_0 \|_{L^\infty(0,T;TV(\mathbb{R}^d))} + \| r \|_{L^1(0,T;TV(\mathbb{R}^d))}. \quad (4.6)$$

4.2.2. Continuous Dependence. Proceeding as in the previous section, we get

$$
\begin{aligned}
\| e \|_{1,T,0}^{+} \le\ & \| e^{+}(0) \|_{L^1(\mathbb{R}^d)} + \int_0^T \int_{\mathbb{R}^d} (r_1 - r_2)^{+} \, dx \, dt \\
& + C \| \mathbf{f}_1'(w) - \mathbf{f}_2'(w) \|_{L^\infty(I(v_{2,0}, r_2))},
\end{aligned}
$$

where

$$C = \| v_{2,0} \|_{L^\infty(0,T;TV(\mathbb{R}^d))} + \| r_2 \|_{L^1(0,T;TV(\mathbb{R}^d))}.$$

4.2.3. An a Posteriori Error Estimate for Smooth u. To obtain an a posteriori error estimate for smooth u, we take the limit as ν goes to zero to get

$$\lim_{\nu \to 0} \| e \|_{1,\epsilon_x,\nu,\hat{\nu},T}^+ \leq \| e^+(T) \|_{L^1(\mathbb{R}^d)} + C(u) \sqrt{\hat{\nu}}$$
$$+ \int_0^T \!\! \int_{\mathbb{R}^d} \left(r - u_t - \nabla_{x'} \cdot \mathbf{f}(u) + \hat{\nu} \Delta_{x'} u \right)^+ dx \, dt, \tag{4.7}$$

where

$$C^2(u) = 8 \, |u|_{L^\infty(0,T;TV(\mathbb{R}^d))} \, |u|_{L^1(0,T;TV(\mathbb{R}^d))}.$$

4.2.4. An a Posteriori Error Estimate for Non-Smooth u. When the approximate solution u is not smooth, we proceed in a similar way. For example, if we choose to pass to the limit in the estimate (3.16), we obtain

$$\lim_{\nu,\hat{\nu} \to 0} \| e \|_{1,\epsilon_t,\epsilon_x,\nu,\hat{\nu},T}^+ \leq \| e^+(0) \|_{L^1(\mathbb{R}^d)}$$

$$+ \left(|v|_{L^\infty(0,T;TV(\mathbb{R}^d))} + |u|_{L^\infty(0,T;TV(\mathbb{R}^d))} \right) \epsilon_x \tag{4.8}$$

$$+ \left(\| v_t \|_{L^\infty(0,T;L^1(\mathbb{R}^d))} + \| u_{t'} \|_{L^\infty(0,T;L^1(\mathbb{R}^d))} \right) \epsilon_t,$$

$$+ \mathrm{RES}_{\epsilon_t,\epsilon_x,T}(v,u)/\mathbb{N}(T),$$

and if we choose to pass to the limit in the estimate (3.18), we obtain

$$\lim_{\nu,\hat{\nu} \to 0} \lim_{\eta \to \frac{1}{2}\chi(-1,1)} \| e \|_{1,\epsilon_t,\epsilon_x,\nu,\hat{\nu},T}^+ \leq 2\| e^+(0) \|_{L^1(\mathbb{R}^d)}$$

$$+ 4 \, |v|_{L^\infty(0,T;TV(\mathbb{R}^d))} \, \epsilon_x + 4 \, \| (v)_t \|_{L^\infty(0,T;L^1(\mathbb{R}^d))} \, \epsilon_t, \tag{4.9}$$

$$+ 4 \lim_{\eta \to \frac{1}{2}\chi(-1,1)} \mathrm{RES}_{\epsilon_t,\epsilon_x,T}(v,u)/\mathbb{N}(T).$$

where the residual form $\mathrm{RES}_{\epsilon_t,\epsilon_x,T}(v,u)$ is given by

$$\mathrm{RES}_{\epsilon_t,\epsilon_x,T}(v,u) = \int_0^T \!\! \int_{\mathbb{R}^d} \Theta(v,u;x,t) \, dx \, dt$$

in which

$$\Theta(c,u;x,t\ ) = \int_0^T \!\! \int_{\mathbb{R}^d} \mathrm{sign}(c-u)^+ \, r \, \varphi \, dx' \, dt'$$

$$+ \int_0^T \!\! \int_{\mathbb{R}^d} \mathrm{sign}(c-u)^+ \, \varphi_{t'} \, dx' \, dt'$$

$$- \int_{\mathbb{R}^d} \mathrm{sign}(c-u(t'=T))^+ \, \varphi(t'=T) \, dx'$$

$$+ \int_{\mathbb{R}^d} \mathrm{sign}(c-u(t'=0))^+ \, \varphi(t'=0) \, dx'$$

$$- \int_0^T \!\! \int_{\mathbb{R}^d} F^+(c,u) \cdot \nabla_{x'} \varphi \, dx' \, dt'.$$

4.3. Concluding Remarks

We have thus shown that the entropy solution inherits the continuous dependence results we obtained for the parabolic solutions in the previous section. We are thus ready to obtain a posteriori error estimates for the entropy solution.

5. A Posteriori Error Estimates for Continuous Approximations

In this section, we apply the a posteriori error estimates obtained in the previous section to approximate solutions u that are taken to be continuous. Then, we perform several numerical experiments with the Engquist-Osher scheme and conclude that the estimate is particularly good when (i) the solution is smooth, even in the nonlinear case, and when (ii) the convective term is linear, even in the presence of discontinuities. We also conclude that when discontinuities are present and $\mathbf{f}$ is strictly nonlinear, the estimate is not useful. We end this section by providing an explanation of this phenomenon and by proposing a new error estimate that does not have this unwanted property.

5.1. The Error Estimate

From the error estimate (4.7), setting $\hat{\nu} = 0$, we easily get the following result.

Theorem 5.1. *Let v be the entropy solution and let u be a continuous approximation. Then,*

$$\| u(T) - v(T) \|_{L^1(\mathbb{R}^d)} \leq \Phi(v_0, u; T),$$

where

$$\Phi(v_0, u; T) = \quad \| u(0) - v_0 \|_{L^1(\mathbb{R}^d)} + \| Residual(u) \|_{L^1((0,T)\times\mathbb{R}^d)}$$

where

$$Residual(u) = u_t + \nabla \cdot f(u), \quad on \ \Omega(T) := (0,T) \times \mathbb{R}^d. \qquad (5.1)$$

If the residual on those smooth solutions is zero — that is, if the approximate solution u is also the entropy solution of the initial value problem (4.1) and (4.2) — then Theorem (3.7) gives the familiar L^1-contraction property of the entropy solutions, namely,

$$\| u(T) - v(T) \|_{L^1(\mathbb{R}^d)} \leq \| u(0) - v_0 \|_{L^1(\mathbb{R}^d)},$$

as expected.

Now, let us apply Theorem 5.1. Suppose that we have a numerical scheme that produces an approximate solution u_h at times t^n, $n = 1, \ldots, N$. Let us denote by $\mathcal{U}$ the set of all the *interpolates* u such that $u(t^n) = u_h(t^n)$ for $n = 1, \ldots, N$. Then a simple application of Theorem 5.1 gives the following a posteriori error estimate:

$$\| u_h(t^n) - v(t^n) \|_{L^1(\mathbb{R}^d)} \leq \inf_{u \in \mathcal{U}} \Phi(v_0, u; t^n), \quad n = 1, \ldots, N. \tag{5.2}$$

In this way, the problem of a posteriori error estimation becomes an interpolation problem. When a front-tracking method is used, the approximate solution is always taken to be smooth except at a finite number of surfaces which contain the discontinuities; this is why the above estimate is very well suited for these methods. Numerical experiments for front-tracking methods have not yet been performed. However, applications of the above estimate to the Engquist-Osher scheme are available. We shall now review those results.

5.2. Application to the Engquist-Osher Scheme

Next, we apply the a posteriori error estimate to the well known Engquist-Osher scheme with the purpose of finding out how good the estimate is.

Thus, we pick $u^\star$ to be a member of $\mathcal{U}$ and study the behavior of the ratio

$$r(u^\star, t^N) = \frac{\Phi(v_0, u^\star; t^N)}{\| u_h(t^N) - v(t^N) \|_{L^1(\mathbb{Z}^d)}},$$

as the discretization parameters go to zero. Of course, the optimal result is when this ratio is equal to unity uniformly with respect to the discretization parameters. In what follows, we present some numerical results that show that this ideal situation can be achieved.

We consider the approximate solution given by the well known Engquist-Osher scheme on uniform grids $\{t^n = n \, \Delta t\}_{n=0}^{N_t} \times \{x_i = i \, \Delta x\}_{i=0}^{N_x}$ defined on the interval $[0, 1)$ with periodic boundary conditions. We use two ways of interpolating the values u_i^n generated by the Engquist-Osher scheme. We denote by $u_1^\star$ the standard piecewise-bilinear function such that $u_1^\star(t^n, x_i) = u_i^n$. We also define a more sophisticated interpolation function $u_2^\star$ which takes into account the nature of the conservation law. We define $u_2^\star$ as follows. First, we take $u_2^\star(n \, \Delta t)$ to be the piecewise-linear function such that $u_2^\star(t^n, x_i) = u_i^n$. We extend $u_2^\star$ to the strips $(n \, \Delta t, (n + 1) \, \Delta t) \times [0, 1)$ as follows. We consider all the trapezoids formed by the lines $t = t^n$, $t = t^{n+1}$ and the approximate characteristics $x = x_i + f'(u(t^n, x_i)) \, (t - t^n)$ and $x = x_i + f'(u(t^{n-1}, x_i)) \, (t - t^{n+1})$. We then define $u_2^\star$ inside of each of these trapezoids as the standard bilinear isoparametric interpolate of the values of u at the vertices. This construction is well defined for the examples considered below.

All the integrals are computed by using the mid-point rule.

Example 1. We take $f(v) = v$, $v_0(x) = 1.0 + 0.5 \sin(\pi(2x - 1))$, and $T = 0.1$. The behavior of the ratio $r(u^\star, T)$ is displayed in Tables 1 and 2.

We can see that with the simple interpolant $u_1^\star$, the ratio $r(u_1^\star, T)$ is not bigger that 1.3 and that with the more sophisticated interpolant $u_2^\star$, the ratio $r(u_2^\star, T)$ differs very little from 1. These results confirm that the error estimate is sharp and that even with a very simple interpolation procedure very good results can be obtained.

Example 2. We take $f(v) = v^2/2$, $v_0(x) = 1.0 + 0.5\sin(\pi(2x - 1))$, and $T = 0.1$. The results obtained for the previous example also hold in this case, as we can see in Tables 3 and 4. (Note that ratios less than 1 might be obtained because we are evaluating the integrals in an approximate way.)

Example 3. We take $f(v) = v$, $T = 1/\sqrt{2}$, and

$$v_0(x) = \begin{cases} 1, & \text{for } x \in (0.4, 0.6), \\ 0, & \text{elsewhere.} \end{cases}$$

The results are displayed in Tables 5 and 6. We see, once more, that the ratio $r(u^\star, T)$ is close to 1 and that, unlike the previous cases, the results with $u_2^\star$ are only slightly better than the results obtained with $u_1^\star$. This might be explained by the fact that (i) the 'smooth region' of the exact solution is flat and so the contribution of the residual in that region is negligible, and by the fact that (ii) the approximate solution is very smooth around the location of the exact discontinuity.

Example 4. We take $f(v) = v^2/2$, $v_0(x) = 1.0 + 0.5\sin(\pi(2x - 1))$, and $T = 0.4$. In this case, the solution does have a single discontinuity and, unfortunately, the ratio blows up as the discretization parameters go to zero!

5.3. Explaining the Numerical Results

It is not difficult to explain the above numerical results with the help of the estimates obtained for the parabolic case. Indeed, the error estimate of Theorem 5.1 is nothing but the limit as the diffusion coefficient ν goes to zero of the first error estimate (3.7) we obtained for the parabolic regularization. To see this, let us recall such estimate:

$$\| e \|_{1,T,\nu} = \| e(0) \|_{L^1(\mathbb{R})} + \int_0^T \int_{\mathbb{R}} \text{sign}(v_1 - v_2)(r_1 - r_2)\, dx\, dt,$$

where

$$\| e \|_{1,T,\nu} = \| e(T) \|_{L^1(\mathbb{R})} + \nu \int_0^T \int_{\mathbb{R}} \nabla e \cdot \nabla \text{sign}(e)\, dx\, dt.$$

If we set

$$r_1 = 0, \qquad r_2 = Residual(u),$$

and note that

$$\| e \|_{1,T,\nu} \geq \| e(T) \|_{L^1(\mathbb{R})},$$

we immediately obtain the estimate of Theorem 5.1.

If we want to study the sharpness of this estimate, it is obvious that we must consider what happens in the limit as ν goes to zero to the quantity

$$\Theta(\nu) = \nu \int_0^T \int_{\mathbb{R}} \nabla e \cdot \nabla \mathrm{sign}(e) \, dx \, dt,$$

where, if we denote by v_ν the exact solution of the parabolic regularization,

$$e = v_\nu - u.$$

It can be seen immediately that when the exact solution v_ν remains smooth as the diffusion coefficient goes to zero, we get

$$\lim_{\nu \downarrow 0} \Theta(\nu) = 0.$$

This explains why we got sharp results in such cases.

In the case in which the entropy solution has discontinuities, we still get that

$$\lim_{\nu \downarrow 0} \Theta(\nu) = 0,$$

in the linear case $\mathbf{f}' \equiv constant$, since in this case,

$$\| \nabla v_\nu \| = O(1/\sqrt{\nu}).$$

To understand the nonlinear case is more delicate. To gain insight into this situation, let us place ourselves in the simple case of the traveling waves that we constructed when studying the traffic flow problem. In this case, we have that

$$v_\nu(x,t) = \phi \left(\frac{x - ct}{\nu} \right),$$

where

$$c = \frac{f(v^+) - f(v^-)}{v^+ - v^-}.$$

and where ϕ was the solution of the following ordinary differential equation:

$$\phi' = f(\phi) - \{ f(v^+) - c(v^+ - \phi) \}.$$

Now, let us *assume* that for each value of the diffusion coefficient ν, the exact solution v_ν and the approximate solution u intersect only on a single curve $(x(t), t)$. Then, a simple *formal* computation gives us that

$$\Theta(\nu) = 2 \int_0^T \nu \, | \, v_{\nu,x}(x(t), t) - u_x(x(t), t) \, | \, dt.$$

Hence,

$$\lim_{\nu \downarrow 0} \Theta(\nu) = 2 \int_0^T | \lim_{\nu \downarrow 0} \nu \, v_\nu \, |(x(t), t) \, dt.$$

But, since v_ν is a traveling wave, we have that

$$\nu\, v_{\nu,x} = \phi',$$

and, noting that on the curve $(x(t), t)$ we have

$$v_{\nu}(x(t), t) = u(x(t), t),$$

we easily obtain that

$$\lim_{\nu \downarrow 0} \Theta(\nu) = 2 \int_0^T |\, f(u) - \{\, f(v^+) - c\,(v^+ - u)\,\}\,|(x(t), t)\, dt.$$

Note that we have assumed that $u(x(t), t)$ lies between $v^+ = v(x(t)+0, t)$ and $v^- = v(x(t) - 0, t)$ and hence, the above limit is different from zero (except. of course, if u coincides with v^+ or with v^- on the discontinuity curve).

If we believe that what happened with the traveling wave solution also happens in the general case, this provides the explanation we were seeking since the value of the approximate solution u given by the Engquist-Osher scheme at the discontinuity of the exact solution always lies between the limits v^- and v^-.

5.3.1. Conclusions. From the above discussion, we can draw several conclusions. The first is that to understand the hyperbolic conservation law case, it is crucial to understand the parabolic case: it was after the examination of the parabolic case that we identified the term

$$\Theta(\nu) = \nu \int_0^T \int_{\mathbb{R}} \nabla e \cdot \nabla \mathrm{sign}(e)\, dx\, dt.$$

The second conclusion is that the above term is zero when (i) the exact solution is smooth, and when (ii) the convection is linear; in these cases, the above a posteriori error estimate seems to be sharp for the Engquist-Osher scheme. Otherwise, this term *has to be added* to the L^1-norm and must be part of the norm in which we measure the error.

Finally, our analytical computation for the case of the traveling waves, which gives that

$$\lim_{\nu \downarrow 0} \Theta(\nu) = 2 \int_0^T |\, f(u) - \{\, f(v^+) - c\,(v^+ - u)\,\}\,|(x(t), t)\, dt,$$

suggests that the above term measures how close $u(x(t), t)$ is to being *outside* the discontinuity of the exact solution. Indeed, if the value $u(x(t), t)$ is never inside the *open* interval determined by $v^+ = v(x(t) + 0, t)$ and $v^- = v(x(t) - 0, t)$, the above term is identically zero. On the other hand, if $u(x(t), t)$ is always, say, in the middle of said interval, the term

$$\lim_{\nu \downarrow 0} \Theta(\nu)$$

will never be zero!

5.4. Another Error Estimate

This unfortunate fact renders impractical the use of the a posteriori error we have considered so far. To try to remedy this situation, we could try to apply the error estimate (4.7), this time with $\hat{\nu} \neq 0$; we easily get the following result.

Theorem 5.2. *Let v be the entropy solution and let u be a smooth approximation. Then,*

$$\| u(T) - v(T) \|_{L^1(\mathbb{R}^d)} \leq \Phi(v_0, u; T),$$

where

$$\Phi(v_0, u; T) = \| u(0) - v_0 \|_{L^1(\mathbb{R}^d)} + \| Residual(u) \|_{L^1((0,T) \times \mathbb{R}^d)} + C(u) \sqrt{\hat{\nu}},$$

where

$$Residual(u) = u_t + \nabla \cdot f(u) - \hat{\nu} \, \Delta u \quad on \ \Omega(T) = (0, T) \times \mathbb{R}^d, \qquad (5.3)$$

and

$$C^2(u) = 8 \, | \, u \, |_{L^\infty(0,T;TV(\mathbb{R}))} \, | \, u \, |_{L^1(0,T;TV(\mathbb{R}))}.$$

Note that in the above result, the quantity $Residual(u)$ involves the term

$$\hat{\nu} \, \Delta u.$$

Accordingly, the numerical approximation should then be devised to make this *new* residual tend to zero. The diffusion parameter $\hat{\nu}$ is nothing but a simple case of the so-called *artificial viscosity* coefficient.

The exploration of this idea constitutes the subject of ongoing research.

Table 1
Example 1: Behavior of $r(u_1^*, T)$ with respect to Δx and $\lambda = \Delta t / \Delta x$.

$1/\Delta x$	$\lambda = .25$	$\lambda = .50$	$\lambda = .75$
10	1.149	1.157	1.146
20	1.184	1.180	1.155
40	1.206	1.202	1.183
80	1.220	1.216	1.204
160	1.227	1.225	1.218
320	1.230	1.229	1.226
640	1.232	1.231	1.230

Table 2
Example 1: Behavior of $r(u_2^*, T)$ with respect to Δx and $\lambda = \Delta t / \Delta x$.

$1/\Delta x$	$\lambda = .25$	$\lambda = .50$	$\lambda = .75$
10	1.02357	1.03430	1.04607
20	1.00642	1.00997	1.00633
40	1.00169	1.00275	1.00225
80	1.00044	1.00073	1.00043
160	1.00011	1.00019	1.00013
320	1.00003	1.00005	1.00004
640	1.00002	1.00002	1.00003

Table 3
Example 2: Behavior of $r(u_1^*, T)$ with respect to Δx and $\lambda = 1.5\Delta t / \Delta x$.

$1/\Delta x$	$\lambda = .25$	$\lambda = .50$	$\lambda = .75$
10	1.177	1.161	1.165
20	1.197	1.187	1.175
40	1.232	1.218	1.197
80	1.249	1.238	1.215
160	1.257	1.249	1.225
320	1.262	1.253	1.231
640	1.264	1.257	1.234

Table 4
Example 2: Behavior of $r(u_2^*, T)$ with respect to Δx and $\lambda = 1.5\Delta t / \Delta x$.

$1/\Delta x$	$\lambda = .25$	$\lambda = .50$	$\lambda = .75$
10	1.04012	1.03034	1.0504
20	1.00009	1.00025	1.01097
40	1.00071	0.99700	1.00065
80	1.00055	0.99875	0.99878
160	0.99947	0.99931	0.99895
320	1.00014	0.99968	0.99935
640	1.00019	0.99990	0.99964

Table 5

Example 3: Behavior of $r(u_1^*, T)$ with respect to Δx and $\lambda = \Delta t / \Delta x$.

$1/\Delta x$	$\lambda = .25$	$\lambda = .50$	$\lambda = .75$
10	1.385	1.347	1.361
20	1.233	1.212	1.248
40	1.155	1.149	1.186
80	1.116	1.112	1.135
160	1.087	1.081	1.096
320	1.064	1.059	1.069
640	1.045	1.042	1.049

Table 6

Example 3: Behavior of $r(u_2^*, T)$ with respect to Δx and $\lambda = \Delta t / \Delta x$.

$1/\Delta x$	$\lambda = .25$	$\lambda = .50$	$\lambda = .75$
10	1.301	1.265	1.287
20	1.159	1.141	1.184
40	1.088	1.085	1.132
80	1.057	1.059	1.094
160	1.038	1.040	1.066
320	1.028	1.029	1.048
640	1.020	1.021	1.034

6. A Posteriori Error Estimates for Discontinuous Approximations

In this section, we apply the a posteriori error estimates obtained in Section 4 to approximate solutions u that are continuous. First, we consider the case in which the approximate solution has a finite number of discontinuities lying on smooth curves. Then, we consider the case in which the approximate solution is piecewise-constant. We end with some concluding remarks.

6.1. The Case of a Finite Number of Smooth Discontinuity Curves

In this section, we assume that

$$u \in C^0(0, T; L^1(\mathbb{R}^d)),$$

is a smooth function, except on a finite number of $(d-1)$-dimensional surfaces $C_i, i = 1, \ldots, I$ and develop an a posteriori error estimate for this type of approximate solution. To state the result, we need to introduce some notation.

We denote by n_{C_i} the unit normal to C_i such that its component on the t-axis is negative and define the jump of $G(u)$ at the point P on surface C_i. $[G(u)](P)$, to be the following quantity:

$$[G(u)](P) = \lim_{\rho \downarrow 0} \left(G(u(P + \rho\, n_{C_i})) - G(u(P - \rho\, n_{C_i})) \right).$$

Finally, we set

$$C_i(T) = (0,T) \times \mathbb{R}^d \cap C_i,$$
$$C(T) = \cup_{i=1}^I C_i(T),$$
$$\Omega(T) = (0,T) \times \mathbb{R}^d \setminus C(T).$$

With this notation, we have the following approximation result.

Theorem 6.1. *Let u be as above and let v be the entropy solution. Then.*

$$\| u(T) - v(T) \|_{L^1(\mathbb{R}^d)} \leq \Phi(v_0, u; T),$$

where

$$\Phi(v_0, u; T) = \quad \| u(0) - v_0 \|_{L^1(\mathbb{R}^d)} + \| Residual(u) \|_{L^1(\Omega(T))}$$
$$+ \sum_{i=1}^I \| Residual(u) \|_{L^1(C_i(T))},$$

where

$$Residual(u) = u_t + \nabla \cdot f(u), \quad on\ \Omega(T), \tag{6.1}$$

and

$$Residual(u) = \max \left\{ 0, \sup_{c \in [a,b]} \left\{ ([F(u,c)], [U(u-c)]) \cdot n_{C_i} \right\} \right\} \quad on\ C_i(T). \tag{6.2}$$

and $U(u-c) = |u-c|$, $F(u,c) = \text{sign}(u-c)\big(f(u) - f(c)\big)$. The interval $[a,b]$ is the range of the initial data v_0.

This result follows from setting ν and $\hat{\nu}$ equal to zero in the estimate (4.8), making a couple of integrations by parts in the term

$$\text{RES}_{\epsilon_t, \epsilon_x, T}(v, u),$$

and then letting the parameters ϵ_t and ϵ_x go to zero.

It is clear that the term $\| Residual(u) \|_{L^1(\Omega(T))}$ is nothing but the truncation error on the smooth regions of the approximation defined by u. The interpretation of the term $\| Residual(u) \|_{L^1(C_i(T))}$, however, requires a more detailed discussion: it measures how close to satisfying the entropy condition are the discontinuities of u on the curve $C_i(T)$.

To see this, let us consider the one-dimensional case, $d = 1$, for the sake of simplicity. If we write $C_i(T) = \{(x,t), x = x_i(t), 0 \leq t_{i1} \leq t \leq t_{i2} \leq T\}$, we have that if

$$([F(u,c)] - \frac{d\,x_i}{dt}\,[U(u-c)])(s, x_i(s)) \leq 0 \quad \text{for } s \in (0,T),$$

for all values of the real constant c, then the discontinuities of u on the curve $C_i(T)$ are entropy-satisfying discontinuities. In this case, we have that

$$
\begin{aligned}
\Theta_i &= \| Residual(u) \|_{L^1(C_i(T))} \\
&= \int_{t_{i1}}^{t_{i2}} \max\left\{0, \sup_{c\in[a,b]} \left\{([F(u,c)] - \frac{d\,x_i}{dt}\,[U(u-c)])(s, x_i(s))\right\}\right\} ds \\
&\leq 0.
\end{aligned}
$$

This means that if all the discontinuity curves of the approximate solution u are entropy-satisfying discontinuities, the Theorem 3.7 becomes:

$$\| u(T) - v(T) \|_{L^1(\mathbb{R}^d)} \leq \| u(0) - v_0 \|_{L^1(\mathbb{R}^d)} + \| Residual(u) \|_{L^1(\Omega(T))}.$$

In other words, we only have to worry about the size of the residual on the smooth regions of the approximate solution.

If the residual on those smooth solutions is zero, that is, if the approximate solution u is also the entropy solution of the initial value problem (4.1) and (4.2), then the theorem above gives the familiar L^1-contraction property of the entropy solutions, namely,

$$\| u(T) - v(T) \|_{L^1(\mathbb{R}^d)} \leq \| u(0) - v_0 \|_{L^1(\mathbb{R}^d)}.$$

as expected.

When a front-tracking method is used, the approximate solution is always taken to be smooth except at a finite number of surfaces which contain the discontinuities; this is why the above estimate is very well suited for these methods. Numerical experiments for front-tracking methods remain to be done. However, we do expect to have problems similar to the ones we encountered in the last section.

6.2. The Case of a Piecewise-Constant Approximation

To try to remedy this situation, we apply the estimate (4.9), but this time we work out the term

$$\mathbb{RES}_{\epsilon_t, \epsilon_x, T}(v, u),$$

in a very different way. Without entering into technical details, let us just say that this time, since the approximate solution is piecewise-constant, we cannot define the term $Residual(u)$ as before; instead, we obtain a *weaker* version of this quantity which can be expressed solely in terms of the discontinuities of the fluxes. To be able to do that, the auxiliary function φ must *absorb a derivative*, so to speak, and, as a consequence, we obtain an estimate of the form

$$\mathbb{RES}_{\epsilon_t,\epsilon_x,T}(v,u) \le C_1/\epsilon_t + C_2/\epsilon_x.$$

As we can see, the parameters ϵ_t and ϵ_x cannot longer be set to zero; instead, the optimal choice for them turns out to be the following:

$$\epsilon_t = O(\sqrt{C_1}), \qquad \epsilon_x = O(\sqrt{C_2}).$$

We carry out this idea for the Engquist-Osher method defined in general grids. Since the approximate solution is discontinuous, we allow ourselves to use different triangulations in different time intervals. Thus, for each partition of the interval $[0, T]$, $\{t_n\}_{0 \le n \le N}$, we define

$$S_n = \mathbb{R}^d \times (t_n, t_{n+1}), \quad n = 0, \dots, N-1.$$

We set $\Delta t_n = t_{n+1} - t_n$. For each value of n, we define a triangulation $T_{h,n}$ of simplices. No compatibility at $t = t_n$ between the meshes of two consecutive slabs S_n and S_{n+1}, $n = 0, \dots, N-2$, is required. We denote by T_e the simplex that shares the face e with the simplex T and write $n_{e,T}$ to denote the unit outer normal to T along its face e. Finally, we denote by h_T the diameter of the element T.

We assume that the approximate solution u has the following form:

$$u(x,t) = u^{T_n} \qquad \forall (x,t) \in T \times \in T_{h,n},$$

and determine these values by using the Engquist-Osher scheme. For $t = t_n$, we let $[u](x, t_n) := u(x, t_n + 0) - u(x, t_n - 0)$.

For this scheme, we have the following a posteriori error estimate.

Theorem 6.2. *Let u be as above and let v be the entropy solution. Then, we have*

$$\|v(T) - u(T)\|_{L^1(\mathbb{R}^d)} \le \Phi(v_0, u; T),$$

where

$$\Phi(v_0, u; T) = 2\|v_0 - u_0\|_{L^1(\mathbb{R})} + C\{|u_0|_{TV(\mathbb{R})} \Theta_0(u)\}^{1/2},$$

where C is a numerical constant, and

$$
\begin{aligned}
\Theta_0(u) &= \sum_{n=0}^{N_\tau - 1} \sum_{T \in T_{h,n}} h_T \sum_{e \in \partial T} \int_e |(f_{e,T}^{EO}(u^T, u^{T_e}) - \tilde{f}(u^T)) \cdot n_{e,T}| \, d\lambda \\
&\quad + \sum_{n=0}^{N_\tau - 1} \Delta t_n \int_{\mathbb{R}^d} |[u]| \, dx.
\end{aligned}
$$

When all the triangulations coincide, it has been proven that the nonlinear functional $\Phi(v_0, u; T)$ goes to zero as the discretization parameters go to zero. As a consequence, the problem we had with the previous error estimate is not present anymore. The question of the sharpness of the estimate, however, remains an open question. Let us just say that the norm in which we measure the error is *not* the L^1-norm but the norm

$$\| e \, \|_{1,\epsilon_t,\epsilon_x,0,0,T}^{+},$$

where the parameters ϵ_t and ϵ_x have been chosen as above! This question remains to be addressed.

7. Concluding Remarks

7.1. Some Bibliographical Remarks

The modeling of the traffic flow in Section 2 was extracted from Witham's book [20]. I could not find any reference for the discussion on traveling waves for the traffic flow equations, although this seems to be unbelievable.

The main part of these notes in contained in Section 3. This section is based on the paper [8], devoted to continuous dependence for nonlinear, degenerate convection-diffusion equations, and on the paper [9], devoted to a posteriori error estimation of convection-diffusion equations.

The whole paper [6] is contained in Section 5 and in the first section of Section 6. The explanation of the numerical results of Section 5 is new and will most probably appear in [9]. The second part of Section 6 is contained in [7] which is devoted to a posteriori error estimates for the discontinuous Galerkin and the streamline-diffusion methods.

7.2. Open Problems

Let us end these notes by pointing out two important problems that could be of interest to researchers in this field.

The first is how to use *already existing* error estimates to define adaptive strategies. Although there have been many a posteriori error estimates for the scalar nonlinear conservation law, the number of papers (see [12]) in which they have been used for adaptivity purposes is unbelievably small. Examples of papers containing these estimates are [13], [14], [5], [19], [6], and [7] among others. The study of the *sharpness* of these estimates and the application of these estimates to define adaptive strategies would be very interesting and has not been done.

The second problem is related to the so-called adjoint equation. In these notes, the a posteriori error estimates we have considered were obtained *without* solving an adjoint equation. As we have seen, the price we had to pay was to loose control on the norm we measured the error. On the other hand, there are methods for which the adjoint equation is solved. For example, in [17], a complete analysis of the solution of the adjoint equation is done for the one-dimensional case and strictly convex nonlinearities f; unfortunately, a more general case cannot be analyzed with that technique. In [11] the case of systems is treated, but the result holds only for a parabolic perturbation of the system of conservation laws and for the streamline-diffusion method. The papers [3], [4] and [10] show how to solve the adjoint equation numerically and how to use this information in the adaptation. This method gives very impressive results and can be applied to quite general problems. However, it seems to be computationally very intensive and restricted to Galerkin methods.

The question is now if there is a *new* technique that would give us more control on the norm in which we measure the error, can be applied to general situations, and is not computationally intensive.

Bibliography

1. R. Abeyaratne and J.K Knowles, *Implications of viscosity and strain gradient effects for the kinetics of propagating phase boundaries in solids*, SIAM J. on Appl. Math. **51** *(1991), 1205–1221.*
2. M. Affouf and R. Caflish, A numerical study of riemann problem solutions and stability for a system of viscous conservation laws of mixed type, SIAM J. Appl. Math. (1991), 605–634.
3. R. Becker and R. Rannacher, *A feed-back approach to error control in finite element methods: Basic analysis and examples*, East-West J. Numer. Math., **4** (1996),237–264.
4. K. Bottcher and R. Rannacher, *Adaptive error control in solving ordinary differential equations by the discontinuous Galerkin method*, Preprint University of Heidelberg, 1996.
5. B. Cockburn, F. Coquel, and P LeFloch *An error estimate for finite-volume methods for conservations laws*, Math. Comp. **63** (1994), 77–103.
6. B. Cockburn and H. Gau, *A posteriori error estimates for general numerical methods for scalar conservation laws*, Mat. Aplic. Comp., **14** (1995), 37–47.
7. B. Cockburn and P.-A. Gremaud, *Error estimates for finite element methods for scalar conservation laws*, SIAM J. Numer. Anal., **33** (1996), 522–554.
8. B. Cockburn and G. Gripenberg, *Continuous dependence on the nonlinearities of solutions of degenerate parabolic equations*, to appear in J. Diff. Equa.
9. B. Cockburn and J. X. Yang, *A posteriori error estimates for nonlinear convection-diffusion equations*, in preparation.
10. R. Furer F.-K. Hebeker and R. Rannacher, *An adaptive finite element method for unsteady convection-dominated flows with stiff source terms*, Preprint University of Heidelberg, 1996.
11. C. Johnson and A. Szepessy, *Adaptive finite element methods for conservation laws based on a posteriori estimates*, Comm. Pure Appl. Math., **48** (1995) 199–243.
12. B. J. Lucier, *A stable adaptive scheme for hyperbolic conservation laws*, SIAM J. Numer. Anal., **22** (1985), 180–203.

13. B. Lucier, *A moving mesh numerical method for hyperbolic conservation laws*, Math. Comp., **46** (1986), 59–69.
14. R. Sanders, *On convergence of monotone finite difference schemes with variable spacing differencing, Math.Comp.* **40** *(1983), 91-106.*
15. *M. Slemrod,* Dynamic phase transitions in a van der waals fluid, J. Diff. Equ. **52** (1984), 1–23.
16. _____, *Lax-Friedrichs and the viscosity-capillarity criterion, Physical Partial Differential Equations (J. Lightbourne and S. Rankin, eds.), Marcel Dekker, New York, 1984, pp. 75-84.*
17. *E. Tadmor,* Local error estimates for discontinuous solutions of nonlinear hyperbolic equations, *SIAM J. Numer. Anal.* **28** *(1991), 891-906.*
18. *L. Truskinovsky,* Equilibrium phase interfaces, Dokl. Akad. Nauk. SSSR **256** (1982), 306–310.
19. J.-P Vila, *Convergence and error estimates for finite volume schemes for general multidimensional scalar conservation laws, Model. Math. Anal. Numér.,* **28** *(1994), 267-295.*
20. *G. B. Witham,* Linear and nonlinear waves, John Wiley& Sons (1974).

Notes on Accuracy and Stability of Algorithms in Numerical Linear Algebra

Nicholas J. Higham*

Department of Mathematics, University of Manchester, Manchester, M13 9PL, England.

1. Introduction

The effects of rounding errors on algorithms in numerical linear algebra have been much-studied for over fifty years, since the appearance of the first digital computers. The subject continues to occupy researchers, for several reasons. First, not everything is known about established algorithms. Second, new algorithms are continually being derived, and their behaviour in finite precision arithmetic needs to be understood. Third, new error analysis techniques lead to different ways of looking at and comparing algorithms, requiring a reassessment of conventional wisdom.

The main purpose of these notes is to describe some up to date results in rounding error analysis in a way accessible to non-experts. We have chosen to analyse several practically important algorithms that are not thoroughly treated in numerical linear algebra textbooks. Section 3 considers block LDL$^\mathrm{T}$ factorization and Aasen's method for symmetric indefinite systems, Section 4 QR factorization and the solution of constrained least squares problems, and Section 5 Jacobi's method for the singular value decomposition.

These notes can be regarded as a supplement to the book [32]. They include mathematical problems and algorithms not treated therein.

2. Preliminaries

We make use of the standard model of floating point arithmetic:

$$fl(x \text{ op } y) = (x \text{ op } y)(1 + \delta), \qquad |\delta| \leq u, \quad \text{op} = +, -, *, /, \qquad (2.1)$$

where u is the unit roundoff. In applying the model, the following lemma is frequently invoked.

Lemma 2.1. *If $|\delta_i| \leq u$ and $\rho_i = \pm 1$ for $i = 1{:}n$, and $nu < 1$, then*

$$\prod_{i=1}^{n}(1 + \delta_i)^{\rho_i} = 1 + \theta_n,$$

* This work was supported by Engineering and Physical Sciences Research Council grant GR/L76532.

48   Nicholas J. Higham

where

$$|\theta_n| \le \frac{nu}{1-nu} =: \gamma_n.$$ (2.2)

Proof. See [32, Problem 3.1].

A useful property is [32, Lemma 3.3]

$$\gamma_j + \gamma_k + \gamma_j\gamma_k \le \gamma_{j+k}.$$ (2.3)

The following lemma [32, Lemma 8.4] helps to simplify the analysis of elimination methods. As usual, a hat denotes a computed quantity.

Lemma 2.2. *If* $y = (c - \sum_{i=1}^{k-1} a_i b_i)/b_k$ *is evaluated in floating point arithmetic, then, no matter what the order of evaluation,*

$$b_k\widehat{y}(1 + \theta_k^{(0)}) = c - \sum_{i=1}^{k-1} a_i b_i(1 + \theta_k^{(i)}),$$

where $|\theta_k^{(i)}| \le \gamma_k$ *for all* i. *If* $b_k = 1$, *so that there is no division, then* $|\theta_k^{(i)}| \le \gamma_{k-1}$ *for all* i.

All our error bounds will be expressed in terms of γ_n in (2.2) and the related constant

$$\tilde{\gamma}_k = \frac{cku}{1-cku},$$

in which c denotes a small integer constant whose exact value is unimportant. Thus we can write, for example, $3\tilde{\gamma}_k = \tilde{\gamma}_k$, and $m\tilde{\gamma}_n = n\tilde{\gamma}_m = \tilde{\gamma}_{mn}$.

Absolute values and inequalities are interpreted componentwise.

The following lemma is useful in the analysis of the application of Householder matrices.

Lemma 2.3. *If* $X_j + \Delta X_j \in \mathbb{R}^{n \times n}$ *satisfies* $\|\Delta X_j\|_F \le \delta_j \|X_j\|_2$ *for all* j, *then*

$$\left\| \prod_{j=0}^{m}(X_j + \Delta X_j) - \prod_{j=0}^{m} X_j \right\|_F \le \left(\prod_{j=0}^{m}(1+\delta_j) - 1 \right) \prod_{j=0}^{m} \|X_j\|_2.$$

Proof. Assume the result is true for $m-1$. Now

$$\prod_{j=0}^{m}(X_j + \Delta X_j) = (X_m + \Delta X_m) \prod_{j=0}^{m-1}(X_j + \Delta X_j),$$

so, using $\|ABC\|_F \le \|A\|_2\|B\|_F\|C\|_2$,

$$\left\| \prod_{j=0}^{m}(X_j + \Delta X_j) - \prod_{j=0}^{m} X_j \right\|_F = \left\| X_m \left[\prod_{j=0}^{m-1}(X_j + \Delta X_j) - \prod_{j=0}^{m-1} X_j \right] \right.$$

$$\left. + \Delta X_m \prod_{j=0}^{m-1}(X_j + \Delta X_j) \right\|_F$$

$$\leq \|X_m\|_2 \left(\prod_{j=0}^{m-1}(1 + \delta_j) - 1 \right) \prod_{j=0}^{m-1} \|X_j\|_2$$

$$+ \delta_m \|X_m\|_2 \prod_{j=0}^{m-1} \left(\|X_j\|_2 (1 + \delta_j) \right)$$

$$= \left(\prod_{j=0}^{m}(1 + \delta_j) - 1 \right) \prod_{j=0}^{m} \|X_j\|_2.$$

$$\square$$

3. Symmetric Indefinite Systems

A symmetric matrix $A \in \mathbb{R}^{n \times n}$ is indefinite if $(x^T A x)(y^T A y) < 0$ for some $x, y \in \mathbb{R}^n$, or, equivalently, if A has both positive and negative eigenvalues. Linear systems with symmetric indefinite coefficient matrices arise in many applications, including least squares problems, optimization and discretized incompressible Navier–Stokes equations.

For solving dense symmetric indefinite linear systems $Ax = b$ two types of factorization are used. The first is the block LDL^T factorization (or symmetric indefinite factorization)

$$PAP^T = LDL^T, \tag{3.1}$$

where P is a permutation matrix, L is unit lower triangular and D is block diagonal with diagonal blocks of dimension 1 or 2. The second factorization is that produced by Aasen's method,

$$PAP^T = LTL^T,$$

where P is a permutation matrix, L is unit lower triangular with first column e_1 and T is tridiagonal. Block LDL^T factorization is much more widely used than Aasen's factorization, but since both factorizations are mathematically interesting we describe them both.

3.1. Block LDL^T Factorization

If the symmetric matrix $A \in \mathbb{R}^{n \times n}$ is nonzero, we can find a permutation Π and an integer $s = 1$ or 2 so that

$$\Pi A \Pi^T = \begin{array}{c} s \\ n-s \end{array} \overset{\displaystyle \overset{s \qquad n-s}{}}{\begin{bmatrix} E & C^T \\ C & B \end{bmatrix}},$$

with E nonsingular. Having chosen such a Π we can factorize

$$\Pi A \Pi^T = \begin{bmatrix} I_s & 0 \\ CE^{-1} & I_{n-s} \end{bmatrix} \begin{bmatrix} E & 0 \\ 0 & B - CE^{-1}C^T \end{bmatrix} \begin{bmatrix} I_s & E^{-1}C^T \\ 0 & I_{n-s} \end{bmatrix}. \qquad (3.2)$$

This process is repeated recursively on the $(n-s) \times (n-s)$ Schur complement

$$\widetilde{A} = B - CE^{-1}C^T,$$

yielding the factorization (3.1) on completion. This factorization method is sometimes called the diagonal pivoting method, and it costs $n^3/3$ operations (the same cost as Cholesky factorization of a positive definite matrix) plus the cost of determining the permutations Π.

There are various ways to choose the permutations. Bunch and Parlett [12] proposed a complete pivoting strategy that requires $O(n^3)$ comparisons. Bunch and Kaufman [11] subsequently proposed a partial pivoting strategy requiring only $O(n^2)$ comparisons, and it is this strategy that is used in LINPACK [20] and LAPACK [2].

To define the Bunch–Kaufman (BK) pivoting strategy it suffices to describe the pivot choice for the first stage of the factorization.

Algorithm 1. (Bunch–Kaufman pivoting strategy) This algorithm determines the pivot for the first stage of the symmetric indefinite factorization applied to a symmetric matrix $A \in \mathbb{R}^{n \times n}$ using the partial pivoting strategy of Bunch and Kaufman.

$\alpha := (1 + \sqrt{17})/8 \ (\approx 0.64)$
$\gamma_1 :=$ maximum magnitude of any subdiagonal entry in column 1.
If $\gamma_1 = 0$ there is nothing to do on this stage of the factorization.
if $|a_{11}| \geq \alpha \gamma_1$
      use a_{11} as a 1×1 pivot $(s = 1, \Pi = I)$.
else
      $r :=$ row index of first (subdiagonal) entry of maximum
         magnitude in column 1.
      $\gamma_r := \left\| \begin{bmatrix} A(1:r-1,r) \\ A(r+1:n,r) \end{bmatrix} \right\|_\infty$

(2)
      if $|a_{11}|\gamma_r \geq \alpha \gamma_1^2$
         use a_{11} as a 1×1 pivot $(s = 1, \Pi = I)$.
      else if $|a_{rr}| \geq \alpha \gamma_r$
         use a_{rr} as a 1×1 pivot $(s = 1, \Pi$ swaps rows and
         columns 1 and r).
      else

use $\begin{bmatrix} a_{ii} & a_{ri} \\ a_{ri} & a_{rr} \end{bmatrix}$ as a 2×2 pivot ($s = 2$, Π swaps
rows and columns 1 and i, and 2 and r).

        end

    end

To understand the BK strategy it helps to consider the matrix

$$\begin{bmatrix} a_{11} & \cdots & a_{r1}(\gamma_1) & \cdots & \cdots & \cdots \\ \vdots & & \vdots & & & \\ a_{r1}(\gamma_1) & \cdots & a_{rr} & \cdots & a_{ir}(\gamma_r) & \cdots \\ \vdots & & \vdots & & & \\ & & a_{ir}(\gamma_r) & & & \\ \vdots & & \vdots & & & \end{bmatrix},$$

and to note that the pivot is one of a_{11}, a_{rr} and $\begin{bmatrix} a_{11} & a_{r1} \\ a_{r1} & a_{rr} \end{bmatrix}$.

The BK pivoting strategy searches at most two columns of the Schur complement at each stage of the factorization, so requires only $O(n^2)$ comparisons in total.

The parameter α is derived by element growth considerations. It is not hard to show (see [11] or [32, Section 10.4]) that for $s = 1$ the elements of the Schur complement satisfy

$$|\tilde{a}_{ij}| \le \left(1 + \frac{1}{\alpha}\right) \max_{i,j} |a_{ij}|,$$

while for $s = 2$,

$$|\tilde{a}_{ij}| \le \left(1 + \frac{2}{1-\alpha}\right) \max_{i,j} |a_{ij}|.$$

Thus over two $s = 1$ steps the elements grow by a factor at most $(1 + 1/\alpha)^2$, while for one $s = 2$ step the growth is by a factor at most $1 + 2/(1-\alpha)$. Equating these growth bounds leads to the quadratic equation $4\alpha^2 - \alpha - 1 = 0$, whose positive root is taken in Algorithm 1.

The numerical stability of the block LDL^T factorization method is described by the following result of Higham [34].

Theorem 3.1. *Let block LDL^T factorization with any pivoting strategy be applied to a symmetric matrix $A \in \mathbb{R}^{n \times n}$ as described above to yield the computed factorization $PAP^T \approx \widehat{L}\widehat{D}\widehat{L}^T$, where P is a permutation matrix and D has diagonal blocks of dimension 1 or 2. Let $\widehat{x}$ be the computed solution to $Ax = b$ obtained using the factorization. Assume that for all linear systems $Ey = f$ involving 2×2 pivots E the computed solution $\widehat{x}$ satisfies*

$$(E + \Delta E)\widehat{y} = f, \qquad |\Delta E| \le (cu + O(u^2))|E|, \tag{3.3}$$

where c is a constant. Then

$$P(A + \Delta A_1)P^T = \widehat{L}\widehat{D}\widehat{L}^T, \qquad (A + \Delta A_2)\widehat{x} = b,$$

where

$$|\Delta A_i| \le p(n)u(|A| + \Pi^T|\widehat{L}||\widehat{D}||\widehat{L}^T|\Pi) + O(u^2), \qquad i = 1:2,$$

with p a linear polynomial.

For the BK pivoting strategy the condition (3.3) can be shown to hold for the two most natural ways of solving the 2×2 systems: Gaussian elimination with partial pivoting (GEPP) and by use of the explicit inverse (as is done in LINPACK and LAPACK). Thus Theorem 3.1 is applicable to the BK pivoting strategy, and the question is whether it implies backward stability, that is, whether the matrix $|\widehat{L}||\widehat{D}||\widehat{L}^T|$ is suitably bounded relative to A. If the elements of L were bounded by a constant then the inequality $\||L||D||L^T|\|_\infty \le \|L\|_\infty\|D\|_\infty\|L^T\|_\infty$ would immediately yield a satisfactory bound. However, for the BK pivoting strategy L is unbounded, as we now show by example. For $\epsilon > 0$, the BK pivoting strategy produces the factorization, with $P = I$,

$$A = \begin{bmatrix} 0 & \epsilon & 0 \\ \epsilon & 0 & 1 \\ 0 & 1 & 1 \end{bmatrix} = \begin{bmatrix} 1 & & \\ 0 & 1 & \\ 1/\epsilon & 0 & 1 \end{bmatrix} \begin{bmatrix} 0 & \epsilon & \\ \epsilon & 0 & \\ & & 1 \end{bmatrix} \begin{bmatrix} 1 & 0 & 1/\epsilon \\ & 1 & 0 \\ & & 1 \end{bmatrix} = LDL^T.$$

As $\epsilon \to 0$, $\|L\|_\infty\|D\|_\infty\|L^T\|_\infty/\|A\|_\infty \to \infty$. Nevertheless, it can be shown in general that the matrix $|L||D||L^T|$ satisfies the bound

$$\||L||D||L^T|\|_M \le 36n\rho_n\|A\|_M,$$

where $\|A\|_M = \max_{i,j} |a_{ij}|$ and

$$\rho_n = \frac{\max_{i,j,k} |a_{ij}^{(k)}|}{\max_{i,j} |a_{ij}|},$$

where the $a_{ij}^{(k)}$ are the elements of the Schur complements arising during the factorization; see Higham [34]. The term ρ_n is a growth factor analogous to that for GEPP; it is bounded by $(1 + \alpha^{-1})^{n-1} = (2.57)^{n-1}$, as can be seen from the derivation of α above, but it is an open problem whether this bound is attainable. The normwise stability can be described as follows.

Theorem 3.2. *Let $A \in \mathbb{R}^{n \times n}$ be symmetric and let $\widehat{x}$ be a computed solution to the linear system $Ax = b$ produced by block LDL^T factorization the partial pivoting strategy of Bunch and Kaufman, where linear systems involving 2×2 pivots are solved by GEPP or by use of the explicit inverse. Then*

$$(A + \Delta A)\widehat{x} = b, \qquad \|\Delta A\|_M \le p(n)\rho_n u\|A\|_M + O(u^2),$$

where p is a quadratic.

For solving linear systems, Theorem 3.2 shows that block LDL$^\mathrm{T}$ factorization with the BK pivoting strategy has satisfactory backward stability. But for certain other applications the possibly large L factor makes the factorization unsuitable. An example is a modified Cholesky factorization algorithm of Cheng and Higham [14] in which a block LDL$^\mathrm{T}$ factorization of a symmetric A is computed and then the D factor perturbed to make it positive definite; for the perturbation of D to correspond to a perturbation of A of similar size it is necessary that $\|L\|$ is not too large. In [14] this problem was overcome by using a new pivoting strategy of Ashcraft, Grimes and Lewis [4] which does guarantee a bounded L in the block LDL$^\mathrm{T}$ factorization. This "bounded Bunch–Kaufman" pivoting strategy is broadly similar to the BK strategy, but it has an iterative phase.

Algorithm 2. (bounded Bunch–Kaufman pivoting strategy) This algorithm determines the pivot for the first stage of the symmetric indefinite factorization applied to a symmetric matrix $A \in \mathbb{R}^{n \times n}$.

> $\alpha := (1 + \sqrt{17})/8 \ (\approx 0.64)$
> $\gamma_1 :=$ maximum magnitude of any subdiagonal entry in column 1.
> If $\gamma_1 = 0$ there is nothing to do on this stage of the factorization.
> if $|a_{11}| \geq \alpha\gamma_1$
> > use a_{11} as a 1×1 pivot ($s = 1$, $\Pi = I$).
>
> else
> > $i := 1;\ \gamma_i := \gamma_1$
> > repeat
> > > $r :=$ row index of first (subdiagonal) entry of maximum magnitude in column i.
> > > $\gamma_r :=$ maximum magnitude of any off-diagonal entry in column r.
> > > if $|a_{rr}| \geq \alpha\gamma_r$
> > > > use a_{rr} as a 1×1 pivot ($s = 1$, Π swaps rows and columns 1 and r).
> > >
> > > else if $\gamma_i = \gamma_r$
> > > > use $\begin{bmatrix} a_{ii} & a_{ri} \\ a_{ri} & a_{rr} \end{bmatrix}$ as a 2×2 pivot ($s = 2$, Π swaps rows and columns 1 and i, and 2 and r).
> > >
> > > else
> > > > $i := r,\ \gamma_i := \gamma_r$.
> > > end
> > until a pivot is chosen
> end

The repeat loop in Algorithm BBK searches for an element a_{ri} that is simultaneously the largest in magnitude in the rth row and the ith column, and it uses this element to build a 2×2 pivot; the search terminates prematurely if a suitable 1×1 pivot is found. Note that the pivot choice in case (2) of the BK strategy (Algorithm 1) can never arise with the BBK strategy.

Since the value of γ_i increases strictly from one pivot step to the next, the search in Algorithm BBK takes at most n steps. The overall cost of the searching is therefore between $O(n^2)$ and $O(n^3)$ comparisons. Matrices are known for which the entire Schur complement must be searched at each step, in which case the cost is $O(n^3)$ comparisons. However, probabilistic results and experimental evidence suggest that usually only $O(n^2)$ comparisons are required [4].

The following properties noted in [4] are readily verified, using the property that any 2×2 pivot satisfies

$$\left| \begin{bmatrix} a_{ii} & a_{ri} \\ a_{ri} & a_{rr} \end{bmatrix}^{-1} \right| \leq \frac{1}{\gamma_r(1-\alpha^2)} \begin{bmatrix} \alpha & 1 \\ 1 & \alpha \end{bmatrix}.$$

1. Every entry of L is bounded by $\max\{1/(1-\alpha), 1/\alpha\} \approx 2.78$.
2. Every 2×2 pivot block D_{ii} satisfies $\kappa_2(D_{ii}) \leq (1+\alpha)/(1-\alpha) \approx 4.56$.
3. The growth factor for the factorization satisfies the same bound as for the BK pivoting strategy.

At the cost of a worst case $O(n^3)$ searching overhead, the BBK pivoting strategy thus gives an L factor with elements of order 1 and produces well conditioned 2×2 blocks of D.

The work of Ashcraft, Grimes and Lewis [4] was motivated by an optimization problem in which solving symmetric linear systems using the BK pivoting strategy led to convergence difficulties, which were traced to the fact that $\|L\|$ is unbounded. The theme of [4] is that pivoting strategies such as the BBK strategy that bound $\|L\|$ lead to higher accuracy. A class of linear systems is given in [4] where the BBK pivoting strategy provides more accurate solutions than the BK strategy. However, a theoretical comparison by Cheng [13] of normwise and componentwise backward and forward stability of the two strategies does not identify clear superiority of the BBK strategy. Therefore with the available evidence it is not possible to conclude that the BBK strategy has superior accuracy or stability to the BK strategy for solving general symmetric indefinite linear systems.

3.2. Aasen's Method

Aasen's method [1] factorizes a symmetric matrix $A \in \mathbb{R}^{n \times n}$

$$PAP^T = LTL^T,$$

where L is unit lower triangular with first column e_1,

$$T = \begin{bmatrix} \alpha_1 & \beta_1 & & & & \\ \beta_1 & \alpha_2 & \beta_2 & & & \\ & \ddots & \ddots & \ddots & & \\ & & \ddots & \ddots & \beta_{n-1} \\ & & & \beta_{n-1} & \alpha_n \end{bmatrix}$$

is tridiagonal, and P is a permutation matrix.

To derive Aasen's method, we initially ignore interchanges and assume that the first $i-1$ columns of T and the first i columns of L are known. We show how to compute the ith column of T and the $(i+1)$st column of L. A key role is played by the matrix

$$H = TL^T, \tag{3.4}$$

which is easily seen to be upper Hessenberg. Equating ith columns in (3.4) we obtain

$$\begin{bmatrix} h_{1i} \\ h_{2i} \\ \vdots \\ h_{i-1,i} \\ \underline{h_{ii}} \\ \underline{h_{i+1,i}} \\ 0 \\ \vdots \\ 0 \end{bmatrix} = T \begin{bmatrix} l_{i1} \\ l_{i2} \\ \vdots \\ l_{i,i-1} \\ 1 \\ 0 \\ \vdots \\ 0 \end{bmatrix} = \begin{bmatrix} \alpha_1 l_{i1} + \beta_1 l_{i2} \\ \beta_1 l_{i1} + \alpha_2 l_{i2} + \beta_2 l_{i3} \\ \vdots \\ \beta_{i-2} l_{i,i-2} + \alpha_{i-1} l_{i,i-1} + \underline{\beta_{i-1}} \\ \underline{\beta_{i-1}} l_{i,i-1} + \underline{\alpha_i} \\ \underline{\beta_i} \\ 0 \\ \vdots \\ 0 \end{bmatrix}. \tag{3.5}$$

We use an underline to denote an unknown quantity to be determined.

The first $i-1$ equations in (3.5) are used to compute $h_{1i}, \ldots, h_{i-1,i}$. The next two equations contain two unknowns each so cannot yet be used. The (i,i) and $(i+1,i)$ elements of the equation $A = LH$ give

$$a_{ii} = \sum_{j=1}^{i-1} l_{ij} h_{ji} + \underline{h_{ii}}, \tag{3.6}$$

$$a_{i+1,i} = \sum_{j=1}^{i} l_{i+1,j} h_{ji} + \underline{h_{i+1,i}}. \tag{3.7}$$

which we solve for h_{ii} and $h_{i+1,i}$. Now we can return to the last two nontrivial equations of (3.5) to obtain α_i and β_i. Finally, the ith column of the equation $A = LH$ yields

$$a_{ki} = \sum_{j=1}^{i+1} l_{kj} h_{ji}, \qquad k = i+2 : n,$$

which yields the elements below the diagonal in the $(i+1)$st column of L:

$$l_{k,i+1} = \frac{a_{ki} - \sum_{j=1}^{i} l_{kj} h_{ji}}{h_{i+1,i}}, \qquad k = i+2\colon n. \tag{3.8}$$

The factorization has thereby been advanced by one step.

The operation count for Aasen's method is the same as for block LDL^T factorization.

To ensure that the factorization does not break down, and to improve its numerical stability, interchanges are incorporated. Before the evaluation of (3.7) and (3.8) we compute $v_k = \alpha_{ki} - \sum_{j=1}^{i} l_{kj} h_{ji}$, $k = i+1\colon n$, find r such that $|v_r| = \max\{\,|v_k| : k = i+1\colon n\,\}$, and then swap v_k and v_r and make corresponding interchanges in A and L. This partial pivoting strategy ensures that $|l_{ij}| \leq 1$ for $i > j$.

3.2.1. Rounding Error Analysis. Aasen [1] states without proof a backward error bound for the factorization. We give a detailed analysis of the factorization and the subsequent solution of a linear system. We will ignore pivoting (or, equivalently, assume that A is "pre-pivoted").

The Factorization

We wish to bound the residual $A - \widehat{L}\widehat{T}\widehat{L}^T$ of the computed factors $\widehat{L}$ and $\widehat{T}$. This can be done in two steps. First, note that applying Lemma 2.2 to (3.8), and using the fact that $\widehat{l}_{k1} = 0$ for $k = 2\colon n$, we obtain

$$\widehat{l}_{k,i+1} \widehat{h}_{i+1,i}(1 + \theta_i^{(0)}) = a_{ki} - \sum_{j=2}^{i} \widehat{l}_{kj} \widehat{h}_{ji}(1 + \theta_i^{(j)}), \qquad k = i+2\colon n,$$

where $|\theta_i^{(j)}| \leq \gamma_i$ for all j. Similar equations are obtained from (3.6) and (3.7). By collecting all these equations together for $i = 1\colon n$ and expressing the result in matrix form we find that

$$|A - \widehat{L}\widehat{H}| \leq \gamma_n |\widehat{L}||\widehat{H}|. \tag{3.9}$$

Similar analysis applied to (3.5) shows that

$$|\widehat{H}(:,i) - \widehat{T}\widehat{L}^T(:,i)| \leq \gamma_3 |\widehat{T}||\widehat{L}^T(:,i)|, \qquad i = 1\colon n,$$

so that

$$|\widehat{H} - \widehat{T}\widehat{L}^T| \leq \gamma_3 |\widehat{T}||\widehat{L}^T|. \tag{3.10}$$

Combining (3.9) and (3.10) we obtain

$$\begin{aligned}
|A - \widehat{L}\widehat{T}\widehat{L}^T| &= |A - \widehat{L}\widehat{H} + \widehat{L}(\widehat{H} - \widehat{T}\widehat{L}^T)| \\
&\leq \gamma_n |\widehat{L}||\widehat{H}| + \gamma_3 |\widehat{L}||\widehat{T}||\widehat{L}^T| \\
&\leq \gamma_n |\widehat{L}|(1 + \gamma_3)|\widehat{T}||\widehat{L}^T| + \gamma_3 |\widehat{L}||\widehat{T}||\widehat{L}^T| \\
&= (\gamma_n + \gamma_n\gamma_3 + \gamma_3)|\widehat{L}||\widehat{T}||\widehat{L}^T| \\
&\leq \gamma_{n+3} |\widehat{L}||\widehat{T}||\widehat{L}^T|,
\end{aligned}$$

using (2.3) for the last inequality. We summarize our bound in the following theorem.

Theorem 3.3. *If Aasen's method applied to $A \in \mathbb{R}^{n \times n}$ runs to completion then the computed factors $\widehat{L}$ and $\widehat{T}$ satisfy*

$$\widehat{L}\widehat{T}\widehat{L}^T = A + \Delta A, \qquad |\Delta A| \leq \gamma_{n+3}|\widehat{L}||\widehat{T}||\widehat{L}^T|.$$

Solution of a Linear System

To solve a linear system $Ax = b$ using the factorization $A = LTL^T$ we solve in turn

$$Lz = b, \quad Ty = z, \quad L^T x = y. \tag{3.11}$$

The system $Ty = z$ has a symmetric tridiagonal coefficient matrix that is indefinite in general. For stability reasons it is usually solved by LU factorization with partial pivoting, which unfortunately destroys the symmetry and so cannot be used to determine the inertia. A factorization $PT = MU$ is obtained, where M has at most one nonzero below the diagonal in each column and U has upper bandwidth 2 ($u_{ij} = 0$ for $j > i + 2$). Straightforward error analysis (cf. [32, Ch. 9]) shows that for the computed factors of the computed $\widehat{T}$ we have

$$P\widehat{T} = \widehat{M}\widehat{U} + \Delta T, \qquad |\Delta T| \leq \gamma_2 |\widehat{M}||\widehat{U}|.$$

We solve the triangular systems $Mw = Pz$ and $Uy = w$. The computed solutions satisfy

$$\begin{aligned}
(\widehat{M} + \Delta M)\widehat{w} &= P\widehat{z}, & |\Delta M| &\leq \gamma_1 |\widehat{M}|, \\
(\widehat{U} + \Delta U)\widehat{y} &= \widehat{w}, & |\Delta U| &\leq \gamma_3 |\widehat{U}|.
\end{aligned}$$

Hence

$$\widehat{z} = P^T(\widehat{M} + \Delta M)(\widehat{U} + \Delta U)\widehat{y} = (\widehat{T} + \Delta T_1)\widehat{y},$$

where

$$\begin{aligned}
|\Delta T_1| &= |P^T(-\Delta T + \Delta M\widehat{U} + \widehat{M}\Delta U + \Delta M \Delta U)| \\
&\leq (\gamma_2 + \gamma_1 + \gamma_3 + \gamma_1\gamma_3)P^T|\widehat{M}||\widehat{U}| \\
&\leq \gamma_6 P^T|\widehat{M}||\widehat{U}|,
\end{aligned}$$

using (2.3). For the triangular solves involving L in (3.11) we have, using Lemma 2.2,

$$\begin{aligned}
(\widehat{L} + \Delta L_1)\widehat{z} &= b, & |\Delta L_1| &\leq \gamma_{n-1}|\widehat{L}|, \\
(\widehat{L}^T + \Delta L_2^T)\widehat{x} &= \widehat{y}, & |\Delta L_2| &\leq \gamma_{n-1}|\widehat{L}|.
\end{aligned}$$

Overall, then,

$$b = (\widehat{L} + \Delta L_1)(\widehat{T} + \Delta T_1)(\widehat{L}^T + \Delta L_2^T)\widehat{x} = (A + \Delta A_1)\widehat{x},$$

where, using Theorem 3.3,

$$
\begin{aligned}
|\Delta A_1| &= |\Delta A + \Delta L_1(\widehat{T} + \Delta T_1)(\widehat{L}^T + \Delta L_2^T) + \widehat{L}\Delta T_1(\widehat{L}^T + \Delta L_2^T) + \widehat{L}\widehat{T}\Delta L_2^T| \\
&\leq \gamma_{n+3}|\widehat{L}||\widehat{T}||\widehat{L}^T| + \gamma_{n-1}(1 + \gamma_{n-1})|\widehat{L}|(|\widehat{T}| + \gamma_6 P^T|\widehat{M}||\widehat{U}|)|\widehat{L}^T| \\
&\quad + \gamma_6(1 + \gamma_{n-1})|\widehat{L}|P^T|\widehat{M}||\widehat{U}||\widehat{L}^T| + \gamma_{n-1}|\widehat{L}||\widehat{T}||\widehat{L}^T| \\
&\leq (\gamma_{n+3} + 2\gamma_{n-1} + \gamma_{n-1}^2)|\widehat{L}||\widehat{T}||\widehat{L}^T| \\
&\quad + \gamma_6(1 + 2\gamma_{n-1} + \gamma_{n-1}^2)|\widehat{L}|P^T|\widehat{M}||\widehat{U}||\widehat{L}^T| \\
&\leq \gamma_{3n+1}|\widehat{L}||\widehat{T}||\widehat{L}^T| + \gamma_{2n+4}|\widehat{L}|P^T|\widehat{M}||\widehat{U}||\widehat{L}^T|,
\end{aligned}
$$

using (2.3). We summarize the analysis in a theorem.

Theorem 3.4. *Let $A \in \mathbb{R}^{n \times n}$ be symmetric and suppose Aasen's method produces computed factors $\widehat{L}$, $\widehat{T}$ and a computed solution $\widehat{x}$ to $Ax = b$. Then*

$$
(A + \Delta A)\widehat{x} = b, \qquad |\Delta A| \leq \gamma_{3n+1}|\widehat{L}||\widehat{T}||\widehat{L}^T| + \gamma_{2n+4}|\widehat{L}|P^T|\widehat{M}||\widehat{U}||\widehat{L}^T|,
$$

where $P\widehat{T} \approx \widehat{M}\widehat{U}$ is the computed factorization produced by LU factorization with partial pivoting. Moreover,

$$
\|\Delta A\|_\infty \leq (n-1)^2 \gamma_{15n+25}\|\widehat{T}\|_\infty.
$$

Proof. We just have to verify the bound for $\|\Delta A\|_\infty$. Every element of $\widehat{L}$ and $\widehat{M}$ is bounded by 1, so, since the first column of L is e_1, $\|\widehat{L}\|_\infty \leq n - 1$ and $\|\widehat{M}\|_\infty \leq 2$. Careful consideration of LU factorization with partial pivoting on a tridiagonal matrix shows that $\|\widehat{U}\|_\infty \leq 3\|\widehat{T}\|_\infty$ (we ignore the trivial effects of rounding error on this bound). The bound follows readily.

3.2.2. The Growth Factor. Theorem 3.4 shows that Aasen's method is a backward stable method for solving $Ax = b$ provided that the growth factor

$$
\rho_n = \frac{\max_{i,j}|t_{ij}|}{\max_{i,j}|a_{ij}|}
$$

is not too large. (Here, we are making the reasonable assumption that $\max_{ij}|t_{ij}| \approx \max_{ij}|\widehat{t}_{ij}|$.)

Using the fact that the multipliers in Aasen's method are bounded by 1, it is straightforward to show that if $\max_{i,j}|a_{ij}| = 1$ then T has a bound illustrated for $n = 5$ by

$$
|T| \leq \begin{bmatrix} 1 & 1 & & & \\ 1 & 1 & 2 & & \\ & 2 & 4 & 8 & \\ & & 8 & 16 & 32 \\ & & & 32 & 64 \end{bmatrix}.
$$

Hence

$$
\rho_n \leq 4^{n-2}.
$$

Whether this bound is attainable for $n \geq 4$ is an open question. Cheng [13] reports experiments using direct search in which he obtained growth of 7.99 for $n = 4$ and 14.61 for $n = 5$, which are to be compared with the corresponding bounds of 16 and 64.

3.3. Aasen's Method Versus Block LDL$^\text{T}$ Factorization

While block LDL$^\text{T}$ of a symmetric matrix using the BK pivoting strategy is implemented in all the major program libraries, the only library we know to contain Aasen's method is the IMSL Fortran 90 library [28]. A comparison of the two methods in the mid 1970s found little to choose between them in speed [7], but no thorough comparison on modern computer architectures has been published. See [4] for some further comments. The greater popularity of block LDL$^\text{T}$ factorization may be due to the fact that it is generally easier to work with a block diagonal matrix with blocks of size at most 2 than with a tridiagonal one.

Note that since $|l_{ij}| \leq 1$ for Aasen's method with pivoting, the method is superior to block LDL$^\text{T}$ factorization with the BK pivoting strategy for applications in which a bounded L is required.

3.4. Tridiagonal Matrices

In the previous section we noted that solving a symmetric tridiagonal system by LU factorization with partial pivoting does not take advantage of the symmetry of A. On the other hand, any attempt to compute the symmetry-preserving factorization $PAP^T = LDL^T$ with a diagonal D can fail, since the factorization does not always exist. Bunch [10] suggested a way to avoid both difficulties: compute a **block** LDL$^\text{T}$ factorization without interchanges (in the same way as in Section 3.1) with a particular strategy for choosing the pivot size (1 or 2) at each stage of the factorization. Bunch's strategy [10] for choosing the pivot size is fully defined by describing the choice of the first pivot.

Algorithm 3 (Bunch's pivoting strategy). This algorithm determines the pivot size, s, for the first stage of block LDL$^\text{T}$ factorization applied to a symmetric tridiagonal matrix $A \in \mathbb{R}^{n \times n}$.

> $\sigma := \max\{\,|a_{ij}| : i, j = 1{:}n\,\}$ (compute once, at the start of the
>   factorization)
> $\alpha := (\sqrt{5} - 1)/2 \approx 0.62$
> if $\sigma|a_{11}| \geq \alpha a_{21}^2$
> $s = 1$
> else
> $s = 2$
> end

The result is a factorization

$$A = LDL^T, \tag{3.12}$$

where L is unit lower triangular and D is block diagonal with each diagonal block having dimension 1 or 2. The value of α is derived in a similar way as for the Bunch–Kaufman pivoting strategy, by equating growth bounds. The inertia of A is the same as that of D, which can be read from the (block) diagonal of D, since any 2×2 block can be shown to have one negative and one positive eigenvalue.

The following stability result is proved by Higham [33].

Theorem 3.5. *Let block* LDL^T *factorization with the pivoting strategy of Algorithm 3 be applied to a symmetric tridiagonal matrix* $A \in \mathbb{R}^{n \times n}$ *to yield the computed factorization* $A \approx \widehat{L}\widehat{D}\widehat{L}^T$, *and let* $\widehat{x}$ *be the computed solution to* $Ax = b$ *obtained using the factorization. Assume that all linear systems* $Ey = f$ *involving* 2×2 *pivots* E *are solved by GEPP or by use of the explicit inverse. Then*

$$A + \Delta A_1 = \widehat{L}\widehat{D}\widehat{L}^T, \qquad (A + \Delta A_2)\widehat{x} = b,$$

where

$$\|\Delta A_i\|_M \le cu\|A\|_M + O(u^2), \qquad i = 1:2, \tag{3.13}$$

with c *a constant.*

Theorem 3.5 shows that block LDL^T factorization with the pivoting strategy of Algorithm 3 is a normwise backward stable way to factorize a symmetric tridiagonal matrix A and to solve a linear system $Ax = b$. Block LDL^T factorization therefore provides an attractive alternative to LU factorization with partial pivoting for solving such linear systems.

4. QR Factorization and Constrained Least Squares Problems

The QR factorization is perhaps the most important factorization in numerical linear algebra. It plays a vital role in the solution of linear systems (well-, under- and overdetermined), and in eigenvalue problems and singular value problems. One recent textbook treats QR factorization before Gaussian elimination [48].

In this section we investigate the accuracy and stability properties of QR factorization and then describe some methods based on QR factorization for solving the constrained least squares problem.

A QR factorization of $A \in \mathbb{R}^{m \times n}$ with $m \ge n$ is a factorization

$$A = QR = [\, Q_1 \quad Q_2 \,] \begin{bmatrix} R_1 \\ 0 \end{bmatrix} = Q_1 R_1,$$

where $Q \in \mathbb{R}^{m \times m}$ is orthogonal and $R_1 \in \mathbb{R}^{n \times n}$ is upper triangular. Whether Q and R or the smaller Q_1 and R_1 are defined as the QR factors depends on the application. A quick proof of existence of the QR factorization for full rank A is obtained by taking R as the Cholesky factor of $A^T A$ and $Q = AR^{-1}$.

We begin with an example to illustrate the versatility of QR factorization.

Any matrix $A \in \mathbb{R}^{m \times n}$ with $m \geq n$ has a polar decomposition $A = UH$, where U has orthonormal columns and H is symmetric positive semidefinite. The polar decomposition has various applications [31] and can be computed using the Newton iteration[1]

$$X_{k+1} = 2X_k(I + X_k^T X_k)^{-1}, \qquad X_0 = A, \tag{4.1}$$

whose iterates converge to U quadratically. By adapting an idea from [51], we can use QR factorization to avoid the explicit matrix inverse. Let

$$B = \begin{matrix} n \\ m \end{matrix} \begin{bmatrix} I \\ A \end{bmatrix} = \begin{matrix} n \\ m \end{matrix} \begin{bmatrix} Q_1 \\ Q_2 \end{bmatrix} R = QR$$

be a QR factorization. Then $I = Q_1 R$ and $A = Q_2 R$. Since Q_1 and R are square, $Q_1 = R^{-1}$, and so

$$Q_2 Q_1^T = AR^{-1}R^{-T} = A(R^T R)^{-1} = A(I + A^T A)^{-1}.$$

Thus $2Q_2 Q_1^T = X_1$, where X_1 is the first Newton iterate from (4.1). It follows that we can implement the Newton iteration in inverse-free form as follows: with $X_0 = A$, for $k = 1, 2, \ldots$

$$\begin{matrix} n \\ m \end{matrix} \begin{bmatrix} I \\ X_k \end{bmatrix} = \begin{matrix} n \\ m \end{matrix} \begin{bmatrix} Q_1^{(k)} \\ Q_2^{(k)} \end{bmatrix} R_k \quad \text{(QR factorization)},$$

$$X_{k+1} = 2Q_2^{(k)} Q_1^{(k)^T}.$$

Inverse-free iterations are potentially attractive for stability reasons [5].

4.1. Householder QR Factorization

A QR factorization can be computed in three main ways. The Gram–Schmidt process, which sequentially orthogonalizes the columns of A, is the oldest method and is described in most linear algebra textbooks. Givens transformations are preferred when A has a special sparsity structure, such as band or Hessenberg structure. Householder transformations provide the most generally useful way to compute the QR factorization and are the subject of the rest of this section.

A Householder matrix (Householder transformation) is a symmetric, orthogonal matrix of the form

---

[1] For square matrices, iteration (4.1) is related to the iteration $Y_{k+1} = (Y_k + Y_k^{-T})/2$, $Y_0 = A$, by $Y_k = X_k^{-T}$. Iteration (4.1) is the more expensive but has the advantage that it applies to rectangular matrices.

$$P = I - \frac{2}{v^T v} v v^T, \qquad 0 \neq v \in \mathbb{R}^m.$$

Its key property, which is easily verified, is that if $\|x\|_2 = \|y\|_2$ but $x \neq y$, then $Px = y$ where $v = x - y$. We typically choose P to transform x into a vector $y = \sigma e_1$ with $\sigma = \pm \|x\|_2$. Most textbooks recommend choosing the sign to avoid cancellation in the computation of $v_1 = x_1 - \sigma$:

$$\mathrm{sign}(\sigma) = -\,\mathrm{sign}(x_1). \qquad (4.2)$$

This has led to the myth that the other choice of sign is unsuitable. In fact, the other sign is perfectly satisfactory provided that the formula for v_1 is suitably rearranged [43], [44, Section 6.3.1]:

$$v_1 = x_1 - \mathrm{sign}(x_1)\|x\|_2 = \frac{x_1^2 - \|x\|_2^2}{x_1 + \mathrm{sign}(x_1)\|x\|_2} = \frac{-(x_2^2 + \cdots + x_m^2)}{x_1 + \mathrm{sign}(x_1)\|x\|_2}. \qquad (4.3)$$

The computation of a QR factorization by Householder transformations can be described as follows.

Algorithm 4. Given $A \in \mathbb{R}^{m \times n}$ with $m \geq n$ this algorithm computes the QR factorization $A = QR$ where $Q \in \mathbb{R}^{m \times m}$ is orthogonal and $R \in \mathbb{R}^{m \times n}$ is upper trapezoidal.

$A_1 = A$
for $k = 1: \min(m - 1, n)$
$\quad \% \; A_k = \begin{bmatrix} R_{k-1} & z_k & B_k \\ 0 & x_k & C_k \end{bmatrix}, \quad R_{k-1} \in \mathbb{R}^{(k-1) \times (k-1)}, \quad x_k \in \mathbb{R}^{m-k+1}.$
$\quad$ Construct a Householder matrix $\widetilde{P}_k \in \mathbb{R}^{(m-k+1) \times (m-k+1)}$
$\quad$ such that $\widetilde{P}_k x_k = \sigma e_1$ and define $P_k = \begin{bmatrix} I_{k-1} & 0 \\ 0 & \widetilde{P}_k \end{bmatrix} \in \mathbb{R}^{m \times m}.$
$\quad$ Let $A_{k+1} = P_k A_k$
end
$R = A_{\min(m, n+1)}$

Cost: $2n^2(m - n/3)$ flops.

Of course, the Householder matrix is never formed in practice; storage and computations use solely the Householder vector v.

By taking σ nonnegative and switching between the formulae $v_1 = x_1 - \sigma$ and (4.3) according as x_1 is nonpositive and positive, respectively, we can obtain from Algorithm 4 an R factor with nonnegative diagonal elements; this is done in [25, Algs. 5.1, 5.2.1], for example. However, this approach is not recommended for badly row scaled matrices, for reasons explained after Theorem 4.2 below.

Now we consider the numerical stability of Householder QR factorization. As is often the case in rounding error analysis, the analysis can be made quite short and instructive if approached in the right way. The following lemma is the key to the analysis. For notational convenience we now write Householder matrices in the form

$$P = I - vv^T, \qquad v^T v = 2.$$

Lemma 4.1. *Let a Householder matrix* $P = I - vv^T \in \mathbb{R}^{m \times m}$ *such that* $Px = \sigma e_1$ *be constructed with either choice of sign of* σ*, as described above. The computed* $\widehat{v}$ *can be written*

$$\widehat{v} = v + \Delta v, \qquad |\Delta v| \leq \tilde{\gamma}_m |v|. \tag{4.4}$$

Proof. The proof consists of straightforward algebra using the model (2.1). See [32, Lemma 18.1]. ∎

For the subsequent analysis we will introduce extra generality by considering v and $\widehat{v}$ satisfying (4.4) without requiring that $Pv = \sigma e_1$.

Lemma 4.2. *Let* $b \in \mathbb{R}^m$ *and consider the computation of* $y = \widehat{P}b = (I - \widehat{v}\widehat{v}^T)b = b - \widehat{v}(\widehat{v}^T b)$*, where* $\widehat{v} \in \mathbb{R}^m$ *satisfies (4.4). The computed* $\widehat{y}$ *satisfies*

$$\widehat{y} = (P + \Delta P)b, \qquad \|\Delta P\|_F \leq \tilde{\gamma}_m,$$

where $P = I - vv^T$*.*

Proof. Again, the proof is straightforward using the model (2.1). See [32, Lemma 18.2]. ∎

In practice we invariably apply a sequence of Householder transformations $P_r P_{r-1} \ldots P_1 A$ and the question is how the errors from each step combine. Since the P_j are applied to the columns of A, columnwise error bounds are to be expected, and these are provided by the next lemma.

We will assume that

$$r\tilde{\gamma}_m < \frac{1}{2}, \tag{4.5}$$

where r is the number of Householder transformations. We will write the jth column of A variously as $A(:,j)$ and a_j.

Lemma 4.3. *Consider the sequence of transformations*

$$A_{k+1} = P_k A_k, \qquad k = 1{:}r,$$

where $A_1 = A \in \mathbb{R}^{m \times n}$ *and* $P_k = I - v_k v_k^T \in \mathbb{R}^{m \times m}$ *is a Householder matrix. Assume that the transformations are performed using computed Householder vectors* $\widehat{v}_k \approx v_k$ *that satisfy (4.4). The computed matrix* $\widehat{A}_{r+1}$ *satisfies*

$$\widehat{A}_{r+1} = Q^T(A + \Delta A), \tag{4.6}$$

where $Q^T = P_r P_{r-1} \ldots P_1$ *and*

$$\|\Delta A(:,j)\|_2 \leq r\tilde{\gamma}_m \|A(:,j)\|_2.$$

Proof. The jth column of A undergoes the transformations

$$a_j^{(r+1)} = P_r \dots P_1 a_j.$$

By Lemma 4.2 we have

$$\widehat{a}_j^{(r+1)} = (P_r + \Delta P_r) \dots (P_1 + \Delta P_1) a_j, \qquad (4.7)$$

where each ΔP_k depends on j and satisfies $\|\Delta P_k\|_F \leq \tilde{\gamma}_m$. Using Lemma 2.3 we obtain

$$\widehat{a}_j^{(r-1)} = Q^T(a_j + d_j),$$

$$\|d_j\|_2 \leq ((1 + \tilde{\gamma}_m)^r - 1)\|a_j\|_2 \leq \frac{r\tilde{\gamma}_m}{1 - r\,\tilde{\gamma}_m}\|a_j\|_2 = r\tilde{\gamma}_m'\|a_j\|_2, \quad (4.8)$$

using Lemma 2.1 and assumption (4.5).

Lemma 4.3 yields almost immediately the standard backward error result for Householder QR factorization.

Theorem 4.1. *Let $\widehat{R} \in \mathbb{R}^{m \times n}$ be the computed upper trapezoidal QR factor of $A \in \mathbb{R}^{m \times n}$ $(m \geq n)$ obtained via the Householder QR algorithm. Then there exists an orthogonal $Q \in \mathbb{R}^{m \times m}$ such that*

$$A + \Delta A = Q\widehat{R},$$

where

$$\|\Delta A(:,j)\|_2 \leq \tilde{\gamma}_{mn}\|A(:,j)\|_2. \qquad (4.9)$$

The matrix Q is given explicitly as $Q = (P_n P_{n-1} \dots P_1)^T$, where P_k is the Householder matrix that corresponds to the exact application of the kth step of the algorithm to $\widehat{A}_k$.

We note that for Householder QR factorization $\Delta P_k = 0$ for $k > j$ in (4.7), and consequently the factor $\tilde{\gamma}_{mn}$ in (4.9) can be reduced to $\tilde{\gamma}_{mj}$.

Theorem 4.1 is often stated in the weaker form $\|\Delta A\|_F \leq \tilde{\gamma}_{mn}\|A\|_F$ that is implied by (4.9) (see, e.g., [25, Section 5.2.1]). For a matrix whose columns vary widely in norm this normwise bound on ΔA is much weaker than (4.9). For an alternative way to express this backward error result define B by $A = BD_C$, where $D_C = \text{diag}(\|A(:,j)\|_2)$; then the result states that there exists an orthogonal $Q \in \mathbb{R}^{m \times m}$ such that

$$(B + \Delta B)D_C = Q\widehat{R}, \qquad \|\Delta B(:,j)\|_2 \leq \tilde{\gamma}_{mn}, \qquad (4.10)$$

so that $\|\Delta B\|_2/\|B\|_2 = O(u)$.

It is natural to ask whether a small row-wise backward error bound holds for Householder QR factorization, since matrices A for which the rows vary widely in norm occur commonly in weighted least square problems, for example (see (4.17) below). In general, the answer is no. However, if column

pivoting is used together with row pivoting or row sorting, and the choice
of sign (4.2) is used, then such a bound does hold. Recall that the column
pivoting strategy exchanges columns at the start of the kth stage of the factorization to ensure that

$$\|a_k^{(k)}(k{:}m)\|_2 = \max_{j \geq k} \|a_j^{(k)}(k{:}m)\|_2. \tag{4.11}$$

In other words, it maximizes the norm of the active part of the pivot column.
With row pivoting, after the column interchange has taken place at the start
of the kth stage we interchange rows to ensure that

$$|a_{kk}^{(k)}| = \max_{i \geq k} |a_{ik}^{(k)}|,$$

where $A_k = (a_{ij}^{(k)})$. The alternative strategy of row sorting reorders the rows
prior to carrying out the factorization so that

$$\|A(i,:)\|_\infty = \max_{j \geq i} \|A(j,:)\|_\infty, \qquad i = 1{:}m.$$

Theorem 4.2. *Let $\widehat{R} \in \mathbb{R}^{m \times n}$ be the computed upper trapezoidal QR factor
of $A \in \mathbb{R}^{m \times n}$ $(m \geq n)$ obtained via the Householder QR algorithm with
column pivoting, with the choice of sign (4.2). Then there exists an orthogonal
$Q \in \mathbb{R}^{m \times m}$ such that*

$$(A + \Delta A)\Pi = Q\widehat{R},$$

*where Π is a permutation matrix that describes the overall effect of the column
interchanges and*

$$|\Delta a_{ij}| \leq j^2 \tilde{\gamma}_m \alpha_i \max_s |a_{is}|,$$

where

$$\alpha_i = \frac{\max_{j,k} |\widehat{a}_{ij}^{(k)}|}{\max_j |a_{ij}|}.$$

The matrix Q is defined as in Theorem 4.1.

Theorem 4.2 was originally proved under some additional assumptions by
Powell and Reid [45]. The result as stated is proved by Cox and Higham [17];
it also follows from a more general result of Higham [35] that sheds more light
on why the theorem holds. In general, the row-wise growth factors α_i can be
arbitrarily large. If row sorting or row pivoting is used it can be shown that
$\alpha_i \leq \sqrt{m}(1 + \sqrt{2})^{n-1}$ for all i [17], [45], with α_i usually small in practice.
Therefore the α_i are somewhat analogous to the growth factor for Gaussian
elimination with partial pivoting. For the alternative choice of sign (4.3) in
the Householder vectors, the α_i are unbounded even if row pivoting or row
sorting is used and so row-wise stability is lost; see [17] for an illustrative
example.

Note that the matrix Q in Theorem 4.1 is not computed by the QR factorization algorithm and is of purely theoretical interest. It is the fact that Q is

exactly orthonormal that makes the result so useful. In most applications of QR factorization (such as solving the least squares problem) it is unnecessary to form the Q factor explicitly. When Q is explicitly formed, two questions arise:

1. How close is the computed $\widehat{Q}$ to being orthonormal?
2. How large is $A - \widehat{Q}\widehat{R}$?

Both questions are easily answered using the analysis above.

First, note that there are two ways to form $Q = (P_n P_{n-1} \dots P_1)^T = P_1 P_2 \dots P_n$: from left to right or from right to left. The right to left evaluation is the more efficient, because the effective dimension of the intermediate products grows from $m - n$ to m, whereas with the left to right order it is m at each stage. The right to left evaluation requires $4(m^2 n - mn^2 + n^3/3)$ flops. Lemma 4.3 gives (with $A_1 = I$)

$$\widehat{Q} = Q(I_m + \Delta I), \qquad \|\Delta I\|_F \leq \sqrt{n}\,\tilde{\gamma}_{mn}.$$

The factor $\sqrt{n}$ is unnecessary, as we can show using a variation of Lemma 4.3. We can write (4.7) with $r = n$ as

$$
\begin{aligned}
\widehat{a}_j^{(n+1)} &= (P_n + \Delta P_n) \dots (P_1 + \Delta P_1) a_j \\
&\equiv (P_n \dots P_1 + \Delta I P_n \dots P_1) a_j \\
&= (I_m + \Delta I) Q^T a_j,
\end{aligned}
$$

where, by Lemma 2.3,

$$\|\Delta I\|_F \leq (1 + \tilde{\gamma}_m)^n - 1 = n\tilde{\gamma}_m'.$$

With $A_1 = I$ we deduce that

$$\widehat{Q} = (I_m + \Delta I) Q^T, \qquad \|\Delta I\|_F \leq \tilde{\gamma}_{mn}.$$

Hence $\|\widehat{Q} - Q\|_F = \|\Delta I Q^T\|_F \leq \tilde{\gamma}_{mn}$, showing that $\widehat{Q}$ is very close to an orthonormal matrix. Moreover, using Theorem 4.1

$$
\begin{aligned}
\|(A - \widehat{Q}\widehat{R})(:,j)\|_F &= \|(A - Q\widehat{R})(:,j) + ((Q - \widehat{Q})\widehat{R}))(:,j)\|_F \\
&\leq \tilde{\gamma}_{mn}\|A(:,j)\|_2 + \|Q - \widehat{Q}\|_F\|\widehat{R}(:,j)\|_2 \\
&\leq \tilde{\gamma}_{mn}'\|A(:,j)\|_2.
\end{aligned}
$$

Thus Theorem 4.1 remains true with Q replaced by $\widehat{Q}$.

One of the main uses of QR factorization is to solve the least squares problem $\min_x \|b - Ax\|_2$. Analogues of Theorems 4.1 and 4.2 hold for the computed LS solution [32, Theorem 19.3], [17].

4.2. The Constrained Least Squares Problem

We now consider the least squares problem with equality constraints

$$\text{LSE}: \quad \min_{Bx=d} \|b - Ax\|_2, \tag{4.12}$$

where $A \in \mathbb{R}^{m \times n}$ and $B \in \mathbb{R}^{p \times n}$, with $m+p \geq n \geq p$. Note that the condition $m \geq n - p$ ensures that the LSE problem is overdetermined. We will assume that

$$\text{rank}(B) = p, \quad \text{null}(A) \cap \text{null}(B) = \{0\}. \tag{4.13}$$

The assumption that B is of full rank ensures that the system $Bx = d$ is consistent and hence that the LSE problem has a solution. The second condition in (4.13), which is equivalent to the condition that the matrix $[A^T, B^T]^T$ has full rank n, then guarantees that there is a unique solution [8, Section 5.1].

The LSE problem arises in various applications, including the analysis of large-scale structures [6] and the solution of the inequality constrained least squares problem [38, Chap. 23].

There are two main classes of methods for solving the LSE problem: null space methods and elimination methods, with more than one variation of method within each class. A basic difference between the classes is that one QR factorizes the constraint matrix B while the other QR factorizes B^T.

We first describe the null space methods, so-called because they employ an orthogonal basis for the null space of the constraint matrix. We begin with a version based on the generalized QR factorization. The generalized QR factorization was introduced by Hammarling [27] and Paige [42] and further analyzed by Anderson, Bai and Dongarra [3] and is of interest in its own right.

Theorem 4.3 (generalized QR factorization). *Let $A \in \mathbb{R}^{m \times n}$ and $B \in \mathbb{R}^{p \times n}$ with $m + p \geq n \geq p$. There are orthogonal matrices $Q \in \mathbb{R}^{n \times n}$ and $U \in \mathbb{R}^{m \times m}$ such that*

$$U^T A Q = \begin{matrix} m-n+p \\ n-p \end{matrix} \begin{bmatrix} \overset{p}{L_{11}} & \overset{n-p}{0} \\ L_{21} & L_{22} \end{bmatrix}, \qquad B Q = p \begin{bmatrix} \overset{p}{S} & \overset{n-p}{0} \end{bmatrix}, \tag{4.14}$$

where L_{22} and S are lower triangular. More precisely, we have

$$U^T A Q = \begin{cases} \begin{matrix} m-n \\ n \end{matrix} \begin{bmatrix} \overset{n}{0} \\ L \end{bmatrix} & \text{if } m \geq n, \\[2em] m \begin{bmatrix} \overset{n-m}{X} & \overset{m}{L} \end{bmatrix} & \text{if } m < n, \end{cases} \tag{4.15}$$

where L is lower triangular. The assumptions (4.13) are equivalent to S and L_{22} being nonsingular.

Proof. Let

$$Q^T B^T = \begin{bmatrix} S^T \\ 0 \end{bmatrix}$$

be a QR factorization of B^T. We can determine an orthogonal U so that $U^T(AQ)$ has the form (4.15), where L is lower triangular (for example, we can construct U as a product of suitably chosen Householder transformations). Clearly, B has full rank if and only if S is nonsingular. Partition $Q = [Q_1\ Q_2]$ conformably with $[S\ 0]$ and assume S is nonsingular. Then, clearly, null$(B) =$ range(Q_2). We can write

$$A[Q_1\ \ Q_2] = [U_1\ \ U_2]\begin{bmatrix} L_{11} & 0 \\ L_{21} & L_{22} \end{bmatrix},$$

so that $AQ_2 = U_2 L_{22}$. It follows that null$(A) \cap$ null$(B) = \{0\}$ is equivalent to L_{22} being nonsingular.

While (4.15) is needed to define the generalized QR factorization precisely, the partitioning of $U^T AQ$ in (4.14) enables us to explain the application to the LSE problem without treating the cases $m \geq n$ and $m < n$ separately.

Using (4.14) the constraint $Bx = d$ may be written

$$Sy_1 = [S\ \ 0]\begin{bmatrix} y_1 \\ y_2 \end{bmatrix} = d, \qquad y = Q^T x.$$

Hence the constraint determines $y_1 \in \mathbb{R}^p$ as the solution of the triangular system $Sy_1 = d$ and leaves $y_2 \in \mathbb{R}^{n-p}$ arbitrary. Since

$$\|b - Ax\|_2 = \|c - U^T AQy\|_2, \qquad c = U^T b,$$

we see that we have to find

$$\min_{y_2}\left\|\begin{bmatrix} c_1 \\ c_2 \end{bmatrix} - \begin{bmatrix} L_{11} & 0 \\ L_{21} & L_{22} \end{bmatrix}\begin{bmatrix} y_1 \\ y_2 \end{bmatrix}\right\|_2 = \min_{y_2}\left\|\begin{bmatrix} c_1 - L_{11}y_1 \\ (c_2 - L_{21}y_1) - L_{22}y_2 \end{bmatrix}\right\|_2.$$

Therefore y_2 is the solution to the triangular system $L_{22}y_2 = (c_2 - L_{21}y_1)$. The solution x is recovered from $x = Qy$. We refer to this solution process as the GQR method. It is the method used by the LAPACK driver routine xgglse.f [2].

The stability of the GQR method is summarized by the following result [15].

Theorem 4.4. *Suppose the LSE problem (4.12) is solved using the GQR method, where the generalized QR factorization is computed using Householder transformations and let the assumptions (4.13) be satisfied. Let $\hat{x}$ denote the computed solution.*

1. $\hat{x} = \bar{x} + \Delta\bar{x}$, where $\bar{x}$ solves $\min\{\|b + \Delta b - (A + \Delta A)x\|_2 : (B + \Delta B)x = d\}$, with

$$\|\Delta\bar{x}\|_2 \;\leq\; \tilde{\gamma}_{np}\|\bar{x}\|_2, \qquad \|\Delta b\|_2 \leq \tilde{\gamma}_{mn}\|b\|_2,$$
$$\|\Delta A\|_F \;\leq\; \tilde{\gamma}_{mn}\|A\|_F, \qquad \|\Delta B\|_F \leq \tilde{\gamma}_{np}\|B\|_F.$$

2. $\hat{x}$ *solves* $\min\{\|b + \Delta b - (A + \Delta A)x\|_2 : (B + \Delta B)x = d + \Delta d\}$, *where*

$$\|\Delta b\|_2 \;\leq\; \tilde{\gamma}_{mn}\|b\|_2 + \tilde{\gamma}_{np}\|A\|_F\|\hat{x}\|_2, \qquad \|\Delta A\|_F \leq \tilde{\gamma}_{mn}\|A\|_F,$$
$$\|\Delta B\|_F \;\leq\; \tilde{\gamma}_{np}\|B\|_F, \qquad\qquad\quad \|\Delta d\|_2 \leq \tilde{\gamma}_{np}\|B\|_F\|\hat{x}\|_2,$$

The first part of the theorem says that $\hat{x}$ is very close to the exact solution of a slightly different LSE problem; this is a mixed form of stability. The second part says that $\hat{x}$ exactly solves a perturbed LSE problem in which the perturbations to A and B are tiny but those to b and d can be relatively large when x is large-normed. It is an open problem whether genuine backward stability holds. For the unconstrained least squares problem Gu [26] proves that mixed stability implies backward stability and it would be interesting to know whether this result can be extended to the LSE problem. In any case, the stability of the GQR method can be regarded as quite satisfactory.

The GQR method can be modified to reduce the amount of computation and the modified versions have the same stability properties [15].

The second main way of solving the LSE problem is by elimination. First, we use QR factorization with column pivoting to factorize

$$B\varPi = Q\,[\,R_1 \quad R_2\,], \qquad R_1 \in \mathbb{R}^{p \times p} \text{ upper triangular, nonsingular.} \quad (4.16)$$

Note that column pivoting is essential here in order to obtain a nonsingular R_1. Then, partitioning $\varPi^T x = [\tilde{x}_1^T,\ \tilde{x}_2^T]^T$, $\tilde{x}_1 \in \mathbb{R}^p$ and substituting the factorization (4.16) into the constraints yields

$$R_1\tilde{x}_1 = Q^T d - R_2\tilde{x}_2.$$

By solving for $\tilde{x}_1$ and partitioning $A\varPi = [\tilde{A}_1,\ \tilde{A}_2]$, $\tilde{A}_1 \in \mathbb{R}^{m \times p}$ we reduce the LSE problem to the unconstrained problem

$$\min_{\tilde{x}_2}\|(\tilde{A}_2 - \tilde{A}_1 R_1^{-1} R_2)\tilde{x}_2 - (b - \tilde{A}_1 R_1^{-1} Q^T d)\|_2.$$

Solving this unconstrained problem by QR factorization completes the elimination method as originally presented by Björck and Golub [9] (see also [38, Chapter 21]). It is instructive to think of the method in terms of transformations on the matrix "B-over-A":

$$\begin{bmatrix} B \\ A \end{bmatrix} = \begin{matrix} p \\ m \end{matrix} \begin{matrix} \overset{p}{} \quad \overset{n-p}{} \\ \begin{bmatrix} B_1 & B_2 \\ A_1 & A_2 \end{bmatrix} \end{matrix} \to \begin{bmatrix} R_1 & R_2 \\ \tilde{A}_1 & \tilde{A}_2 \end{bmatrix}$$

$$\to \begin{bmatrix} R_1 & R_2 \\ 0 & \tilde{A}_2 - \tilde{A}_1 R_1^{-1} R_2 \end{bmatrix} \to \begin{bmatrix} R_1 & R_2 \\ 0 & R_3 \\ 0 & 0 \end{bmatrix},$$

where $R_3 \in \mathbb{R}^{(n-p) \times (n-p)}$ is upper triangular. Note that the penultimate transformation is simply the annihilation of $\tilde{A}_1$ by Gaussian elimination. The B-over-A matrix also arises in the method of weighting for solving the LSE problem, which is based on the observation that the LSE solution is the limit of the solution of the unconstrained problem

$$\min_x \left\| \begin{bmatrix} \mu d \\ b \end{bmatrix} - \begin{bmatrix} \mu B \\ A \end{bmatrix} x \right\|_2 \qquad (4.17)$$

as the weight μ tends to infinity.

A row-wise backward error result is available for the elimination method [16]. The computed solution exactly solves a perturbed LSE problem for which row-wise backward error bounds hold that involve row-wise growth factors; if row sorting or row pivoting is used (separately on A and B) then the growth factors have similar behaviour to the α_i in Theorem 4.2.

5. The Singular Value Decomposition and Jacobi's Method

> Jacobi used as a computing system his student Ludwig Seidel, apparently operating in eight-digit decimal arithmetic!
> *J. C. NASH, 'Compact Numerical Methods for Computers' (1990)*

One of the oldest methods for solving the symmetric eigenvalue problem and computing the singular value decomposition (SVD) is Jacobi's method. The method's publication date of 1846 predates the development of matrix theory (the term "matrix" was first used in 1850 by Sylvester). Although the QR algorithm has always been favoured because of its lower operation count, Jacobi's method has attracted renewed interest since the 1980s because of its suitability for parallel implementation and its high accuracy properties. We describe Jacobi's method for the SVD, concentrating on its accuracy and stability.

The perturbation theory and error analysis in this section is based on that in the book of Demmel [18] and gives a relatively accessible explanation of the accuracy properties of Jacobi methods. The research literature should be consulted for further details, of which there are many. The best starting points are Demmel and Veselić [19] and Mathias [39].

Recall that the SVD of $A \in \mathbb{R}^{m \times n}$, where $m \geq n$, has the form

$$A = U \Sigma V^T, \qquad \Sigma = \operatorname{diag}(\sigma_i) \in \mathbb{R}^{m \times n}, \qquad \sigma_1 \geq \sigma_2 \geq \cdots \geq \sigma_n \geq 0,$$

where $U \in \mathbb{R}^{m \times m}$ and $V \in \mathbb{R}^{n \times n}$ are orthogonal. The σ_i are the singular values and the columns of U and V contain the left and right singular vectors, respectively. Jacobi's method computes the reduced-size SVD in which $U \in \mathbb{R}^{m \times n}$ has orthonormal columns and $\Sigma \in \mathbb{R}^{n \times n}$. We will use this form of the SVD throughout the section. The last $m - n$ columns of U are arbitrary subject to completing an orthogonal matrix.

5.1. Jacobi's Method

First, consider the following orthogonal similarity transformation of a symmetric 2×2 matrix:

$$\begin{bmatrix} h'_{jj} & h'_{kj} \\ h'_{kj} & h'_{kk} \end{bmatrix} = J^T C J = \begin{bmatrix} c & -s \\ s & c \end{bmatrix}^T \begin{bmatrix} h_{jj} & h_{kj} \\ h_{kj} & h_{kk} \end{bmatrix} \begin{bmatrix} c & -s \\ s & c \end{bmatrix}, \quad (5.1a)$$

$$c = \cos\theta, \quad s = \sin\theta. \quad (5.1b)$$

We wish to choose c and s to make h'_{kj} zero. If $h_{kj} = 0$ then we can set $c = 1$, $s = 0$. Otherwise, multiplying out we find that

$$h'_{kj} = sc(h_{kk} - h_{jj}) + h_{kj}(c^2 - s^2),$$

and so $h'_{kj} = 0$ if $t = \tan\theta$ satisfies

$$t^2 + 2\tau t - 1 = 0, \quad \tau = \frac{h_{jj} - h_{kk}}{2h_{kj}}.$$

We take the smaller of the two roots which can be expressed as

$$t = \frac{\text{sign}(\tau)}{|\tau| + \sqrt{1 + \tau^2}}$$

(where $\text{sign}(0) = 1$), and then obtain c and s from

$$c = \frac{1}{\sqrt{1 + t^2}}, \quad s = tc. \quad (5.2)$$

The corresponding rotation angle θ satisfies $|\theta| \leq \pi/4$; choosing a small rotation angle is essential for the convergence theory [44, Chapter 9].

The transformation (5.1) with (5.2) is the basis of Jacobi's method for computing the eigensystem of a symmetric $H \in \mathbb{R}^{n \times n}$. For a sequence of pairs (j, k) the transformation is applied in the (j, k) plane to the matrix H to eliminate the (j, k) and (k, j) elements. Thus effectively we are applying orthogonal similarity transformations to H with matrices of the form

$$J(j, k, \theta) = \begin{matrix} & & & j & & k & \\ & \begin{bmatrix} 1 & & \vdots & & \vdots & \\ & \ddots & \vdots & & \vdots & \\ j & \cdots & \cdots & c & & s & \\ & & & & \ddots & & \\ k & \cdots & \cdots & -s & & c & \\ & & & & & & \ddots \\ & & & & & & & 1 \end{bmatrix} \end{matrix},$$

which are called Jacobi matrices or Jacobi rotations (they are identical to Givens rotations). A rotation will, in general, destroy zeros introduced by an earlier rotation, but the hope is that the norm of the off-diagonal will converge to zero, leaving the eigenvalues on the diagonal.

Returning to the SVD, recall that the singular values of A are the square roots of the eigenvalues of $A^T A$. Therefore we can compute the singular values of A by applying Jacobi's method to $A^T A$. It is undesirable to form $A^T A$ explicitly, because of the potential loss of information when A is ill conditioned. It is preferable to apply Jacobi's method implicitly, at each stage postmultiplying A by the Jacobi transformation defined by Jacobi's method for the symmetric eigenproblem applied to $A^T A$. This idea is encapsulated in the **one-sided Jacobi algorithm** (from the right), which was first proposed by Hestenes [30].

Algorithm 5 (one-sided Jacobi). This algorithm computes the SVD of $A = U \Sigma V^T \in \mathbb{R}^{m \times n}$, $m \geq n$, by the one-sided Jacobi algorithm.

$\quad$ done_rot $=$ true; $V = I$
$\quad$ while done_rot $=$ true
$\quad\quad$ done_rot $=$ false
$\quad\quad$ for $j = 1 : n - 1$
$\quad\quad\quad$ for $k = j + 1 : n$
$\quad\quad\quad\quad h_{jj} = A(:,j)^T A(:,j)$
$\quad\quad\quad\quad h_{kj} = A(:.j)^T A(:,k)$
$\quad\quad\quad\quad h_{kk} = A(:,k)^T A(:,k)$
$\quad\quad\quad\quad$ if $|h_{kj}| > u \sqrt{h_{jj} h_{kk}}$
$\quad\quad\quad\quad\quad$ done_rot $=$ true
$\quad\quad\quad\quad\quad \tau = (h_{jj} - h_{kk})/(2 h_{kj})$
$\quad\quad\quad\quad\quad t = \operatorname{sign}(\tau)/(|\tau| + \sqrt{1 + \tau^2})$
$\quad\quad\quad\quad\quad c = 1/\sqrt{1 + t^2}$, $s = ct$

$$A(:, [j\ k]) = A(:, [j\ k]) \begin{bmatrix} c & -s \\ s & c \end{bmatrix}$$

$$V(:, [j\ k]) = V(:, [j\ k]) \begin{bmatrix} c & -s \\ s & c \end{bmatrix}$$

$\quad\quad\quad$ end
$\quad\quad$ end
$\quad$ end
$\quad$ end
$\quad$ Sort the columns of A in decreasing order of 2-norm
$\quad$ and interchange the columns of V in the same way.
$\quad$ for $j = 1 : n$
$\quad\quad \sigma_j = \|A(:,j)\|_2$
$\quad\quad U(:,j) = A(:,j)/\sigma_j$
$\quad$ end

Denoting the matrix on a particular step of Algorithm 5 by A, the 2×2 principal submatrix of $A^T A$ whose off-diagonal element is to be zeroed is formed explicitly in order to determine the Jacobi rotation. However, the rotation is applied to A (on the right) and not $A^T A$ (from both sides), which leads to excellent stability properties, as we will see.

The criterion $|h_{kj}| > u\sqrt{h_{jj}h_{kk}}$ for deciding when to carry out a rotation can be expressed as

$$u < \frac{|A(:,j)^T A(:,k)|}{\|A(:,j)\|_2 \|A(:,k)\|_2} = \cos(\theta_{jk}),$$

where θ_{jk} is the angle between $A(:,j)$ and $A(:,k)$. A rotation is therefore carried out whenever the jth and kth columns are not already orthogonal to working precision. Further justification for this criterion is given in Section 5.2.

The convergence test for Algorithm 5 is to stop when a sweep has been completed without carrying out any rotations, where a sweep denotes that all the elements in the upper triangle of $A^T A$ have been eliminated in turn. Note that an alternative (and expensive) convergence test is to stop when $\mathrm{off}(A^T A)/\|A\|_F^2$ is less than some modest multiple of u, where $\mathrm{off}(B) = (\sum_{i \neq j} b_{ij}^2)^{1/2}$. However, with this weaker criterion the computed matrices of singular vectors may not be numerically orthogonal. A numerical example illustrates this point. With A the floating point approximation $\tilde{H}_{10}$ to the 10×10 Hilbert matrix we applied Algorithm 5 as stated and then again using the stopping criterion $\mathrm{off}(A^T A)/\|A\|_F^2 < u$. The computations were done in MATLAB, with $u = 2^{-53} \approx 1.1 \times 10^{-16}$. The results are shown in Table 5.1. (As the exact singular values of $\tilde{H}_{10}$ we took the values computed in 50 digit arithmetic by MATLAB's Symbolic Math Toolbox [40]). Notice that the off stopping criterion results in fewer iterations but provides less accurate computed singular values and a U completely lacking orthogonality. A bizarre stopping criterion for Jacobi's method for the symmetric eigenproblem is used in [46], based on testing whether $\mathrm{off}(A)$ underflows to zero. Proper choice of stopping criterion is vital if a reliable Jacobi code is to be produced.

Table 5.1. Effect of stopping criterion in Algorithm 5.

	Original stopping criterion	off stopping criterion
Number of sweeps	9	5
$\max_i |\sigma_i - \hat{\sigma}_i|/\sigma_i$	4.7e-5	6.6e0
$\|U^T U - I\|_2$	5.2e-16	9.9e-1
$\|V^T V - I\|_2$	3.0e-15	2.9e-15

Convergence of the algorithm corresponds to $A^T A$ being diagonal, which means that the columns of A are orthonormal with 2-norms equal to the

74   Nicholas J. Higham

singular values. The left singular vectors that make up the columns of U are
then obtained by scaling the columns by these singular values. Algorithm 5
does converge, as proved by Forsythe and Henrici [24], and the asymptotic
rate of convergence is quadratic; see [29], [44, Chapter 9] for details.

The overall cost of applying Algorithm 5 can be reduced by computing a
QR factorization of A or A^T and then applying the algorithm to the square
upper triangular factor.

Finally, we note that the code in Algorithm 5 is very compact by comparison with that for methods based on reduction to bidiagonal form. The
ease of coding of Jacobi methods was an advantage on computers that had
little memory [41], but is less of an issue nowadays. Indeed there are many
subtleties in the implementation of Jacobi methods—even how to construct
and apply a single rotation in an accurate and robust fashion is a nontrivial
question [22].

5.2. Relative Perturbation Theory

In order to explain the accuracy properties of the one-sided Jacobi algorithm
we need a new style of perturbation theory known as relative perturbation
theory. Whereas traditional perturbation theory for eigenvalues and singular
values provides absolute error bounds that are the same for each eigenvalue
or singular value, relative perturbation theory provides relative error bounds
that are the same for each eigenvalue or singular value (these bounds are
therefore stronger than what is obtained by direct conversion of the absolute bounds). We will treat the perturbations in multiplicative rather than
additive form, in order to match the error analysis of the next section. For
a survey of relative perturbation theory and additive versus multiplicative
representation of perturbations see Ipsen [36].

The result that we need for singular values is obtained via eigenvalue
perturbation theory, so we first consider eigenvalues. We denote the ith largest
eigenvalue of a symmetric matrix $A \in \mathbb{R}^{n \times n}$ by $\lambda_i(A)$, so that the eigenvalues
are ordered $\lambda_n \leq \lambda_{n-1} \leq \cdots \leq \lambda_1$. Similarly, $\sigma_i(B)$ denotes the ith largest
singular value of B.

We need two standard results. The first is an immediate consequence of
the Courant–Fischer characterization of eigenvalues [47, p. 201] and is also a
special case of Weyl's inequality [47, p. 203].

Lemma 5.1. If $A \in \mathbb{R}^{n \times n}$ and $E \in \mathbb{R}^{n \times n}$ are symmetric then $|\lambda_i(A + E) - \lambda_i(A)| \leq \|E\|_2$.

The next two results are from Eisenstat and Ipsen [23, Theorems 2.1, 3.3].

Theorem 5.1. Let the symmetric matrices A and $\widetilde{A} = X^T A X$ have eigenvalues λ_i and $\widetilde{\lambda}_i$, respectively, and assume X is nonsingular. Then

$$|\lambda_i - \widetilde{\lambda}_i| \leq |\lambda_i|\epsilon,$$

where $\epsilon = \|X^TX - I\|_2$.

Proof. Sylvester's inertia theorem tells us that the matrices $A - \lambda_i I$ and $X^T(A - \lambda_i I)X$ have the same number of negative, zero and positive eigenvalues. Since the ith eigenvalue of $A - \lambda_i I$ is zero, so is the ith eigenvalue of

$$X^T(A - \lambda_i I)X = (X^TAX - \lambda_i I) + \lambda_i(I - X^TX) \equiv C + E.$$

From Lemma 5.1 it follows that $|\lambda_i(C) - 0| \leq \|E\|_2$, which gives the result since $\|E\|_2 \leq |\lambda_i|\|X^TX - I\|_2$.

Our desired singular value result now follows as a corollary.

Corollary 5.1. *Let* $B \in \mathbb{R}^{m \times n}$ *and* $\widetilde{B} = Y^TBX$ *have singular values* σ_i *and* $\widetilde{\sigma}_i$, *respectively, where* $Y \in \mathbb{R}^{m \times m}$ *and* $X \in \mathbb{R}^{n \times n}$, *and assume* X *and* Y *have full rank. Then*

$$|\sigma_i - \widetilde{\sigma}_i| \leq |\sigma_i|\epsilon,$$

where $\epsilon = \max(\|X^TX - I\|_2, \|Y^TY - I\|_2)$.

Proof. Since X and Y have full rank, B and $\widetilde{B}$ have the same number of zero singular values. Recall that the nonzero singular values of B are plus and minus the nonzero eigenvalues of

$$A = \begin{bmatrix} 0 & B \\ B^T & 0 \end{bmatrix}.$$

Similarly, for $\widetilde{B}$ we can write

$$\widetilde{A} = \begin{bmatrix} 0 & Y^TBX \\ X^TB^TY & 0 \end{bmatrix} = \text{diag}(Y, X)^T \begin{bmatrix} 0 & B \\ B^T & 0 \end{bmatrix} \text{diag}(Y, X).$$

Applying Theorem 5.1 to A and $\widetilde{A}$ yields the result.

Corollary 5.1 makes precise the intuitive notion that a transformation $B \to Y^TBX$ will have little effect on the singular values only if X and Y are nearly unitary.

With the aid of Theorem 5.1 we can justify the convergence test in Algorithm 5. Consider $H = A^TA$ and assume without loss of generality that the diagonal elements of H are sorted in decreasing order. Suppose $|h_{kj}| \leq \epsilon\sqrt{h_{jj}h_{kk}}$ for all k and j. Let $D = \text{diag}(h_{jj}^{1/2})$ and define Δ by

$$\delta_{ij} = \begin{cases} 0, & i = j, \\ \dfrac{h_{ij}}{\sqrt{h_{ii}h_{jj}}}, & i \neq j, \end{cases}$$

so that $H = D(I + \Delta)D$. Write $I + \Delta = X^2$, where X is the symmetric positive definite square root. Then

$$H = DX^2D = (XD)^{-1}XD^2X(XD).$$

Thus H is similar to XD^2X, and so by Theorem 5.1

$$|\lambda_i(H) - h_{ii}| \leq \|X^TX - I\|_2\lambda_i(H) = \|\Delta\|_2\lambda_i(H).$$

But $\lambda_i(H) = \sigma_i^2(A)$ and $\|\Delta\|_2 \leq (n-1)\epsilon$, so

$$\begin{aligned}
|\sigma_i(A) - h_{ii}^{1/2}| &\leq (n-1)\epsilon\frac{\sigma_i^2(A)}{\sigma_i(A) + h_{ii}^{1/2}} \\
&\leq (n-1)\epsilon\sigma_i(A).
\end{aligned}$$

Hence the singular values of A agree with the square roots of the (sorted) diagonal elements of H to high relative accuracy and so Algorithm 5 terminates the iteration when the desired accuracy has been obtained.

5.3. Error Analysis

For the error analysis of the one-sided Jacobi algorithm we will assume that A is square, which will be the case if a preliminary QR factorization has been used.

Standard error analysis for SVD algorithms based on orthogonal transformations says that the computed singular values $\widehat{\sigma}_i$ of $A \in \mathbb{R}^{m \times n}$ are the exact ones of $A + \Delta A$, where $\|\Delta A\|_2 \leq p(m,n)u\|A\|_2$ for some polynomial p. This yields the forward error bound for the singular values, from an analogue for singular values of Lemma 5.1,

$$|\sigma_i - \widehat{\sigma}_i| \leq p(m,n)u\sigma_1.$$

Hence the large singular values are guaranteed to be computed to high accuracy but the small ones (if there are any) are not. Algorithm 5 can provide better relative accuracy—a fact that has only come to be widely appreciated in the 1990s. The following result from [18, Theorem 5.15], [21, Theorem 6.1] provides a relative error bound of the same size for each singular value.

Theorem 5.2. Let $A \in \mathbb{R}^{n \times n}$ be nonsingular and write $A = DX$, where D is diagonal. Let $\widehat{A}$ be the matrix obtained from Algorithm 5 after applying p Jacobi rotations. Then the singular values σ_i of A and $\widehat{\sigma}_i$ of $\widehat{A}$ satisfy

$$\frac{|\sigma_i - \widehat{\sigma}_i|}{\sigma_i} \leq \kappa_2(X)\sqrt{n}\tilde{\gamma}_p. \tag{5.3}$$

Proof. It is straightforward to show that the construction and application of a single Jacobi rotation, $y = Jx$, results in a computed $\widehat{y}$ for which $\widehat{y} = (J+\Delta J)x$, with $\|\Delta J\|_F \leq \tilde{\gamma}_1$, where J is exactly orthogonal (the proof is very similar to that for application of a Givens rotation in QR factorization [32, Lemma 18.7]). In Algorithm 5 we apply p Jacobi rotations from the right and so the final matrix $\widehat{A}$ satisfies

$$\widehat{A}(i,:) = A(i,:)(J_1 + \Delta J_1)\dots(J_p + \Delta J_p), \qquad \|\Delta J_i\|_F \le \tilde{\gamma}_1.$$

From Lemma 2.3 it follows that

$$\widehat{A}(i,:) = A(i,:)Q + \Delta A(i,:),$$

where $Q = J_1 \dots J_p$ and

$$\|\Delta A(i,:)\|_2 \le ((1 + \tilde{\gamma}_1)^p - 1)\|A(i,:)\|_2 = \tilde{\gamma}_p\|A(i,:)\|_2.$$

Hence $\|D^{-1}\Delta A\|_2 \le \sqrt{n}\tilde{\gamma}_p\|X\|_2$. Thus

$$
\begin{aligned}
\widehat{A} &= AQ + \Delta A = AQ(I + Q^T A^{-1}\Delta A) = AQ(I + Q^T X^{-1}D^{-1}\Delta A) \\
&\equiv AQ(I + E),
\end{aligned}
$$

where

$$\|E\|_2 \le \kappa_2(X)\sqrt{n}\tilde{\gamma}_p. \tag{5.4}$$

Applying Corollary 5.1 we deduce that, with $\widehat{\sigma}_i = \sigma_i(\widehat{A})$,

$$
\begin{aligned}
\frac{|\sigma_i - \widehat{\sigma}_i|}{\sigma_i} &\le \|(I + E)^T(I + E) - I\|_2 = \|E + E^T + E^T E\|_2 \le 3\|E\|_2 \\
&\le \kappa_2(X)\sqrt{n}\tilde{\gamma}_p,
\end{aligned}
$$

where we have assumed that $\|E\|_2 < 1$.

Theorem 5.2 shows that Jacobi transformations from the right introduce only small relative errors in the singular values in floating point arithmetic provided that X is well conditioned. It is natural to ask what is the minimum value of $\kappa_2(X)$ over all nonsingular diagonal D. A result of van der Sluis [49] shows that

$$\kappa_2(D_R A) \le \sqrt{n}\min_D \kappa_2(DA),$$

where $D_R = \text{diag}(\|A(i,:)\|_2)^{-1}$. Therefore $X = D_R A$, which has rows of unit 2-norm, gives approximately the best bound (5.3).

Error bounds can also be obtained for the singular vectors. They show that the absolute error in the singular vectors is bounded by a multiple of the reciprocal of the relative gap between the singular values; see [19] for details. For standard SVD methods it is the reciprocal of the absolute gap that appears in the bounds, and the absolute gap can be much smaller than the relative gap for small clustered singular values.

We give a numerical example to illustrate Theorem 5.2. We chose $A = DX \in \mathbb{R}^{10 \times 10}$, where $D = \text{diag}(10^{-10}, 10^{-9}, \dots, 1)$ and X is a random matrix from the normal $(0,1)$ distribution, with $\kappa_2(X) = 1.4 \times 10^2$. Table 5.2 shows the relative errors in the singular values of A computed by Algorithm 5 and by MATLAB's SVD routine, which implements the Golub–Reinsch algorithm (bidiagonalization followed by the implicit QR algorithm). As Theorem 5.2 predicts, the one-sided Jacobi algorithm provides high relative accuracy in

Table 5.2. Relative errors in computed singular values.

| | Relative Error | |
True singular value	Algorithm 5	MATLAB's SVD
1.9e+00	2.3e-16	1.2e-16
1.4e-01	1.9e-16	3.8e-16
1.0e-02	6.6e-16	0.0e+00
9.2e-04	2.3e-16	1.1e-15
8.5e-05	8.0e-16	6.4e-16
7.0e-06	1.2e-16	2.4e-16
7.0e-07	1.5e-16	2.0e-13
6.6e-08	2.0e-16	5.3e-12
1.6e-09	3.8e-15	1.2e-08
2.1e-10	3.4e-15	1.2e-07

all the singular values, but MATLAB's SVD loses up to half the digits in the small singular values.

The representation $A = DX$ in Theorem 5.2 factors out the row scaling of A. It is natural to as whether a similar result holds if we factor out the column scaling: $A = XD$. The answer is yes, and in fact this is the result originally proved by Demmel and Veselić [19, Corollary 4.2]. Mathias [39] calls the case where the orthogonal transformations are applied on the same side as the scaling the "harder case" and explains the differences with the "easy case" that we have considered here. Repeating the numerical example above with A defined as $A = XD$ rather than $A = DX$, we found that the relative errors were very similar to those in Table 5.2.

5.4. Other Issues

Several variations of the Jacobi method are possible. For the SVD we can apply one-sided Jacobi from the left, thereby acting on the rows (which is not recommended in Fortran, since arrays are stored columnwise). We can apply two-sided Jacobi transformations, in which in each 2×2 subproblem a pair of rotations is applied from the left and the right to diagonalize the submatrix; this is known as the Kogbetliantz algorithm [8, Section 2.6.6], [37].

The one-sided Jacobi algorithm can also be used to compute the eigensystem of a symmetric positive definite matrix A [19], [39], [50]. The idea is to compute a Cholesky factorization $A = R^T R$ and then apply Algorithm 5 to R^T. The SVD $R^T = U\Sigma V^T$ yields the eigendecomposition $A = U\Sigma^2 U^T$. This approach has high accuracy, as we now show using error analysis for Cholesky factorization. The computed Cholesky factor $\widehat{R}$ satisfies [32, Theorem 10.5]

$$A + \Delta A = \widehat{R}^T \widehat{R}, \qquad |\Delta A| \leq \tilde{\gamma}_n dd^T,$$

where $d_i = a_{ii}^{1/2}$. Define $Y = \widehat{R}^{-T} R^T$, so that

$$RR^T = Y^T \widehat{R}\widehat{R}^T Y. \tag{5.5}$$

Now

$$
\begin{aligned}
Y^T Y &= R\widehat{R}^{-1}\widehat{R}^{-T}R^T = R(A + \Delta A)^{-1}R^T \\
&\approx R(A^{-1} - A^{-1}\Delta A A^{-1})R^T \\
&= I - R^{-T}\Delta A R^{-1} \tag{5.6}
\end{aligned}
$$

and, with $D = \mathrm{diag}(d_i)$ and $e = [1, 1, \dots, 1]^T$,

$$
\begin{aligned}
\|R^{-T}\Delta A R^{-1}\|_2 &= \|R^{-1}D \cdot D^{-1}\Delta A D^{-1} \cdot DR^{-1}\|_2 \\
&\leq \|DR^{-1}\|_2^2 \|D^{-1}\Delta A D^{-1}\|_2 \\
&\leq \|DR^{-1}\|_2^2 \, \tilde{\gamma}_n \, \|ee^T\|_2 \\
&= n\tilde{\gamma}_n \|DR^{-1}\|_2^2.
\end{aligned}
$$

Now $R^T R = A =: DCD$, where $c_{ii} \equiv 1$, which gives $DR^{-1} \cdot R^{-T}D = C^{-1}$ and hence $\|DR^{-1}\|_2^2 = \|C^{-1}\|_2$. Therefore

$$\|R^{-T}\Delta A R^{-1}\|_2 \leq n\tilde{\gamma}_n \kappa_2(C), \tag{5.7}$$

since $1 \leq \|C\|_2 \leq n$. From (5.5)–(5.7) and Theorem 5.1 we deduce that (to first order) the eigenvalues of RR^T and $\widehat{R}\widehat{R}^T$ differ by a relative amount at most $n\tilde{\gamma}_n \kappa_2(C)$. If we write $R^T = DX^T$, then $X^T X = C$, so $\kappa_2(C) = \kappa_2(X)^2$. Thus the eigenvalues of A and the squared singular values of $\widehat{R}$ differ by at most $O(\kappa_2(X)^2 u)$. From Theorem 5.2 we know that the Jacobi algorithm applied to R^T introduces errors of order $\kappa_2(X)u$ and we see that these are no larger and possibly much smaller than the errors introduced by the Cholesky factorization.

Finally, we note that although Jacobi methods have typically been slower than the QR algorithm for the symmetric eigenproblem, a sophisticated implementation using recent "preconditioning" techniques can compete with and sometimes beat the QR algorithm for speed (Drmač, private communication).

Acknowledgements

I am grateful to Zlatko Drmač for suggesting several improvements to Section 5.

Bibliography

1. Jan Ole Aasen. On the reduction of a symmetric matrix to tridiagonal form. *BIT*, 11:233–242, 1971.

2. E. Anderson, Z. Bai, C. H. Bischof, J. W. Demmel, J. J. Dongarra, J. J. Du Croz, A. Greenbaum, S. J. Hammarling, A. McKenney, S. Ostrouchov, and D. C. Sorensen. *LAPACK Users' Guide, Release 2.0*. Second edition, Society for Industrial and Applied Mathematics, Philadelphia, PA, USA, 1995. xix+325 pp. ISBN 0-89871-345-5.

3. E. Anderson, Z. Bai, and J. Dongarra. Generalized QR factorization and its applications. *Linear Algebra and Appl.*, 162-164:243–271, 1992.

4. Cleve Ashcraft, Roger G. Grimes, and John G. Lewis. Accurate symmetric indefinite linear equation solvers. To appear in SIAM J. Matrix Anal. Appl., September 1998. 51 pp.

5. Zhaojun Bai, James W. Demmel, and Ming Gu. Inverse free parallel spectral divide and conquer algorithms for nonsymmetric eigenproblems. *Numer. Math.*, 76:279–308, 1997.

6. J. L. Barlow, N. K. Nichols, and R. J. Plemmons. Iterative methods for equality-constrained least squares problems. *SIAM J. Sci. Stat. Comput.*, 9(5):892–906, 1988.

7. Victor Barwell and Alan George. A comparison of algorithms for solving symmetric indefinite systems of linear equations. *ACM Trans. Math. Software*, 2(3):242–251, 1976.

8. Åke Björck. *Numerical Methods for Least Squares Problems*. Society for Industrial and Applied Mathematics, Philadelphia, PA, USA, 1996. xvii+408 pp. ISBN 0-89871-360-9.

9. Åke Björck and Gene H. Golub. Iterative refinement of linear least squares solutions by Householder transformation. *BIT*, 7:322–337, 1967.

10. James R. Bunch. Partial pivoting strategies for symmetric matrices. *SIAM J. Numer. Anal.*, 11(3):521–528, 1974.

11. James R. Bunch and Linda Kaufman. Some stable methods for calculating inertia and solving symmetric linear systems. *Math. Comp.*, 31(137):163–179, 1977.

12. James R. Bunch and Beresford N. Parlett. Direct methods for solving symmetric indefinite systems of linear equations. *SIAM J. Numer. Anal.*, 8(4):639–655, 1971.

13. Sheung Hun Cheng. *Symmetric Indefinite Matrices: Linear System Solvers and Modified Inertia Problems*. PhD thesis, University of Manchester, Manchester, England, January 1998. 150 pp.

14. Sheung Hun Cheng and Nicholas J. Higham. A modified Cholesky algorithm based on a symmetric indefinite factorization. Numerical Analysis Report No. 289, Manchester Centre for Computational Mathematics, Manchester, England, April 1996. 18 pp. To appear in SIAM J. Matrix Anal. Appl.

15. Anthony J. Cox and Nicholas J. Higham. Accuracy and stability of the null space method for solving the equality constrained least squares problem. Numerical Analysis Report No. 306, Manchester Centre for Computational Mathematics, Manchester, England, August 1997. 20 pp. To appear in BIT, 39(1): 1999.

16. Anthony J. Cox and Nicholas J. Higham. Row-wise backward stable elimination methods for the equality constrained least squares problem. Numerical Analysis Report No. 319, Manchester Centre for Computational Mathematics, Manchester, England, March 1998. 18 pp. To appear in SIAM J. Matrix Anal. Appl.

17. Anthony J. Cox and Nicholas J. Higham. Stability of Householder QR factorization for weighted least squares problems. In *Numerical Analysis 1997, Proceedings of the 17th Dundee Biennial Conference*, D. F. Griffiths, D. J. Higham, and

G. A. Watson, editors, volume 380 of *Pitman Research Notes in Mathematics*, Addison Wesley Longman, Harlow, Essex, UK, 1998, pages 57–73.

18. James W. Demmel. *Applied Numerical Linear Algebra*. Society for Industrial and Applied Mathematics, Philadelphia, PA, USA, 1997. xi+419 pp. ISBN 0-89871-389-7.

19. James W. Demmel and Krešimir Veselić. Jacobi's method is more accurate than QR. *SIAM J. Matrix Anal. Appl.*, 13(4):1204–1245, 1992.

20. J. J. Dongarra, J. R. Bunch, C. B. Moler, and G. W. Stewart. *LINPACK Users' Guide*. Society for Industrial and Applied Mathematics, Philadelphia, PA, USA, 1979. ISBN 0-89871-172-X.

21. Zlatko Drmač. *Computing the Singular and the Generalized Singular Values*. PhD thesis, Lehrgebiet Mathematische Physik, Fernuniversität Hagen, 1994. 193 pp.

22. Zlatko Drmač. Implementation of Jacobi rotations for accurate singular value computation in floating point arithmetic. *SIAM J. Sci. Comput.*, 18(4):1200–1222, 1997.

23. Stanley C. Eisenstat and Ilse C. F. Ipsen. Relative perturbation techniques for singular value problems. *SIAM J. Numer. Anal.*, 32(6):1972–1988, 1995.

24. G. E. Forsythe and P. Henrici. The cyclic Jacobi method for computing the principal values of a complex matrix. *Trans. Amer. Math. Soc.*, 94:1–23, 1960.

25. Gene H. Golub and Charles F. Van Loan. *Matrix Computations*. Third edition, Johns Hopkins University Press, Baltimore, MD, USA, 1996. xxvii+694 pp. ISBN 0-8018-5413-X (hardback), 0-8018-5414-8 (paperback).

26. Ming Gu. Backward perturbation bounds for linear least squares problems. *SIAM J. Matrix Anal. Appl.*, 1998. To appear.

27. Sven J. Hammarling. The numerical solution of the general Gauss–Markov linear model. In *Mathematics in Signal Processing*, T. S. Durrani, J. B. Abbiss, and J. E. Hudson, editors, Oxford University Press, 1987, pages 451–456.

28. Richard J. Hanson. Aasen's method for linear systems with self-adjoint matrices. Visual Numerics, Inc., http://www.vni.com/books/whitepapers/Aasen.html, July 1997.

29. Vjeran Hari. On sharp quadratic convergence bounds for the serial Jacobi methods. *Numer. Math.*, 60:375–406, 1991.

30. Magnus R. Hestenes. Inversion of matrices by biorthogonalization and related results. *J. Soc. Indust. Appl. Math.*, 6(1):51–90, 1958.

31. Nicholas J. Higham. Computing the polar decomposition—with applications. *SIAM J. Sci. Stat. Comput.*, 7(4):1160–1174, October 1986.

32. Nicholas J. Higham. *Accuracy and Stability of Numerical Algorithms*. Society for Industrial and Applied Mathematics, Philadelphia, PA, USA, 1996. xxviii+688 pp. ISBN 0-89871-355-2.

33. Nicholas J. Higham. Stability of block LDL^T factorization of a symmetric tridiagonal matrix. Numerical Analysis Report No. 308, Manchester Centre for Computational Mathematics, Manchester, England, September 1997. 8 pp. To appear in Linear Algebra and Appl.

34. Nicholas J. Higham. Stability of the diagonal pivoting method with partial pivoting. *SIAM J. Matrix Anal. Appl.*, 18(1):52–65, January 1997.

35. Nicholas J. Higham. QR factorization with complete pivoting and accurate computation of the SVD. Numerical Analysis Report 324, Manchester Centre for Computational Mathematics, Manchester, England, August 1998. 26 pp.

36. Ilse C. F. Ipsen. Relative perturbation results for matrix eigenvalues and singular values. In *Acta Numerica*, volume 7, Cambridge University Press, 1998, pages 151–201.

37. E. G. Kogbetliantz. Solution of linear equations by diagonalization of coefficients matrix. *Quart. Appl. Math.*, 13(2):123–132, 1955.
38. Charles L. Lawson and Richard J. Hanson. *Solving Least Squares Problems.* Society for Industrial and Applied Mathematics, Philadelphia, PA, USA, 1995. xii+337 pp. Revised republication of work first published in 1974 by Prentice-Hall. ISBN 0-89871-356-0.
39. Roy Mathias. Accurate eigensystem computations by Jacobi methods. *SIAM J. Matrix Anal. Appl.*, 16(3):977–1003, 1995.
40. Cleve Moler and Peter J. Costa. *Symbolic Math Toolbox Version 2.0: User's Guide.* The MathWorks, Inc., Natick, MA, USA, 1997.
41. J. C. Nash. *Compact Numerical Methods for Computers: Linear Algebra and Function Minimisation.* Second edition, Adam Hilger, Bristol, 1990. xii+278 pp. ISBN 0-85274-319-X.
42. C. C. Paige. Some aspects of generalized QR factorizations. In *Reliable Numerical Computation*, M. G. Cox and S. J. Hammarling, editors, Oxford University Press, 1990, pages 73–91.
43. Beresford N. Parlett. Analysis of algorithms for reflections in bisectors. *SIAM Review*, 13(2):197–208, 1971.
44. Beresford N. Parlett. *The Symmetric Eigenvalue Problem.* Society for Industrial and Applied Mathematics, Philadelphia, PA, USA, 1998. xxiv+398 pp. Unabridged, amended version of book first published by Prentice-Hall in 1980. ISBN 0-89871-402-8.
45. M. J. D. Powell and J. K. Reid. On applying Householder transformations to linear least squares problems. In *Proc. IFIP Congress 1968*, North-Holland, Amsterdam, The Netherlands, 1969, pages 122–126.
46. William H. Press, Saul A. Teukolsky, William T. Vetterling, and Brian P. Flannery. *Numerical Recipes in FORTRAN: The Art of Scientific Computing.* Second edition, Cambridge University Press, 1992. xxvi+963 pp. ISBN 0 521 43064 X.
47. G. W. Stewart and Ji-guang Sun. *Matrix Perturbation Theory.* Academic Press, London, 1990. xv+365 pp. ISBN 0-12-670230-6.
48. Lloyd N. Trefethen and David Bau III. *Numerical Linear Algebra.* Society for Industrial and Applied Mathematics, Philadelphia, PA, USA, 1997. xii+361 pp. ISBN 0-89871-361-7.
49. A. van der Sluis. Condition numbers and equilibration of matrices. *Numer. Math.*, 14:14–23, 1969.
50. Krešimar Veselić and Vjeran Hari. A note on a one-sided Jacobi algorithm. *Numer. Math.*, 56:627–633, 1989.
51. Hongyuan Zha and Zhenyue Zhang. Fast parallelizable methods for the Hermitian eigenvalue problem. Technical Report CSE-96-041, Department of Computer Science and Engineering, Pennsylvania State University, University Park, PA, May 1996. 19 pp.

Numerical Analysis of Semilinear Parabolic Problems

Stig Larsson

Department of Mathematics, Chalmers University of Technology and Göteborg University, SE–412 96 Göteborg, Sweden

Summary. In these lectures I discuss error analysis techniques for finite element methods for systems of reaction-diffusion equations with applications in dynamical systems theory. The emphasis is on pedagogical aspects and analysis techniques rather than on results. The list of techniques discussed include: analytic semigroup, parabolic smoothing, non-smooth data error estimate, a priori error estimate, a posteriori error estimate, exponential dichotomy, shadowing.

1. The Continuous Problem

We consider the following initial-boundary value problem for a system of reaction-diffusion equations,

$$
\begin{aligned}
u_t - \epsilon \Delta u &= \tilde{f}(u), & x \in \Omega,\ t > 0, \\
u &= 0, & x \in \partial\Omega,\ t > 0, \\
u(\cdot, 0) &= u_0, & x \in \Omega,
\end{aligned}
\tag{1.1}
$$

where Ω is a bounded domain in $\mathbb{R}^d$, $d = 1, 2, 3$, $u = u(x, t) \in \mathbb{R}^p$. $u_t = \partial u/\partial t$, $\Delta u = \sum_{i=1}^{d} \partial^2 u/\partial x_i^2$, $\epsilon = \mathrm{diag}(\epsilon_1, \ldots, \epsilon_p)$ is a diagonal matrix of constant coefficients $\epsilon_j > 0$, and $\tilde{f} : \mathbb{R}^p \to \mathbb{R}^p$ is twice continuously differentiable. If $d = 2, 3$ we assume, in addition, that the derivatives of $\tilde{f}$ satisfy the growth condition

$$
|\tilde{f}^{(l)}(\xi)| \leq C(1 + |\xi|^{\delta+1-l}), \quad \xi \in \mathbb{R}^p,\ l = 1, 2,
\tag{1.2}
$$

where $|\cdot|$ denotes the Euclidean norm on $\mathbb{R}^p$ and the induced operator norms, and where $\delta = 2$ if $d = 3$, $\delta \in [1, \infty)$ if $d = 2$.

We assume that Ω is a convex polygon, so that we have access to the elliptic regularity theory and so that finite element meshes can be fitted exactly to the domain.

In the sequel we use the Hilbert space $H = (L_2(\Omega))^p$, with its standard norm and inner product

$$
\|v\| = \left(\int_\Omega |v|^2\, dx \right)^{1/2}, \quad (v, w) = \int_\Omega v \cdot w\, dx.
\tag{1.3}
$$

The norms in the Sobolev spaces $(H^m(\Omega))^p$, $m \geq 0$, are denoted by

$$
\|v\|_m = \left(\sum_{|\alpha| \leq m} \|D^\alpha v\|^2 \right)^{1/2}.
\tag{1.4}
$$

The space $V = (H_0^1(\Omega))^p$, with norm $\|\cdot\|_1$, consists of the functions in $(H^1(\Omega))^p$ that vanish on $\partial\Omega$. $V^* = (H^{-1}(\Omega))^p$ is the dual space of V with norm

$$\|v\|_{-1} = \sup_{\chi \in V} \frac{|(v, \chi)|}{\|\chi\|_1}. \tag{1.5}$$

If X, Y are Banach spaces, then $\mathcal{L}(X, Y)$ denotes the space of bounded linear operators from X into Y, $\mathcal{L}(X) = \mathcal{L}(X, X)$, and $B_X(x, R)$ denotes the closed ball in X with center x and radius R. In particular, we let $B_R = B_V(0, R)$ denote the the closed ball of radius R in V:

$$B_R = \{v \in V : \|v\|_1 \le R\}.$$

We also use the notation

$$\|v\|_{L_\infty([0,T], X)} = \sup_{t \in [0,T]} \|v(t)\|_X.$$

We set the problem up in the framework of [6]. We define the unbounded operator $A = -\epsilon\Delta$ on H with domain of definition $\mathcal{D}(A) = (H^2(\Omega) \cap H_0^1(\Omega))^p$. Then A is a closed, densely defined, and self-adjoint positive definite operator in H with compact inverse. Moreover, our assumption (1.2) guarantees that the mapping $\tilde{f}$ induces a nonlinear operator $f : V \to H$ through $f(v)(x) = \tilde{f}(v(x))$, see Lemma 1.1 below. The initial-boundary value problem (1.1) may then be formulated as an initial value problem in V: find $u(t) \in V$ such that

$$u' + Au = f(u), \ t > 0; \quad u(0) = u_0. \tag{1.6}$$

The operator $-A$ is the infinitesimal generator of the analytic semigroup $E(t) = \exp(-tA)$ defined by

$$E(t)v = \sum_{j=1}^{\infty} e^{-t\lambda_j}(v, \varphi_j)\varphi_j, \quad v \in H, \tag{1.7}$$

where λ_j and φ_j denote the eigenvalues and an orthonormal basis of eigenvectors of A. The semigroup $E(t)$ is the solution operator of the initial value problem for the homogeneous equation,

$$u' + Au = 0, \ t > 0; \quad u(0) = u_0.$$

Its solution is thus given by $u(t) = E(t)u_0$. By Duhamel's principle it follows that solutions of (1.6) satisfy

$$u(t) = E(t)u_0 + \int_0^t E(t - s)f(u(s))\, ds, \quad t \ge 0. \tag{1.8}$$

Conversely, appropriately defined solutions of the nonlinear integral equation (1.8) are solutions of the differential equation (1.6); see Theorem 1.2 below. We shall mainly work with (1.8) and discretized variants of it.

An important ingredient in our framework is that f can be controlled by fractional powers of A. We define the fractional powers of A by means of the spectral theorem and we have

$$\begin{aligned}
\|A^\alpha v\| &= \left(\sum_{j=1}^\infty \left(\lambda_j^\alpha (v, \varphi_j)\right)^2\right)^{1/2}, \quad \alpha \in \mathbb{R}, \\
\mathcal{D}(A^\alpha) &= \{v : \|A^\alpha v\| < \infty\}, \quad \alpha \in \mathbb{R}.
\end{aligned} \tag{1.9}$$

Using also the elliptic regularity estimate,

$$\|v\|_2 \le C \|Av\|, \quad v \in (H^2(\Omega) \cap H_0^1(\Omega))^p, \tag{1.10}$$

and the trace inequality,

$$\|v\|_{L_2(\partial\Omega)} \le C \|v\|_1, \quad v \in (H^1(\Omega))^p,$$

we obtain $\mathcal{D}(A^{l/2}) = (H^l(\Omega) \cap H_0^1(\Omega))^p$, $l = 1, 2$, with the equivalence of norms

$$c \|v\|_l \le \left\|A^{l/2} v\right\| \le C \|v\|_l, \quad v \in \mathcal{D}(A^{l/2}), \ l = 1, 2; \tag{1.11}$$

cf. [12]. A simple exercise using the special case $l = 1$ of (1.11) shows

$$c \|v\|_{-1} \le \left\|A^{-1/2} v\right\| \le C \|v\|_{-1}; \tag{1.12}$$

cf. (2.7) below.

The analyticity of the semigroup is reflected in the inequalities ($D_t = \partial/\partial t$)

$$\left\|D_t^l E(t) v\right\| = \left\|A^l E(t) v\right\| \le C_l t^{-l} \|v\|, \quad t > 0, \ v \in H, \ l \ge 0. \tag{1.13}$$

These follow easily from (1.7) and Parseval's identity. Using also the norm equivalences (1.11), (1.12) we obtain the smoothing property

$$\left\|D_t^l E(t) v\right\|_\beta \le C_l t^{-l-(\beta-\alpha)/2} \|v\|_\alpha, \quad \begin{array}{l} t > 0, \ v \in \mathcal{D}(A^{\alpha/2}), \\ -1 \le \alpha \le \beta \le 2, \ l = 0, 1. \end{array} \tag{1.14}$$

We need the fact that the operator $f : V \to H$ satisfies a local Lipschitz condition. This is contained in our first lemma. We also prove that the Fréchet derivative $f' : V \to \mathcal{L}(V, H)$ satisfies a local Lipschitz condition. The proof is based on the assumption (1.2) and the Sobolev inequality (where $p = 6$ if $d \le 3$, $p < \infty$ if $d = 2$, and $p = \infty$ if $d = 1$)

$$\|v\|_{L_p} \le C \|v\|_1, \tag{1.15}$$

and on Hölder's inequality

$$\left\|v^\delta w\right\|_{L_r} \le \|v\|_{L_q}^\delta \|w\|_{L_p}, \quad \tfrac{\delta}{q} + \tfrac{1}{p} = \tfrac{1}{r}, \ \delta > 0. \tag{1.16}$$

Lemma 1.1. *For each nonnegative number R there is a constant $C(R)$ such that, for all $u, v, w \in B_R$, $l = 0, 1$,*

$$\|f'(u)\|_{\mathcal{L}(V,H)} \leq C(R), \tag{1.17}$$

$$\|f'(u)\|_{\mathcal{L}(H,V^*)} \leq C(R), \tag{1.18}$$

$$\|f(u) - f(v)\| \leq C(R)\|u - v\|_1, \tag{1.19}$$

$$\|f(u) - f(v)\|_{-1} \leq C(R)\|u - v\|, \tag{1.20}$$

$$\|f'(u) - f'(v)\|_{\mathcal{L}(V,H)} \leq C(R)\|u - v\|_1, \tag{1.21}$$

$$|f'(u) - f'(v)\|_{\mathcal{L}(H,V^*)} \leq C(R)\|u - v\|_1, \tag{1.22}$$

$$\begin{aligned} \|f(u) - f(v) - f'(w)(u - v)\|_{l-1} \\ \leq C(R)\left(\|u - w\|_1 + \|v - w\|_1\right)\|u - v\|_1. \end{aligned} \tag{1.23}$$

If, in addition, $u \in (H^2(\Omega))^p$ and $z \in (H^2(\Omega))^p \cap V$, then

$$\left\|A^{1/2}(f'(u)z)\right\| \leq C(R)\left(\|z\|_2 + \|u\|_2 \|z\|_1\right). \tag{1.24}$$

Proof. We have in view of (1.2) and the Hölder and Sobolev inequalities, for $z \in V$,

$$\begin{aligned} \|f'(u)z\|_{L_2} &\leq C\left(1 + \|u\|_{L_q}^{\delta}\right)\|z\|_{L_p} \\ &\leq C\left(1 + \|u\|_1^{\delta}\right)\|z\|_1, \end{aligned}$$

where $\frac{1}{p} + \frac{\delta}{q} = \frac{1}{2}$ with $p = q = 6$ if $d = 3$, and with arbitrary $p \in (1, \infty)$ if $d \leq 2$. This proves (1.17) and (1.19) follows. Similarly, for any $z \in V$,

$$\begin{aligned} \|(f'(u) - f'(v))z\|_{L_2} &\leq C\left(1 + \|u\|_{L_q}^{\delta-1} + \|v\|_{L_q}^{\delta-1}\right)\|u - v\|_{L_p}\|z\|_{L_q} \\ &\leq C\left(1 + \|u\|_1^{\delta-1} + \|v\|_1^{\delta-1}\right)\|u - v\|_1\|z\|_1, \end{aligned}$$

with the same p and q as before. This proves (1.21).

Moreover, for any $z, \chi \in V$,

$$\begin{aligned} (f'(u)z, \chi) &\leq C\left(1 + \|u\|_{L_q}^{\delta}\right)\|z\|_{L_2}\|\chi\|_{L_p} \\ &\leq C\left(1 + \|u\|_1^{\delta}\right)\|z\|\,\|\chi\|_1, \end{aligned}$$

and

$$\begin{aligned} ((f'(u) - f'(v))z, \chi) &\leq C\left(1 + \|u\|_{L_q}^{\delta-1} + \|v\|_{L_q}^{\delta-1}\right)\|u - v\|_{L_2}\|z\|_{L_q}\|\chi\|_{L_p} \\ &\leq C\left(1 + \|u\|_1^{\delta-1} + \|v\|_1^{\delta-1}\right)\|u - v\|\,\|z\|_1\|\chi\|_1, \end{aligned}$$

where $\frac{\delta}{q} + \frac{1}{2} + \frac{1}{p} = 1$, i.e., with the same p and q as before. This proves (1.18) and (1.22); (1.20) follows from (1.18).

Next, (1.23) is obtained by applying (1.21) and (1.22) to the identity

$$f(u) - f(v) - f'(w)(u - v) = \int_0^1 \left(f'(su + (1 - s)v) - f'(w) \right) ds \, (u - v).$$

Finally, (1.24) is proved in a similar way, by using the equivalence of norms (1.11) and computing the first order partial derivatives of $f'(u)z$. Note that this uses the fact that $f'(u)z \in V$, i.e., $f'(u)z = 0$ on $\partial\Omega$.

We may now prove local existence of solutions of (1.8) and hence of (1.6).

Theorem 1.1. *For any $R_0 > 0$ there is $\tau = \tau(R_0)$ such that (1.8) has a unique solution $u \in C([0, \tau], V)$ for any initial value $u_0 \in V$ with $\|u_0\|_1 \leq R_0$. Moreover, there is c such that $\|u\|_{L_\infty([0,\tau],V)} \leq cR_0$.*

Proof. Let $u_0 \in B_{R_0}$, define

$$\mathcal{S}(u)(t) = E(t)u_0 + \int_0^t E(t - s)f(u(s)) \, ds,$$

and note that (1.8) is a fixed point equation $u = \mathcal{S}(u)$. We shall choose τ and R such that we can apply Banach's fixed point theorem (the contraction mapping theorem) in the closed ball

$$\mathcal{B} = \{u \in C([0, \tau], V) : \|u\|_{L_\infty([0,\tau],V)} \leq R\}$$

in the Banach space $C([0, \tau], V)$.

We must show (i) that $\mathcal{S}$ maps $\mathcal{B}$ into itself, (ii) that $\mathcal{S}$ is a contraction on $\mathcal{B}$. In order to prove (i) we take $u \in \mathcal{B}$ and first note that the Lipschitz condition (1.19) implies that

$$\begin{aligned}
\|f(u(t))\| &\leq \|f(0)\| + \|f(u(t)) - f(0)\| \\
&\leq \|f(0)\| + C(R)\|u(t)\|_1 \\
&\leq \|f(0)\| + C(R)R, \quad 0 \leq t \leq \tau.
\end{aligned} \tag{1.25}$$

Hence, using also (1.14), we get

$$\begin{aligned}
\|\mathcal{S}(u)(t)\|_1 &\leq \|E(t)u_0\|_1 + \int_0^t \|E(t - s)f(u(s))\|_1 \, ds \\
&\leq c_0 \|u_0\|_1 + c_1 \int_0^t (t - s)^{-1/2} \|f(u(s))\| \, ds \\
&\leq c_0 R_0 + 2c_1 \tau^{1/2} \left(\|f(0)\| + C(R)R \right), \quad 0 \leq t \leq \tau.
\end{aligned}$$

This implies

$$\|\mathcal{S}(u)\|_{L_\infty([0,\tau],V)} \leq c_0 R_0 + 2c_1 \tau^{1/2} \left(\|f(0)\| + C(R)R \right).$$

Choose $R = 2c_0 R_0$ and $\tau = \tau(R_0)$ so small that

$$2c_1 \tau^{1/2} \left(\|f(0)\| + C(R)R \right) \leq \tfrac{1}{2} R. \tag{1.26}$$

Then $\|\mathcal{S}(u)\|_{L_\infty([0,\tau],V)} \le R$ and we conclude that $\mathcal{S}$ maps $\mathcal{B}$ into itself.

To show (ii) we take $u, v \in \mathcal{B}$ and note that

$$\|f(u(t)) - f(v(t))\| \le C(R)\, \|u - v\|_{L_\infty([0,\tau],V)}, \quad 0 \le t \le \tau.$$

Hence

$$
\begin{aligned}
\|\mathcal{S}(u)(t) - \mathcal{S}(v(t))\|_1 &\le \int_0^t \|E(t - s)(f(u(s)) - f(v(s)))\|_1\, ds \\
&\le c_1 \int_0^t (t - s)^{-1/2}\, \|f(u(s)) - f(v(s))\|\, ds \\
&\le 2c_1 \tau^{1/2} C(R)\, \|u - v\|_{L_\infty([0,\tau],V)}, \quad 0 \le t \le \tau,
\end{aligned}
$$

so that

$$\|\mathcal{S}(u) - \mathcal{S}(v)\|_{L_\infty([0,\tau],V)} \le 2c_1 \tau^{1/2} C(R)\, \|u - v\|_{L_\infty([0,\tau],V)} \,.$$

It follows from (1.26) that $2c_1 \tau^{1/2} C(R) \le \frac{1}{2}$ and we conclude that $\mathcal{S}$ is a contraction on $\mathcal{B}$. Hence $\mathcal{S}$ has a unique fixed point $u \in \mathcal{B}$.

The initial value problem (1.8) (or (1.6)) thus has a unique local solution for any initial datum $u_0 \in V$. We denote by $S(t, \cdot)$ the corresponding (local) solution operator, so that $u(t) = S(t, u_0)$ is the solution of (1.8).

The following theorem provides regularity estimates for solutions u of (1.8). These will be used in our error analysis, but the theorem also shows that $u'(t) \in H$ and $Au(t) \in H$ for $t > 0$, so that any solution of the integral equation (1.8) is also a solution of the differential equation (1.6).

Theorem 1.2. *Let $R \ge 0$ and $\tau > 0$ be given and let $u \in C([0, \tau], V)$ be a solution of (1.8). If $\|u(t)\|_1 \le R$ for $t \in [0, \tau]$, then*

$$
\begin{aligned}
\|u(t)\|_2 &\le C(R, \tau) t^{-1/2}, & t \in (0, \tau], & \quad\quad (1.27) \\
\|u_t(t)\|_s &\le C(R, \tau) t^{-1-(s-1)/2}, & t \in (0, \tau],\ s = 0, 1, 2. & \quad\quad (1.28)
\end{aligned}
$$

Proof. In view of (1.19) we have $\|f(u(t))\| \le C(R)$ for $t \in [0, \tau]$, cf. (1.25). The proof of (1.27), (1.28) can now obtained by tracing the constants in [6, Theorem 3.5.2]. The argument is based on a generalization of Gronwall's lemma, Lemma 1.2 below. The case $s = 2$ of (1.28) requires that we differentiate (1.6) with respect to t and use (1.24).

Lemma 1.2. *(Generalized Gronwall lemma.) Let the function $\varphi(t, \tau) \ge 0$ be continuous for $0 \le \tau < t \le T$. If*

$$\varphi(t, \tau) \le A\,(t - \tau)^{-1+\alpha} + B \int_\tau^t (t - s)^{-1+\beta}\varphi(s, \tau)\, ds, \quad 0 \le \tau < t \le T,$$

for some constants $A, B \ge 0$, $\alpha, \beta > 0$, then there is a constant $C = C(B, T, \alpha, \beta)$ such that

$$\varphi(t, \tau) \le CA\,(t - \tau)^{-1+\alpha}, \quad 0 \le \tau < t \le T.$$

Proof. Iterating the given inequality $N - 1$ times, using the identity

$$\int_\tau^t (t - s)^{-1+\alpha}(s - \tau)^{-1+\beta}\, ds = C(\alpha, \beta)\,(t - \tau)^{-1+\alpha+\beta}, \quad \alpha, \beta > 0, \quad (1.29)$$

(Abel's integral) and estimating $(t - \tau)^\beta$ by T^β, we obtain

$$\varphi(t, \tau) \le C_1 A\,(t - \tau)^{-1+\alpha} + C_2 \int_\tau^t (t - s)^{-1+N\beta}\varphi(s, \tau)\, ds, \quad 0 \le \tau < t \le T,$$

where $C_1 = C_1(B, T, \alpha, \beta, N)$, $C_2 = C_2(B, \beta, N)$. We now choose the smallest N such that $-1 + N\beta \ge 0$, and estimate $(t-s)^{-1+N\beta}$ by $T^{-1+N\beta}$. If $-1+\alpha \ge 0$ we obtain the desired conclusion by the standard version of Gronwall's lemma. Otherwise we set $\psi(t, \tau) = (t - \tau)^{1-\alpha}\varphi(t, \tau)$ to obtain

$$\psi(t, \tau) \le C_1 A + C_3 \int_\tau^t (s - \tau)^{-1+\alpha}\psi(s, \tau)\, ds, \quad 0 \le \tau < t \le T,$$

and the standard Gronwall lemma yields $\psi(t, \tau) \le CA$ for $0 \le \tau < t \le T$, which is the desired result.

Note that the constant in Gronwall's lemma grows exponentially with the length T of the time interval. Hence, results derived by means of this lemma are often useful only for short time intervals. There is also a discrete version of Lemma 1.2; see [3, Lemma 7.1].

We conclude this section with a remark about the abstract framework. It is tempting to start by saying "Let X be a Banach space and A the generator of an analytic semigroup in X." This approach goes a rather long way with the continuous problem (see [6]) and with pure time discretization. But when it comes to spatial discretization we soon need to know that, for example, $X = L_2(\Omega)$ and A is an elliptic operator of second order. "PDE theory is not a branch of functional analysis" as L. C. Evans puts it in the preface of his textbook [5].

2. Local a Priori Error Estimates

In this section we first discretize (1.1) with respect to the spatial variables by means of a standard piecewise linear finite element method. We then briefly discuss completely discrete approximation by means of the backward Euler time-stepping. This material is taken from [7]. A general reference is [12].

2.1. The Spatially Semidiscrete Problem

The weak formulation of (1.1) is: find $u(t) \in V$ such that

$$
\begin{aligned}
(u', v) + a(u, v) &= (f(u), v), \quad \forall v \in V, \ t > 0, \\
u(0) &= u_0,
\end{aligned}
\tag{2.1}
$$

where $a(u, v) = (\epsilon \nabla u, \nabla v) = (-\epsilon \Delta u, v) = (Au, v)$ is the bilinear form associated with A.

Let $\{V_h\}_{0 < h < 1}$ be a family of finite dimensional subspaces of V, where each V_h consists of continuous piecewise polynomials of degree ≤ 1 with respect to a triangulation $\mathcal{T}$ of Ω with maximal mesh size h, see [2]. The approximate solution $u_h(t) \in V_h$ of (1.1) is defined by

$$
\begin{aligned}
(u_h', \chi) + a(u_h, \chi) &= (f(u_h), \chi), \quad \forall \chi \in V_h, \ t > 0, \\
u_h(0) &= u_{h,0},
\end{aligned}
\tag{2.2}
$$

where $u_{h,0} \in V_h$ is an approximation of u_0.

Introducing the linear operator $A_h : V_h \to V_h$ and the orthogonal projection $P_h : H \to V_h$, defined by

$$
(A_h \psi, \chi) = a(\psi, \chi), \quad (P_h g, \chi) = (g, \chi) \quad \forall \psi, \chi \in V_h, \ g \in H,
\tag{2.3}
$$

we may write (2.2) as

$$
u_h' + A_h u_h = P_h f(u_h), \ t > 0; \quad u_h(0) = u_{h,0}.
\tag{2.4}
$$

The operator A_h is self-adjoint positive definite (uniformly in h); the corresponding semigroup $E_h(t) = \exp(-tA_h) : V_h \to V_h$ therefore satisfies inequalities analogous to (1.13) (uniformly in h):

$$
\left\| D_t^l E_h(t)v \right\| = \left\| A_h^l E_h(t)v \right\| \leq C_l t^{-l} \left\| v \right\|, \quad t > 0, \ v \in V_h, \ l \geq 0.
$$

Moreover, for the operator A_h we have the equivalence of norms (cf. (1.11))

$$
c \left\| v \right\|_1 \leq \left\| A_h^{1/2} v \right\| = \sqrt{a(v, v)} = \left\| A^{1/2} v \right\| \leq C \left\| v \right\|_1, \quad v \in V_h.
\tag{2.5}
$$

We also have

$$
\left\| P_h f \right\| \leq \left\| f \right\|, \quad f \in H,
$$

and (cf. (1.12))

$$
\left\| A_h^{-1/2} P_h f \right\| \leq C \left\| f \right\|_{-1}, \quad f \in H,
\tag{2.6}
$$

which is obtained by using (2.5) in the calculation

$$
\begin{aligned}
\left\| A_h^{-1/2} P_h f \right\| &= \sup_{v_h \in V_h} \frac{|(A_h^{-1/2} P_h f, v_h)|}{\|v_h\|} = \sup_{v_h \in V_h} \frac{|(f, A_h^{-1/2} v_h)|}{\|v_h\|} \\
&= \sup_{w_h \in V_h} \frac{|(f, w_h)|}{\left\| A_h^{1/2} w_h \right\|} \leq C \sup_{w_h \in V_h} \frac{|(f, w_h)|}{\|w_h\|_1} \\
&\leq C \sup_{w \in V} \frac{|(f, w)|}{\|w\|_1} = C \left\| f \right\|_{-1}.
\end{aligned}
\tag{2.7}
$$

Using these inequalities we easily prove the smoothing property of $E_h(t)P_h$:

$$\left\|D_t^l E_h(t)P_h f\right\|_\beta \leq Ct^{-l-(\beta-\alpha)/2}\left\|f\right\|_\alpha, \qquad t > 0, \ f \in \mathcal{D}(A^{\alpha/2}),$$
$$-1 \leq \alpha \leq \beta \leq 1, \ l = 0, 1. \tag{2.8}$$

Note that the upper limit to β is 1, while it is 2 in the continuous case (1.14). The initial-value problem (2.4) is equivalent to the integral equation

$$u_h(t) = E_h(t)u_{h,0} + \int_0^t E_h(t-s)P_h f(u_h(s))\,ds, \quad t \geq 0. \tag{2.9}$$

Because of (2.8) the proof of Theorem 1.1 carries over verbatim to the semidiscrete case. We thus have:

Theorem 2.1. *For any* $R_0 > 0$ *there is* $\tau = \tau(R_0)$ *such that (2.9) has a unique solution* $u_h \in C([0,\tau],V)$ *for any initial value* $u_{h,0} \in V_h$ *with* $\|u_{h,0}\|_1 \leq R_0$. *Moreover, there is* c *such that* $\|u_h\|_{L_\infty([0,\tau],V)} \leq cR_0$.

We denote by $S_h(t,\cdot)$ the corresponding (local) solution operator, so that $u_h(t) = S_h(t, u_{h,0})$ is the solution of (2.9).

We may now estimate the difference between the local solutions $u(t) = S(t,u_0)$ and $u_h(t) = S_h(t,u_{h,0})$ that we have obtained. We refer to the following result as a local a priori error estimate. It is local because the constant $C(R,\tau)$ grows exponentially with the length τ of the time interval and also because it grows with the size R of the solutions. It is a priori because the error is evaluated in terms of derivatives of u, which are estimated a priori in Theorem 1.2.

Theorem 2.2. *Let* $R \geq 0$ *and* $\tau > 0$ *be given. Let* $u(t)$ *and* $u_h(t)$ *be solutions of (2.1) and (2.2) respectively, such that* $u(t), u_h(t) \in B_R$ *for* $t \in [0,\tau]$. *Then*

$$\|u_h(t) - u(t)\|_1 \leq C(R,\tau)t^{-1/2}\big(\|u_{h,0} - P_h u_0\| + h\big), \ t \in (0,\tau], \tag{2.10}$$
$$\|u_h(t) - u(t)\| \leq C(R,\tau)\big(\|u_{h,0} - P_h u_0\| + h^2 t^{-1/2}\big). \ t \in (0,\tau]. \tag{2.11}$$

Proof. We introduce the elliptic projection operator $R_h : V \to V_h$ defined by

$$a(R_h v, \chi) = a(v, \chi), \quad \forall \chi \in V_h. \tag{2.12}$$

Under the usual regularity assumptions on the triangulation and the elliptic regularity estimate (1.10), we have the error bounds (see [2])

$$\|R_h v - v\| + h\|R_h v - v\|_1 \leq Ch^s\|v\|_s, \quad v \in V \cap (H^s(\Omega))^p, \ s = 1, 2, \tag{2.13}$$

Following a standard practice we divide the error into two parts:

$$e(t) \equiv u_h(t) - u(t) = \big(u_h(t) - R_h u(t)\big) + \big(R_h u(t) - u(t)\big) \equiv \theta(t) + \rho(t).$$

In view of (2.13) and (1.27), (1.28) we have, for $j = 0, 1$ and $s = 1, 2$,

$$\|\rho(t)\|_j \le Ch^{s-j} \|u(t)\|_s \le C(R,\tau)h^{s-j}t^{-(s-1)/2}, \quad t \in (0,\tau], \quad (2.14)$$

$$\|\rho_t(t)\| \le Ch^s \|u_t(t)\|_s \le C(R,\tau)ht^{-1-(s-1)/2}, \quad t \in (0,\tau]. \quad (2.15)$$

It remains to estimate $\theta(t)$, which belongs to V_h. In view of (2.2), the identity $A_h R_h = P_h A$ (which follows easily from (2.3) and (2.12)), and (1.6), we find that

$$\theta_t + A_h\theta = P_h\big(f(u_h) - f(u) - \rho_t\big). \quad (2.16)$$

Hence

$$\theta(t) = E_h(t)\theta(0) + \int_0^t E_h(t-\sigma)P_h\left(f(u_h(\sigma)) - f(u(\sigma)) - D_\sigma\rho(\sigma)\right) d\sigma, \quad (2.17)$$

where $D_\sigma = \partial/\partial\sigma$. Integration by parts yields

$$-\int_0^{t/2} E_h(t-\sigma)P_h D_\sigma\rho(\sigma)\, d\sigma \;=\; E_h(t)P_h\rho(0) - E_h(t/2)P_h\rho(t/2)$$
$$+ \int_0^{t/2} \big(D_\sigma E_h(t-\sigma)\big) P_h\rho(\sigma)\, d\sigma.$$

Hence

$$\theta(t) \;=\; E_h(t)P_h e(0) - E_h(t/2)P_h\rho(t/2) + \int_0^{t/2} \big(D_\sigma E_h(t-\sigma)\big) P_h\rho(\sigma)\, d\sigma$$
$$- \int_{t/2}^t E_h(t-\sigma)P_h D_\sigma\rho(\sigma)\, d\sigma$$
$$+ \int_0^t E_h(t-\sigma)P_h \left(f(u_h(\sigma)) - f(u(\sigma))\right) d\sigma.$$

Using the smoothing property (2.8) of $E_h(t)P_h$, the error estimates (2.14), (2.15) with $j = 0$, $s = 1$, and the Lipschitz condition (1.19) we obtain

$$\|\theta(t)\|_1 \;\le\; Ct^{-1/2}\big(\|P_h e(0)\| + \|\rho(t/2)\|\big) + C\int_0^{t/2} (t-\sigma)^{-3/2}\|\rho(\sigma)\|\, d\sigma$$
$$+ C\int_{t/2}^t (t-\sigma)^{-1/2}\|D_\sigma\rho(\sigma)\|\, d\sigma$$
$$+ C\int_0^t (t-\sigma)^{-1/2}\|f(u_h(\sigma)) - f(u(\sigma))\|\, d\sigma$$
$$\le\; C(R,\tau)t^{-1/2}\big(\|P_h e(0)\| + h\big)$$
$$+ C(R,\tau)h\left(\int_0^{t/2} (t-\sigma)^{-3/2}\, d\sigma + \int_{t/2}^t (t-\sigma)^{-1/2}\sigma^{-1}\, d\sigma\right)$$
$$+ C(R)\int_0^t (t-\sigma)^{-1/2}\|e(\sigma)\|_1\, d\sigma$$
$$\le\; C(R,\tau)t^{-1/2}\big(\|P_h e(0)\| + h\big) + C(R)\int_0^t (t-\sigma)^{-1/2}\|e(\sigma)\|_1\, d\sigma,$$

for $t \in (0, \tau]$. Together with (2.14), (2.15) and $e = \theta + \rho$ this yields, for $t \in (0, \tau]$,

$$\|e(t)\|_1 \leq C(R, \tau) t^{-1/2} \big(\|P_h e(0)\| + h \big) + C(R) \int_0^t (t - \sigma)^{-1/2} \|e(\sigma)\|_1 \, d\sigma,$$

and the desired bound follows by the generalized Gronwall lemma, cf. (1.2). This proves (2.10), because $P_h e(0) = u_{h,0} - P_h u_0$.

To prove (2.11) we use the Lipschitz condition (1.20) instead of (1.19). We omit the details.

We now formulate a consequence of the previous theorem, which is the form in which we will apply it later on. (It is convenient to change the notation so that the length of the local interval of existence is now 2τ.)

Theorem 2.3. *For any $R_0 \geq 0$ there are numbers $\tau = \tau(R_0) > 0$ and $C(R_0)$ such that if $v \in B_{R_0}$, $v_h \in V_h \cap B_{R_0}$, then $S(t, v)$ and $S_h(t, v_h)$ exist for $t \in [0, 2\tau]$. Moreover, for $l = 0, 1$,*

$$\|S_h(t, v_h) - S(t, v)\|_l \leq C(R_0) t^{-1/2} \big(\|v_h - P_h v\| + h^{2-l} \big), \quad t \in (0, 2\tau],$$

and

$$\|S_h(t, v_h) - S(t, v)\|_l \leq C(R_0) \big(\|v_h - P_h v\| + h^{2-l} \big), \quad t \in [\tau, 2\tau].$$

Proof. The local existence theorems 1.1 and 2.1 give $S(t, v), S_h(t, v_h) \in B_R$, $t \in [0, 2\tau]$, for some $\tau = \tau(R_0)$ and with $R = cR_0$. We now apply Theorem 2.2.

2.2. A Completely Discrete Scheme

In this subsection we show that the above program can be carried through also for a completely discrete scheme based on the backward Euler method. We replace the time derivative in (2.2) by a backward difference quotient $\partial_t U_j = (U_j - U_{j-1})/k$, where k is a time step and U_j is the approximation of $u_j = u(t_j)$ and $t_j = jk$. The discrete solution $U_j \in V_h$ thus satisfies

$$\partial_t U_j + A_h U_j = P_h f(U_j), \quad t_j > 0; \quad U_0 = u_{h,0}. \tag{2.18}$$

Duhamel's principle yields

$$U_j = E_{kh}^j u_{h,0} + k \sum_{l=1}^{j} E_{kh}^{j-l-1} P_h f(U_l), \quad t_j \geq 0, \tag{2.19}$$

where $E_{kh} = (I + kA_h)^{-1}$. Since A_h is self-adjoint positive definite (uniformly in h), we have the inequality

$$\left\| \partial_t^l E_{kh}^j v \right\| = \left\| A_h^l E_{kh}^j v \right\| \leq C_l t_j^{-l} \|v\|, \quad t_j \geq t_l, \ v \in V_h, \ l \geq 0,$$

which is the discrete analogue of (1.13). In view of the inequalities (2.5) and (2.6) this leads to a smoothing property analogous to (2.8):

$$\left\| \partial_t^l E_{kh}^j P_h f \right\|_\beta \le C t_j^{-l-(\beta-\alpha)/2} \|f\|_\alpha, \qquad t_j > 0, \ f \in \mathcal{D}(A^{\alpha/2}),$$
$$-1 \le \alpha \le \beta \le 1, \ l = 0, 1. \tag{2.20}$$

Again the proof of Theorem 1.1 carries over verbatim to the discrete case. We thus have:

Theorem 2.4. *For any* $R_0 > 0$ *there is* $\tau = \tau(R_0)$ *such that (2.19) has a unique solution* U_j, $t_j \in [0, \tau]$, *for any initial value* $u_{h,0} \in V_h$ *with* $\|u_{h,0}\|_1 \le R_0$. *Moreover, there is* c *such that* $\max_{t_j \in [0,\tau]} \|U_j\|_1 \le cR_0$.

We denote by $S_{hk}(t_j, \cdot)$ the corresponding (local) solution operator, so that $U_j = S_{hk}(t_j, u_{h,0})$ is the solution of (2.19).

We may now estimate the difference between the local solutions $u(t) = S(t, u_0)$ and $U_j = S_{hk}(t_j, u_{h,0})$ that we have obtained. We refer to the following result as a local a priori error estimate. The proof can be found in [7]. It follows the lines of the proof of Theorem 2.2 but is more technical.

Theorem 2.5. *Let* $R \ge 0$ *and* $\tau > 0$ *be given. Let* $u(t)$ *and* U_j *be solutions of (2.1) and (2.19) respectively, such that* $u(t), U_j \in B_R$ *for* $t, t_j \in [0, \tau]$. *Then, for* $k \le k_0(R)$, *we have*

$$\|U_j - u(t_j)\|_1 \le C(R, \tau) \big(\|u_{h,0} - P_h u_0\| t_j^{-1/2} + h t_j^{-1/2} + k t_j^{-1} \big), \ t_j \in (0, \tau],$$
$$\|U_j - u(t_j)\| \le C(R, \tau) \big(\|u_{h,0} - P_h u_0\| + h^2 t_j^{-1/2} + k t_j^{-1/2} \big), \quad t_j \in (0, \tau].$$

There is also an analogue of Theorem 2.3.

Theorem 2.6. *For any* $R_0 \ge 0$ *there are numbers* $\tau = \tau(R_0) > 0$, $k_0(R_0) > 0$ *and* $C(R_0)$ *such that if* $v \in B_{R_0}$, $v_h \in V_h \cap B_{R_0}$, *then* $S(t, v)$ *and* $S_{hk}(t_j, v_h)$ *exist for* $t, t_j \in [0, 2\tau]$. *Moreover, for* $k \le k_0(R_0)$, *we have*

$$\|S_{hk}(t_j, v_h) - S(t_j, v)\|_l \le C(R_0) \big(h^{2-l} + k \big), \quad t_j \in [\tau, 2\tau], \ l = 0, 1.$$

3. Shadowing—First Approach

In this section we present a numerical shadowing result. It is taken essentially from [9] but we present it in a slightly more general form.

3.1. Linearization

Let $\bar{u} \in C([0,T],V)$ with $\|\bar{u}(t)\|_1 \leq R$, $t \in [0,T]$, for some T and R. In order to linearize around $\bar{u}$ we rewrite the differential equation in (1.6),

$$u' + Au + B(t)u = F(t,u), \tag{3.1}$$

where $B(t)$ is the linear part of f and $F(t,\cdot)$ is the nonlinear remainder:

$$\begin{aligned}
B(t) &= -f'(\bar{u}(t)) \in \mathcal{L}(V,H), \\
F(t,v) &= f(v) - f'(\bar{u}(t))v.
\end{aligned} \tag{3.2}$$

Let $L(t,s)$, $0 \leq s \leq t \leq T$, denote the solution operator of the linearized homogeneous problem

$$v' + Av + B(t)v = 0, \ t > s; \quad v(s) = \phi. \tag{3.3}$$

Thus, $v(t) = L(t,s)\phi$ is the solution of (3.3). The operator $w \mapsto -B(t)w = f'(\bar{u}(t))w$ satisfies a local Lipschitz condition from V into H (uniformly in t); cf. Lemma 1.1. We therefore obtain local existence in $C([0,\tau],V)$ by the same argument as for (1.6). The following lemma provides a global a priori bound, which implies that any local solution can be extended to a global solution on the whole interval $[0,T]$. The lemma will also be useful in our analysis below. The bound grows exponentially with the length T of the time interval; this is realistic when the linearized problem is unstable, otherwise it is an overestimate. Note, by the way, that the linearized operator $A + B(t)$ is not self-adjoint in general even though A is.

Lemma 3.1. *Assume that $\bar{u} \in C([0,T],V)$ with $\|\bar{u}(t)\|_1 \leq R$, $t \in [0,T]$, for some T and R. Then any solution $v(t) = L(t,s)\phi$ of (3.3) satisfies*

$$\|L(t,s)\phi\|_m \leq C(R,T)(t-s)^{-(m-l)/2}\|\phi\|_l, \quad 0 \leq s < t \leq T, \ 0 \leq l \leq m \leq 1.$$

Proof. By Duhamel's principle we have

$$v(t) = E(t-s)\phi + \int_s^t E(t-\sigma)f'(\bar{u}(\sigma))v(\sigma)\,d\sigma,$$

so that, in view of (1.14), (1.17) and (1.18),

$$\begin{aligned}
\|v(t)\|_m &\leq C(t-s)^{-(m-l)/2}\|\phi\|_l + C\int_s^t (t-\sigma)^{-1/2}\|f'(\bar{u}(\sigma))v(\sigma)\|_{m-1}\,d\sigma \\
&\leq C(t-s)^{-(m-l)/2}\|\phi\|_l + C(R)\int_s^t (t-\sigma)^{-1/2}\|v(\sigma)\|_m\,d\sigma.
\end{aligned}$$

The desired bound now follows by Gronwall's lemma 1.2.

In order to conveniently take advantage of the smoothing properties of our problems, as reflected in the error bounds of Theorems 2.3 and 2.6, we replace the continuous dynamical system $u(t) = S(t, u_0)$, $t \geq 0$, by a discrete dynamical system $u_n = u(n\tau) = S(\tau, u_{n-1})$, $n = 1, 2, \ldots$, with an appropriately chosen step τ.

Let $\|\bar{u}(t)\|_1 \leq R$, $t \in [0, T]$, and let $\tau = \tau(R + 1)$ be half the time of local existence as in Theorems 2.3 and 2.6, adjusted so that $N\tau = T$ for some positive integer N. This implies that the operator $S = S(\tau, \cdot)$ is defined on the ball $D = B_{R-1}$. We may thus think of S as a nonlinear operator $S : D \subset X \to X$ with both $X = V$ and $X = H$; its domain $D = B_{R+1}$ is a ball in V in both cases.

Let $T_n = n\tau$ and $L_n = L(T_{n+1}, T_n)$. Duhamel's principle applied to (3.1) gives

$$u(t) = S(t - T_n, v) = L(t, T_n)v + \int_{T_n}^t L(t, s)F(s, u(s)) \, ds. \qquad (3.4)$$

With $t = T_{n+1}$ we get a decomposition of $S = S(\tau, \cdot)$,

$$S = L_n + N_n,$$

where

$$N_n(v) = \int_{T_n}^{T_{n+1}} L(t, s)F(s, S(s - T_n, v)) \, ds.$$

We think of L_n as the linear part of the nonlinear time-τ evolution S at time level T_n. In the following lemma we estimate the nonlinear remainder N_n.

Lemma 3.2. *Let $\rho \leq 1$. (a) If*

$$\|S(t - T_n, v_i) - \bar{u}(t)\|_1 \leq \rho, \quad t \in [T_n, T_{n+1}], \ i = 1, 2, \qquad (3.5)$$

then

$$\|N_n(v_1) - N_n(v_2)\|_1 \leq C(R, \tau)\rho \|v_1 - v_2\|_l, \quad l = 0, 1. \qquad (3.6)$$

(b) If $\bar{u}$ is a solution of (1.6) and $\|v_i - \bar{u}(T_n)\|_1 \leq \rho$, $i = 1, 2$, then (3.6) holds.

Proof. Let $u_i(t) = S(t - T_n, v_i)$ and $w_i(t) = \int_{T_n}^t L(t, s)F(s, u_i(s)) \, ds$ for $i = 1, 2$. Then $N_n(v_i) = w_i(T_{n+1})$. (Note that $u_i(t)$ exists for $t \in [T_n, T_{n+1}]$ because $v_i \in D = B_{R+1}$.) Applying Lemma 3.1 over the (short!) time interval $[T_n, T_{n-1}]$ we get, with $e = w_1 - w_2$,

$$\|e(t)\|_1 \leq \int_{T_n}^t \|L(t, s) \left(F(s, u_1(s)) - F(s, u_2(s))\right)\|_1 \, ds$$

$$\leq C(R, \tau) \int_{T_n}^t (t - s)^{-1/2} \|F(s, u_1(s)) - F(s, u_2(s))\| \, ds.$$

In view of (1.23) and (3.5) we have here (note that $u_i(t) \in B_{R+1}$)

$$\|F(s,u_1) - F(s,u_2)\| \le C(R) \sum_{i=1}^{2} \|u_i - \bar{u}\|_1 \|u_1 - u_2\|_1 \le C(R)\rho \|u_1 - u_2\|_1 ,$$

so that

$$\|e(t)\|_1 \le C(R,\tau)\rho \int_{T_n}^{t} (t-s)^{-1/2} \|u_1(s) - u_2(s)\|_1 \, ds.$$

Here, by (3.4) and Lemma 3.1,

$$\begin{aligned}
\|u_1(s) - u_2(s)\|_1 &= \|L(s,T_n)(v_1 - v_2) + w_1(s) - w_2(s)\|_1 \\
&\le C(R,\tau)(s-T_n)^{-(1-l)/2} \|v_1 - v_2\|_l + \|e(s)\|_1 .
\end{aligned}$$

Hence, since

$$\int_{T_n}^{t} (t-s)^{-1/2}(s-T_n)^{-(1-l)/2} \, ds \le C\tau^{l/2},$$

see (1.29), we get

$$\|e(t)\|_1 \le C(R,\tau)\rho \|v_1 - v_2\|_l + C(R,\tau)\rho \int_{T_n}^{t} (t-s)^{-1/2} \|e(s)\|_1 \, ds,$$

and by Gronwall's Lemma (Lemma 1.2),

$$\|e(T_{n+1})\|_1 \le C(R,\tau)\rho \|v_1 - v_2\|_l ,$$

which is (3.6) and part (a) is proved. If $\bar{u}$ is a solution of (1.6), then a similar Gronwall argument gives

$$\|u_i(t) - \bar{u}(t)\|_1 \le C(R,\tau) \|v_i - \bar{u}(T_n)\|_1 , \quad t \in [T_n, T_{n+1}].$$

Combining this with the result of part (a) we obtain the second statement of the lemma.

We have now decomposed the operator S as $S = L_n + N_n$, where. according to Lemma 3.1, the L_n are bounded linear operators on both V and H.

$$\|L_n\|_{\mathcal{L}(X)} \le C(R,\tau), \quad n = 0,\dots,N-1, \; X = V, H,$$

and where the nonlinear remainder $N_n : D \subset X \to V$ has a Lipschitz constant that can be rendered arbitrarily small (uniformly in n) by restricting to a small neighborhood of $\bar{u}$; see Lemma 3.2.

Note that L_n is the linearization of S at $\bar{u}(T_n)$. In fact, the mapping $S : D \subset V \to V$ is Fréchet differentiable with $L_n = S'(\bar{u}(T_n)) \in \mathcal{L}(V)$. However, the mapping $S : D \subset H \to H$ is not differentiable, because $D_n = B_V(\bar{u}(T_n), \rho)$ is not a neighborhood of $\bar{u}(T_n)$ with respect to the topology of H.

3.2. Exponential Dichotomies

We need a condition under which we can prove existence of global solutions of (1.6). Let X denote either V or H. We assume that the linear evolution operator $L(t, s)$ has an exponential dichotomy in X on the interval $J = [0, T]$; see [6]. This means that there are projections $P(t) \in \mathcal{L}(X)$, $t \in J$, and constants $M \geq 1$, $\beta > 0$ such that the following conditions are satisfied for $s, t \in J, t \geq s$.

1. $L(t, s)P(s) = P(t)L(t, s)$.
2. The restriction $L(t, s)|_{\mathcal{R}(I - P(s))} : \mathcal{R}(I - P(s)) \to \mathcal{R}(I - P(t))$ is an isomorphism. We define $L(s, t) : \mathcal{R}(I - P(t)) \to \mathcal{R}(I - P(s))$ to be its inverse.
3. $\|L(t, s)P(s)\|_{\mathcal{L}(X)} \leq Me^{-\beta(t-s)}$.
4. $\|L(s, t)(I - P(t))\|_{\mathcal{L}(X)} \leq Me^{-\beta(t-s)}$.

The ranges $\mathcal{R}(P(t))$ of the projections $P(t)$ are called stable subspaces; the complementary spaces $\mathcal{R}(I - P(t)) = \mathcal{N}(P(t))$ (null spaces of $P(t)$) are called unstable subspaces. Note that, due to parabolic smoothing, $L(t, s)$ is not invertible. Condition (2) means that $L(t, s)$ can be inverted on the unstable subspaces, so that we can evolve the solution backwards there. Condition (1) implies that the stable/unstable subspaces are invariant. Conditions (3) and (4) mean that solutions decay exponentially: forwards in time on the stable subspaces, backwards in time on the unstable subspaces. The projections are not orthogonal in general.

Example 3.1. (Hyperbolic fixed point.) Let $\bar{u} \in V$ be a stationary solution of (1.6); $A\bar{u} = f(\bar{u})$. Assume, for simplicity, that $B = -f'(\bar{u})$ is self-adjoint so that the linearized operator $A + B$ is self-adjoint, too, and has real spectrum. Let μ_j, ϕ_j be the eigenvalues and an orthonormal basis of eigenvectors of $A + B$. The evolution operator $L(t, s) = \exp(-(t - s)(A + B))$ is then given by

$$e^{-t(A+B)}v = \sum_{j=1}^{\infty} e^{-t\mu_j}(v, \phi_j)\phi_j.$$

The fixed point $\bar{u}$ is said to be hyperbolic if the spectrum does not intersect the imaginary axis, i.e., in this case, if all eigenvalues μ_j are non-zero. We then have an unbounded sequence

$$\mu_1 \leq \ldots \leq \mu_q < 0 < \mu_{q+1} \leq \ldots < \infty.$$

It is now easy to show that we have a constant exponential dichotomy with constant stable and unstable projections

$$Pv = \sum_{j=q+1}^{\infty} (v, \phi_j)\phi_j, \quad I - P = \sum_{j=1}^{q}(v, \phi_j)\phi_j.$$

When we linearize around a time-dependent solution, $\bar{u}(t)$, then we cannot use spectral theory in this way. This is why we need the more general concept of exponential dichotomy.

If $L(t, s)$ has an exponential dichotomy on the interval J, then the inhomogeneous boundary value problem, with $u_0 \in \mathcal{R}(P_0)$, $u_T \in \mathcal{R}(I - P_N)$,

$$u' + Au + B(t)u = g(t), \quad t \in J,$$
$$P(0)u(0) = u_0, \quad (I - P(T))u(T) = u_T, \tag{3.7}$$

is well-posed. We leave this as an exercise; see Exercise 3.1 below. Instead we switch the discussion to the discrete case, where we shall prove an analogous result.

Recall the sequence of operators $L_n = L(T_{n+1}, T_n)$, $n = 0, \ldots, N - 1$. Define

$$L_{mn} = L_{m-1} \ldots L_n, \; m > n; \quad L_{nn} = I$$

(note $L_n = L_{n-1,n}$). If $L(t, s)$ has an exponential dichotomy on J, then $\{L_n\}_{n=0}^{N-1}$ has a discrete exponential dichotomy; see [6]. This means that there are projections P_n, $n = 0, \ldots, N$, and constants $M \geq 1$ and $\theta \in (0, 1)$ such that the following conditions are satisfied for $n = 0, \ldots, N - 1$.

1. $L_n P_n = P_{n+1} L_n$.
2. The restriction $L_n|_{\mathcal{R}(I - P_n)} : \mathcal{R}(I - P_n) \to \mathcal{R}(I - P_{n+1})$ is an isomorphism. We define $L_{n,n+1} : \mathcal{R}(I - P_{n+1}) \to \mathcal{R}(I - P_n)$ to be its inverse.
3. $\|L_{mn} P_n\|_{\mathcal{L}(X)} \leq M\theta^{m-n}$, $m \geq n$.
4. $\|L_{nm}(I - P_m)\|_{\mathcal{L}(X)} \leq M\theta^{m-n}$, $m \geq n$.

Note that (1) implies $L_{mn} P_n = P_m L_{mn}$, $m \geq n$.

Let $X_0^S = \mathcal{R}(P_0)$ and $X_N^U = \mathcal{R}(I - P_N)$. Consider the following linear boundary value problem with data $\{y_n\}_{n=0}^{N+1}$, where $y_0 \in X_0^S$, $y_{N+1} \in X_N^U$, and $y_n \in X$ for $n = 1, \ldots, N$.

$$x_{n+1} = L_n x_n + y_{n+1}, \quad n = 0, \ldots, N - 1,$$
$$P_0 x_0 = y_0, \quad (I - P_N)x_N = y_{N-1}. \tag{3.8}$$

A simple calculation gives (Duhamel's principle)

$$x_n = L_{n0} x_0 + \sum_{j=1}^{n} L_{nj} y_j.$$

Hence, by condition (1),

$$P_n x_n = L_{n0} P_0 x_0 + \sum_{j=1}^{n} L_{nj} P_j y_j.$$

Similarly, evolving from T_n to T_N,

$$(I - P_N)x_N = L_{Nn}(I - P_n)x_n + \sum_{j=n+1}^{N} L_{Nj}(I - P_j)y_j,$$

where, by condition (2), $L_{Nn}|_{\mathcal{R}(I-P_n)}$ is invertible, so that

$$(I - P_n)x_n = L_{nN}(I - P_N)x_N - \sum_{j=n+1}^{N} L_{nj}(I - P_j)y_j.$$

Hence, using also the boundary conditions, we get

$$
\begin{aligned}
x_n &= P_n x_n + (I - P_n)x_n \\
&= L_{n0}P_0 y_0 + \sum_{j=1}^{n} L_{nj}P_j y_j \\
&\quad + L_{nN}(I - P_N)y_{N+1} - \sum_{j=n+1}^{N} L_{nj}(I - P_j)y_j.
\end{aligned}
\tag{3.9}
$$

Using (3) and (4) we get

$$
\begin{aligned}
\|x_n\|_X &\leq M\left(\theta^n + \sum_{j=1}^{n} \theta^{n-j} + \theta^{N-n} + \sum_{j=n+1}^{N} \theta^{j-n} \right) \max_j \|y_j\|_X \\
&\leq M\left(1 + \sum_{j=n+1}^{N} \theta^{j-n} + \theta^n + \sum_{j=1}^{n} \theta^{n-j} \right) \max_j \|y_j\|_X \\
&\leq M\left(\sum_{j=n}^{\infty} \theta^{j-n} + \sum_{j=-\infty}^{n} \theta^{n-j} \right) \max_j \|y_j\|_X = \frac{2M}{1-\theta} \max_j \|y_j\|_X .
\end{aligned}
$$

It follows that (3.8) is uniquely solvable and the solution satisfies

$$\max_{0 \leq n \leq N} \|x_n\|_X \leq \frac{2M}{1-\theta} \max_{0 \leq n \leq N+1} \|y_n\|_X . \tag{3.10}$$

We formulate this as a lemma.

Lemma 3.3. *Assume that $\{L_n\}_{n=0}^{N-1}$ has a discrete exponential dichotomy in the Banach space X with projections P_n and constants $M \geq 1$ and $\theta \in (0,1)$. Set $\mathbf{X} = X^{N+1}$, $\mathbf{Y} = X_0^S \times X^N \times X_N^U$, and define a linear operator $\mathbf{A} : \mathbf{X} \to \mathbf{Y}$ by $\mathbf{A} : (x_0, \ldots, x_N) \mapsto (y_0, \ldots, y_{N+1})$, where*

$$
\begin{aligned}
y_{n+1} &= x_{n-1} - L_n x_n, \quad n = 0, \ldots, N-1, \\
y_0 &= P_0 x_0. \quad y_{N-1} = (I - P_N)x_N.
\end{aligned}
$$

Then $\mathbf{A}$ is invertible and

$$\|\mathbf{A}^{-1}\|_{\mathcal{L}(\mathbf{Y},\mathbf{X})} \leq \frac{2M}{1-\theta}.$$

For future reference we note that the solution formula (3.9) may be written

$$x_n = G_{n0}y_0 - G_{nN}y_{N+1} + \sum_{j=1}^{N} G_{nj}y_j, \qquad (3.11)$$

by means of the Green's operator

$$G_{nj} = \begin{cases} L_{nj}P_j, & 0 \le j \le n, \\ -L_{nj}(I - P_j), & n < j \le N. \end{cases} \qquad (3.12)$$

Exercise 3.1. Show that the solution of (3.7) may be written

$$u(t) = G(t, 0)u_0 - G(t, T)u_T + \int_0^T G(t, s)g(s)\, ds, \qquad (3.13)$$

where

$$G(t, s) = \begin{cases} L(t, s)P(s), & 0 \le s \le t, \\ -L(t, s)(I - P(s)), & t < s \le T. \end{cases} \qquad (3.14)$$

Use this to derive an estimate of $\|u(t)\|_X$ similar to (3.10).

3.3. Shadowing

Let $\bar{u} \in C([0, T], V)$ be a solution of (1.6) with $\|\bar{u}(t)\|_1 \le R$, $t \in [0, T]$, for some T and R. Assume that we have numerical solution $u_h(t) = S_h(t, u_{h,0})$, $t \in [0, T]$, which belongs to a neighborhood of $\bar{u}$ in the sense that, for some $\rho_0 \le 1$,

$$\|u_h(t) - \bar{u}(t)\|_1 \le \rho_0, \quad t \in [0, T]. \qquad (3.15)$$

With $T_n = n\tau$ chosen as in §3.1 we form the sequence $u_{h,n} = S_h(n\tau, u_{h,0})$. This means that $u_{h,n}$ is an orbit of the nonlinear mapping $S_h = S_h(\tau, \cdot)$, i.e., $u_{h,n+1} = S_h(u_{h,n})$. Comparing with the mapping $S = S(\tau, \cdot)$ we get, in view of Theorem 2.3,

$$\begin{aligned} \|u_{h,n+1} - S(u_{h,n})\|_l &= \|S_h(\tau, u_{h,n}) - S(\tau, u_{h,n})\|_l \\ &\le C(R + 1)h^{2-l} =: \delta, \quad n = 0, \dots, N - 1, \ l = 0.1. \end{aligned}$$
$$\qquad (3.16)$$

(A similar result holds for a completely discrete sequence via Theorem 2.6.) With the terminology used in shadowing theory this means that the sequence $\{u_{h,n}\}_{n=0}^N$ is a δ-pseudo-orbit of S with $\delta = C(R + \rho)h^{2-l}$. Our aim is to find an orbit $\{u_n\}_{n=0}^N$ of S, i.e., $u_{n+1} = S(u_n)$, such that, for some σ,

$$\max_{0 \le n \le N} \|u_n - u_{h,n}\|_l \le \sigma\delta =: \epsilon, \quad l = 0, 1.$$

This means that $\{u_{h,n}\}_{n=0}^N$ is ϵ-shadowed by the true orbit $\{u_n\}_{n=0}^N$.

In order to achieve this we shall prove the following theorem which is set in a more general context.

Theorem 3.1 (Shadowing theorem). *Consider a nonlinear operator* $S :$ $D \subset X \to X$, *where* X *is a Banach space and* D *a nonempty subset of* X. *Assume that* S *can be decomposed in the form*

$$S = L_n + N_n \tag{3.17}$$

where $\{L_n\}_{n=0}^{N-1}$ *has a discrete exponential dichotomy on* X *with projections* P_n *and constants* $M \geq 1$, $\theta \in (0,1)$, *and where the operators* $N_n : D_n \subset$ $X \to X$ *are Lipschitz continuous with Lipschitz constants that satisfy*

$$\mathrm{Lip}(N_n) \leq \frac{1 - \theta}{4M}, \quad n = 0, \dots, N - 1. \tag{3.18}$$

Set

$$\sigma = \frac{4M}{1 - \theta}. \tag{3.19}$$

(a) *Let* $\{\tilde{x}_n\}_{n=0}^N$ *be a sequence with* $\tilde{x}_n \in D_n$, $n = 0, \dots, N - 1$. *If* $\{x_n\}_{n=0}^N$ *satisfies* $x_n \in D_n$, $n = 0, \dots, N - 1$, *and*

$$\begin{aligned} x_{n+1} &= S(x_n), \quad n = 0, \dots, N - 1, \\ P_0 x_0 &= P_0 \tilde{x}_0, \ (I - P_N)x_N = (I - P_N)\tilde{x}_N, \end{aligned} \tag{3.20}$$

then

$$\max_{0 \leq n \leq N} \|x_n - \tilde{x}_n\|_X \leq \sigma \max_{0 \leq n \leq N-1} \|\tilde{x}_{n+1} - S(\tilde{x}_n)\|_X. \tag{3.21}$$

(b) *Assume that* D_n *contains a closed ball* $B_X(z_n, \rho)$ *for* $n = 0, \dots, N - 1$, *where* $\{z_n\}_{n=0}^N$ *is a sequence that satisfies*

$$\|z_{n+1} - S(z_n)\|_X \leq \rho/\sigma, \quad n = 0, \dots, N - 1. \tag{3.22}$$

Set $\rho_0 = \rho/(\sigma M)$. *Then, for any data* $\tilde{x}_0 \in B_X(z_0, \rho_0)$, $\tilde{x}_N \in B_X(z_N, \rho_0)$, *the boundary value problem (3.20) has unique solution* $\{x_n\}_{n=0}^N$ *with* $x_n \in$ $B_X(z_n, \rho)$.

The proof is based on the following simple lemmas.

Lemma 3.4. *Let* $\mathbf{X}$ *and* $\mathbf{Y}$ *be Banach spaces and let* $\mathbf{A} : \mathbf{X} \to \mathbf{Y}$ *be a linear bijection with bounded inverse* $\mathbf{A}^{-1}$. *Assume that the mapping* $\mathbf{N} : \mathbf{D} \subset \mathbf{X} \to$ $\mathbf{Y}$, *defined in a nonempty subset* $\mathbf{D}$ *of* $\mathbf{X}$, *is Lipschitz continuous with*

$$\alpha := \left\|\mathbf{A}^{-1}\right\|_{\mathcal{L}(\mathbf{Y},\mathbf{X})} \mathrm{Lip}(\mathbf{N}) < 1.$$

Define $\mathbf{F} = \mathbf{A} + \mathbf{N}$ *and set* $\sigma := \left\|\mathbf{A}^{-1}\right\|_{\mathcal{L}(\mathbf{Y},\mathbf{X})} /(1 - \alpha)$. *Then*

$$\|\mathbf{x_1} - \mathbf{x_2}\|_{\mathbf{X}} \leq \sigma \|\mathbf{F}(\mathbf{x_1}) - \mathbf{F}(\mathbf{x_2})\|_{\mathbf{Y}}, \quad \text{for all } \mathbf{x_1}, \mathbf{x_2} \in \mathbf{D}. \tag{3.23}$$

Proof. The bound (3.23) follows readily from the identity

$$\mathbf{x_1} - \mathbf{x_2} = \mathbf{A}^{-1}\big(\mathbf{F}(\mathbf{x_1}) - \mathbf{F}(\mathbf{x_2})\big) - \mathbf{A}^{-1}\big(\mathbf{N}(\mathbf{x_1}) - \mathbf{N}(\mathbf{x_2})\big).$$

Lemma 3.5. *In addition to the hypotheses of Lemma 3.4, assume that the domain* $\mathbf{D}$ *of* $\mathbf{N}$ *contains a closed ball* $B_{\mathbf{X}}(\mathbf{z}, \rho)$. *Then, for each* $\mathbf{y} \in \mathbf{Y}$ *with*

$$\|\mathbf{y} - \mathbf{F}(\mathbf{z})\|_{\mathbf{Y}} \le \rho/\sigma, \tag{3.24}$$

the equation $\mathbf{F}(\mathbf{x}) = \mathbf{y}$ *has a unique solution* $\mathbf{x} \in B_{\mathbf{X}}(\mathbf{z}, \rho)$.

Proof. This is a simple consequence of the contraction mapping theorem. In fact, if we define $\mathbf{T}(\mathbf{x}) = \mathbf{A}^{-1}(\mathbf{y} - \mathbf{N}(\mathbf{x}))$, then $\mathrm{Lip}(\mathbf{T}) = \alpha < 1$. Moreover, the identity

$$\mathbf{T}(\mathbf{x}) - \mathbf{z} = \mathbf{A}^{-1}(\mathbf{y} - \mathbf{F}(\mathbf{z})) - \mathbf{A}^{-1}(\mathbf{N}(\mathbf{x}) - \mathbf{N}(\mathbf{z}))$$

and (3.24) imply that $\mathbf{T}$ maps $B_{\mathbf{X}}(\mathbf{z}, \rho)$ into itself.

Proof (of Theorem 3.1). Set $X_0^S = \mathcal{R}(P_0)$, $X_N^U = \mathcal{R}(I - P_N)$, $\mathbf{X} = X^{N+1}$, $\mathbf{Y} = X_0^S \times X^N \times X_N^U$, $\mathbf{D} = D_0 \times \ldots \times D_{N-1} \times X$, and define $\mathbf{y} \in \mathbf{Y}$, $\mathbf{A} : \mathbf{X} \to \mathbf{Y}$, $\mathbf{N} : \mathbf{D} \subset \mathbf{X} \to \mathbf{Y}$ by

$$\mathbf{y} = \begin{bmatrix} y_0 \\ y_{n+1} \\ y_{N+1} \end{bmatrix} := \begin{bmatrix} P_0 \tilde{x}_0 \\ 0 \\ (I - P_N)\tilde{x}_N \end{bmatrix},$$

$$\mathbf{A}\mathbf{x} := \begin{bmatrix} P_0 x_0 \\ x_{n+1} - L_n x_n \\ (I - P_N)x_N \end{bmatrix}, \quad \mathbf{N}(\mathbf{x}) := \begin{bmatrix} 0 \\ -N_n(x_n) \\ 0 \end{bmatrix},$$

where $n = 0, \ldots, N - 1$, and set $\mathbf{F} = \mathbf{A} + \mathbf{N}$. The boundary value problem (3.20) then becomes

$$\mathbf{y} = \mathbf{F}(\mathbf{x}).$$

The assumption (3.18) implies that $\mathrm{Lip}(\mathbf{N}) \le (1-\theta)/(4M)$. We may therefore apply Lemma 3.4 with $\alpha = 1/2$; this yields a value of $\sigma = \|\mathbf{A}^{-1}\|_{\mathcal{L}(\mathbf{Y}, \mathbf{X})}/(1 - \alpha) = 4M/(1 - \theta)$ that coincides with (3.19). The error bound (3.21) follows from (3.23) since

$$\mathbf{F}(\tilde{\mathbf{x}}) - \mathbf{F}(\mathbf{x}) = \begin{bmatrix} 0 \\ \tilde{x}_{n+1} - S(\tilde{x}_n) \\ 0 \end{bmatrix}.$$

This proves part (a). To prove part (b) we apply Lemma 3.5. Set $\mathbf{z} = (z_0, \ldots, z_N)$. Our assumption implies that $B_{\mathbf{X}}(\mathbf{z}, \rho) \subset \mathbf{D}$. Moreover, with $\mathbf{y}$ as above,

$$\mathbf{y} - \mathbf{F}(\mathbf{z}) = \begin{bmatrix} P_0(\tilde{x}_0 - z_0) \\ -(z_{n+1} - S(z_n)) \\ (I - P_N)(\tilde{x}_N - z_N) \end{bmatrix}.$$

Here, by conditions (3) and (4) in the definition of discrete exponential dichotomy, we have $\|P_0\|_{\mathcal{L}(X)} \le M$, $\|I - P_N\|_{\mathcal{L}(X)} \le M$, so that (3.24) follows by our assumptions.

We now return to the problem of shadowing a numerical solution $u_h(t)$ in a neighborhood of $\bar{u}$.

Theorem 3.2 (Numerical shadowing theorem 1). *Let* $\bar{u} \in C([0,T], V)$ *be a solution of (1.6) with*

$$\max_{t \in [0,T]} \|\bar{u}(t)\|_1 \leq R, \tag{3.25}$$

for some T *and* R. *Assume that the solution operator* $L(t,s)$ *of the linearized problem (3.3) has an exponential dichotomy in* V *and* H *on the interval* $[0,T]$. *Then there are numbers* ρ_0 *and* C *such that, for each solution* $u_h(t) = S_h(t, u_{h,0})$, $t \in [0,T]$, *of (2.2) with*

$$\max_{t \in [0,T]} \|u_h(t) - \bar{u}(t)\|_1 \leq \rho_0, \tag{3.26}$$

there is a solution u *of (1.6) such that*

$$|u_h(t) - u(t)\|_l \leq C(1 + t^{-1/2})h^{2-l}, \quad t \in (0,T], \ l = 0,1. \tag{3.27}$$

Proof. Let $u_{h,n} = u_h(T_n)$, $T_n = n\tau$, and $S = L_n + N_n$ be as above. We apply Theorem 3.1 with $\tilde{x}_n = u_{h,n}$. Set $D = B_{R+1} = B_V(0, R+1)$, $z_n = \bar{u}(T_n)$. and

$$D_n = \{v \in V : \|v - \bar{u}(T_n)\|_1 \leq \rho\} = B_V(z_n, \rho).$$

Lemma 3.2 (b) shows that $N_n : D_n \to X$ has Lipschitz constant

$$\text{Lip}(N_n) \leq C(R)\rho$$

for both $X = V$ and $X = H$. We fix ρ so that (3.18) holds and set $\rho_0 = \rho/(\sigma M)$.

In order to prove existence we use $X = V$. Then D_n is a closed ball in X. Moreover, $z_{n+1} - S(z_n) = 0$ so that (3.22) holds, and (3.26) implies that $u_{h,n} = \tilde{x}_n \in B_V(z_n, \rho_0) \subset D_n$, $n = 0, \ldots, N$. Theorem 3.1 (b) now gives the existence of an orbit u_n, $n = 0, \ldots, N$, of S, satisfying (3.20), and hence, by part (a), (3.21) and (3.16) give the special case $l = 1$ of the inequality

$$\max_{0 \leq n \leq N} \|u_{h,n} - u_n\|_l \leq \sigma \max_{0 \leq n \leq N-1} \|u_{h,n+1} - S(u_{h,n})\|_l \leq Ch^{2-l}, \quad l = 0,1. \tag{3.28}$$

Another application of Theorem 3.1 (a), now with $X = H$, proves the case $l = 0$ of (3.28). Note that D_n does not contain any ball defined using the norm of $X = H$ so part (b) does not apply.

Using the sequence u_n we define $u(t) = S(t - T_n, u_n)$ for $t \in [T_n, T_{n+1}]$. By existence and uniqueness of solutions (Theorem 2.3) this is a solution of (1.6). Error bounds at intermediate times are obtained by combining (3.28) with Theorem 2.3 as follows. For $t \in [0, T_1]$ we have

$$\|u_h(t) - u(t)\|_1 \leq C(R)t^{-1/2} \left(\|u_{h,0} - u_0\|_1 + h\right) \leq Ct^{-1/2}h.$$

Here we used the fact that

$$\|u_{h,0} - P_h u_0\| \leq \|u_{h,0} - u_0\| \leq C\,\|u_{h,0} - u_0\|_1 \leq C\big(\|u_{h,0}\|_1 + \|u_0\|_1\big) \leq CR.$$

For $t \in [T_{n+1}, T_{n+2}]$, $n \geq 0$, we have

$$\|u_h(t) - u(t)\|_1 \leq C(R)\,\big(\|u_{h,n} - u_n\|_1 + h\big) \leq Ch.$$

This proves the special case $l = 1$ of (3.27). The case $l = 0$ is obtained similarly.

Exercise 3.2. State and prove a similar theorem without the assumption that $\bar{u}$ is a solution. (Hint: (3.22) means that z_n needs only be a pseudo-orbit of S.)

Exercise 3.3. State and prove a similar theorem for the backward Euler discretization in §2.2.

Note that the shadowing property is a property of the evolution operator S; properties of S_h (or S_{hk}) enter only through (3.16).

Shadowing in a neighborhood of a hyperbolic fixed point (Example 3.1) was proved in [10] and [9] by the above technique. There we also proved the converse: every exact solution near the fixed point is shadowed by a numerical solution. The proof involves showing that the exponential dichotomy of the evolution operator L carries over to its numerical counterpart and then letting S and S_h change roles. This work was inspired by the paper [1], where similar results were proved in the context of ordinary differential equations.

In [8] we obtained shadowing in a neighborhood of the whole attractor of S. Here we assume that the attractor has the Morse-Smale property, which means, among other things, that S has a finite number of fixed points, all of them hyperbolic, and the attractor consists of the union of their unstable manifolds. After reduction to finite dimension by means of an inertial manifold, we apply a more complicated, global shadowing theorem of Pilyugin. Again properties of S_h (or S_{hk}) enter mainly through (3.16).

4. A Posteriori Error Estimates

In this section we express the error in terms of an appropriately defined residual of the computed solution and the solution of a linearized adjoint problem. The residual is estimated in terms of a posteriori computable residuals. The presentation here is inspired by [4].

4.1. The Error Equation

We will derive an equation for the error by means of a duality argument. We therefore begin by recalling the weak form of (1.1): find $u(t) \in V$ such that

$$(u', v) + a(u, v) - (f(u), v) = 0, \quad \forall v \in V, \ t > 0, \tag{4.1}$$
$$u(0) = u_0,$$

where $a(u, v) = (\epsilon \nabla u, \nabla v) = (-\epsilon \Delta u, v) = (Au, v)$ is the bilinear form associated with A. To allow for time-discretization by means of discontinuous piecewise polynomials in t, we introduce a mesh on the time interval $[0, T]$.

$$0 = t_0 < \ldots < t_{j-1} < t_j < \ldots < t_N = T,$$

with time intervals $I_j = (t_{j-1}, t_j)$ of lengths $k_j = t_j - t_{j-1}$, and we define function spaces

$$\mathcal{V} = C([0, T], V),$$
$$\tilde{\mathcal{V}} = \{v : v \in C(\bar{I}_j, V) \cap C^1(\bar{I}_j, V^*), \ j = 1, \ldots, N\}. \tag{4.2}$$

Functions in $\tilde{\mathcal{V}}$ are piecewise smooth with respect to the mesh, but may be discontinuous at the nodes t_j.

Let $U \in \tilde{\mathcal{V}}$ be arbitrary. (In the end U will be a numerical approximation of u, but this is not essential right now.) We define the local residual of U,

$$r_{t_{j-1}, t}(U, v) = \int_{t_{j-1}}^t \left((U', v) + a(U, v) - (f(U), v) \right) d\sigma \tag{4.3}$$
$$+ ([U]_{t_{j-1}}, v(t_{j-1}^-)), \quad v \in \tilde{\mathcal{V}}, \ t \in (t_{j-1}, t_j],$$

where $[v]_{t_j} = v(t_j^+) - v(t_j^-)$ denotes the jump at t_j.

Let $s, t \in [0, T]$ with $s < t$. We define a punctured interval and two sets of nodes: if $s \in [t_m, t_{m+1})$, $t \in (t_{n-1}, t_n]$, then

$$J_{s,t} = (s, t_{m+1}) \cup I_{m+2} \cup \ldots I_{n-1} \cup (t_{n-1}, t),$$

$$J_{s,t}^- = \{t_j : s \le t_j < t\}, \quad J_{s,t}^+ = \{t_j : s < t_j \le t\}.$$

We can now define a global residual

$$R_{s,t}(U, v) = \int_{J_{s,t}} \left((U', v) + a(U, v) - (f(U), v) \right) d\sigma \tag{4.4}$$
$$+ \sum_{J_{s,t}^-} ([U]_{t_j}, v(t_j^+)), \quad v \in \tilde{\mathcal{V}}.$$

Any solution u of (4.1) satisfies

$$R_{s,t}(u,v) = 0, \quad \forall v \in \tilde{\mathcal{V}}. \tag{4.5}$$

Hence

$$
\begin{aligned}
R_{s,t}(U,v) &= R_{s,t}(U,v) - R_{s,t}(u,v) \\[2mm]
&= \int_{J_{s,t}} ((U'-u',v) + a(U-u,v) - (f(U)-f(u),v))\, d\sigma \\[2mm]
&\quad + \sum_{J_{s,t}^-} ([U-u]_{t_j}, v(t_j^+)), \quad \forall v \in \tilde{\mathcal{V}}.
\end{aligned}
$$

We linearize around U. Thus, let

$$
\begin{aligned}
e &:= U - u, \\
B(t) &:= -f'(U(t)) \in \mathcal{L}(V,H), \\
\eta = \eta(U,u) &:= f(U) - f(u) - f'(U)(U-u);
\end{aligned} \tag{4.6}
$$

cf. (3.2) with $\bar{u}$ replaced by U. Then

$$
\begin{aligned}
R_{s,t}(U,v) &= \int_{J_{s,t}} ((e',v) + a(e,v) + (Be,v) - (\eta,v))\, d\sigma \\[2mm]
&\quad + \sum_{J_{s,t}^-} ([e]_{t_j}, v(t_j^+)), \quad \forall v \in \tilde{\mathcal{V}}.
\end{aligned} \tag{4.7}
$$

We define the bilinear form

$$
\begin{aligned}
B_{s,t}(w,v) &= \int_{J_{s,t}} ((w',v) + a(w,v) + (Bw,v))\, d\sigma \\[2mm]
&\quad + \sum_{J_{s,t}^-} ([w]_{t_j}, v(t_j^+)) + (w(s^-), v(s^+)) \\[4mm]
&= \int_{J_{s,t}} ((w,-v') + a(w,v) + (w,B^*v))\, d\sigma \\[2mm]
&\quad + \sum_{J_{s,t}^+} (w(t_j^-), -[v]_{t_j}) + (w(t^-), v(t^+)), \quad w,v \in \tilde{\mathcal{V}},
\end{aligned} \tag{4.8}
$$

where the second variant is obtained by integration by parts. When restricted to smooth test functions $v \in C_0^\infty$ the relevant terms in (4.8) coincide with the definition of the derivative w' in the sense of distributions. The point is that we extend this definition to piecewise smooth test functions $v \in \tilde{\mathcal{V}}$.

Now (4.7) becomes

$e \in \tilde{\mathcal{V}}$;

$$B_{s,t}(e,v) = (e(s^-), v(s^+)) + R_{s,t}(U,v) + \int_s^t (\eta, v)\, d\sigma, \quad \forall v \in \tilde{\mathcal{V}}. \quad (4.9)$$

We argue by duality. We therefore introduce the linearized adjoint problem with data $\psi \in H$:

$$-z' + Az + B^*(s)z = 0, \quad s < t; \quad z(t) = \psi. \quad (4.10)$$

This is the adjoint of the linearized problem (3.3) (but linearized around $U(t)$ instead of $\bar{u}(t)$). Using the second form of (4.8) we write (4.10) as:

$$z \in \mathcal{V}: \quad B_{s,t}(w,z) = (w(t^-), \psi), \quad \forall w \in \tilde{\mathcal{V}}. \quad (4.11)$$

Let $K(s,t)$ be the solution operator of (4.10), so that its solution is $z(s) = K(s,t)\psi$. Global existence on $[0,T]$ is obtained in the same way as for the evolution operator $L(t,s)$ in §3.1. In fact, the solution $u(t) = L(t,s)\phi$ of (3.3) with data $\phi \in H$ satisfies

$$u \in \mathcal{V}; \quad B_{s,t}(u,v) = (\phi, v(s^+)), \quad \forall v \in \tilde{\mathcal{V}}. \quad (4.12)$$

With $v = z$ in (4.12) and $w = u$ in (4.11) we get

$$(L(t,s)\phi, \psi) = B(u,z) = (\phi, K(s,t)\psi).$$

We conclude that $K(s,t) = L(t,s)^*$.

We now return to the equation (4.9) for e. With $v = z$ in (4.9) and $w = e$ in (4.11) we get

$$(e(t^-), \psi) = (e(s^-) \cdot z(s^-)) + R_{s,t}(U,z) + \int_s^t (\eta, z)\, d\sigma, \quad \forall \psi \in H. \quad (4.13)$$

or, with $z(s) = K(s,t)\psi$,

$$\begin{aligned}
(e(t^-), \psi) &= (e(s^-), K(s,t)\psi) \\
&\quad + \int_{J_{s,t}} \big((U' - f(U), K(\sigma,t)\psi) + a(U, K(\sigma,t)\psi)\big)\, d\sigma \quad (4.14) \\
&\quad + \sum_{J_{s,t}^-} ([U]_{t_j}, K(t_j,t)\psi) + \int_s^t (\eta, K(\sigma,t)\psi)\, d\sigma, \quad \forall \psi \in H.
\end{aligned}$$

Formally, since $K(s,t) = L(t,s)^*$, we may write

$$\begin{aligned}
e(t^-) &= L(t,s)e(s^-) + \int_s^t L(t,\sigma)(U' + AU - f(U))\, d\sigma \\
&\quad + \int_s^t L(t,\sigma)\eta\, d\sigma.
\end{aligned} \quad (4.15)$$

This is the variation of constants formula (Duhamel's principle) for the equation

$$e' + Ae + B(t)e = U' + AU - f(U) + \eta.$$

The first term in the right side of (4.15) describes the evolution of the initial error $e(s^-)$; the second term is the contribution to $e(t^-)$ from the residual $U' + AU - f(U)$. The third term is the remainder from linearization. However, the derivatives $U'(\sigma)$ and $AU(\sigma) = -\epsilon \Delta U(\sigma)$, where $U(\sigma)$ is only in V, have to be understood in the sense of distributions. This is why we argue by duality and use the weak formulation (4.13).

We finish this subsection by stating an estimate of the remainder η defined in (4.6). It is an immediate consequence of (1.23).

Lemma 4.1. *Assume that* $\|U(t)\|_1 \leq R$, $\|u(t)\|_1 \leq R$, *for* $t \in [0,T]$. *Then, for* $s, t \in [0,T]$, $l = 0, 1$,

$$\left| \int_s^t (\eta, v) \, d\sigma \right| \leq C(R) \|e\|_{L_\infty([s,t],H^1)} \|e\|_{L_\infty([s,t],H^l)} \|v\|_{L_1([s,t],H^{1-l})}.$$

4.2. Local Estimates of the Residual

We now assume that U is defined by the discontinuous Galerkin finite element method; see [12]. We define a space of functions that are discontinuous, piecewise constant in time, and piecewise linear in space:

$$\tilde{\mathcal{V}}_h = \{v \in \tilde{\mathcal{V}} : v|_{I_j}(t) = v(t_j^-) \in V_h, \ j = 1, \ldots, N\}.$$

Then $U \in \tilde{\mathcal{V}}_h$ is defined by

$$\int_{I_j} ((U', v) + a(U, v) - (f(U), v)) \, d\sigma + ([U]_{t_{j-1}}, v(t_{j-1}^+)) = 0, \quad \forall v \in \tilde{\mathcal{V}}_h,$$
(4.16)

for $j = 1, \ldots, N$. An initial value $U(t_0^-) = u_{h,0}$ must be provided. Here. of course, $U' = 0$, since U is constant on I_j. Setting $U_j = U(t_j^-)$, $v(t) = \chi \in V_h$ on I_j, we see that (4.16) becomes

$$k_j a(U_j, \chi) - k_j (f(U_j), \chi) + (U_j - U_{j-1}, \chi) = 0, \quad \forall \chi \in V_h,$$
(4.17)

or

$$A_h U_j - P_h f(U_j) + k_j^{-1}(U_j - U_{j-1}) = 0,$$

which is the backward Euler method (2.18).

The equation (4.16) means that

$$r_{t_{j-1}, t_j}(U, v) = 0, \quad \forall v \in \tilde{\mathcal{V}}_h.$$
(4.18)

As a consequence of this "orthogonality" we have the following a posteriori estimate of the local residual.

Lemma 4.2. *Let $U \in \tilde{V}_h$ be a solution of (4.16) and assume that $v \in V$ is sufficiently smooth. Then, for $l = 0, 1$, $m = 1, 2$, $\tilde{I}_j = (t_{j-1}, t) \subset I_j$,*

$$|r_{t_{j-1},t}(U,v)| \leq C \|kR_t\|_{L_\infty(\tilde{I}_j, H^l)} \min\left(\|v'\|_{L_1(\tilde{I}_j, H^{-l})}, \|v\|_{L_\infty(\tilde{I}_j, H^{-l})}\right)$$
$$+ C \|h^m R_x\|_{L_\infty(\tilde{I}_j, H)} \left\|\int_{\tilde{I}_j} v \, d\sigma\right\|_m,$$

where the functions $h = h(x)$, $k = k(t)$, $R_t = R_t(x,t)$, $R_x = R_x(x,t)$, are defined piecewise by

$$
\begin{aligned}
h|_K &= h_K = \operatorname{diam}(K), \quad k|_{I_j} = k_j, \\
R_t|_{K \times I_j} &= k_j^{-1}[U]_{t_{j-1}} = k_j^{-1}(U_j - U_{j-1}), \\
R_x|_{K \times I_j} &= |k_j^{-1}[U]_{t_{j-1}} - f(U_j)| + |K|^{-1/2} h_K^{-1/2} \|\epsilon[\partial U_j/\partial n]\|_{L_2(\partial K)}.
\end{aligned}
\tag{4.19}
$$

Here $[\partial U_j/\partial n]$ is the jump across ∂K in the exterior normal derivative $\partial U_j/\partial n$.

Proof. Since U is constant on I_j we have by (4.17) and (4.3), for any $\chi \in V_h$.

$$
\begin{aligned}
r_{t_{j-1},t}(U,v) &= a(U, V(t)) + (U' - f(U), V(t)) + ([U]_{t_{j-1}}, v(t_{j-1}^+)) \\
&= a(U_j, V(t)) + (k_j^{-1}[U]_{t_{j-1}} - f(U_j), V(t)) \\
&\quad + ([U]_{t_{j-1}}, v(t_{j-1}^+) - k_j^{-1}V(t)) \\
&= \left(a(U_j, V(t) - \chi) + (k_j^{-1}[U]_{t_{j-1}} - f(U_j), V(t) - \chi)\right) \\
&\quad + ([U]_{t_{j-1}}, v(t_{j-1}^+) - k_j^{-1}V(t)) \equiv \text{I} + \text{II},
\end{aligned}
$$

where $V(t) = \int_{t_{j-1}}^t v \, d\sigma = \int_{\tilde{I}_j} v \, d\sigma$. Here, by the definition of R_t and a simple estimation,

$$
\begin{aligned}
|\text{II}| &\leq \|k_j R_t\|_l \left\|v(t_{j-1}^+) - k_j^{-1}V(t)\right\|_{-l} \\
&\leq \|k_j R_t\|_l \min\left(\|v'\|_{L_1(\tilde{I}_j, H^{-l})}, 2\|v\|_{L_\infty(\tilde{I}_j, H^{-l})}\right).
\end{aligned}
$$

We now estimate I. We write $g = -f(U_j) + k_j^{-1}[U]_{t_{j-1}}$ for short. Elementwise integration by parts gives, since $\Delta U_j = 0$ on K,

$$
\begin{aligned}
a(U_j, w) + (g, w) &= \sum_K ((\epsilon \nabla U_j, \nabla w)_K + (g, w)_K) \\
&= \sum_K ((-\epsilon \Delta U_j + g, w)_K + (\epsilon \partial U_j/\partial n, w)_{\partial K}) \\
&= \sum_K \left((g, w)_K - \tfrac{1}{2}(\epsilon[\partial U_j/\partial n], w)_{\partial K \setminus \partial \Omega}\right), \quad \forall w \in V.
\end{aligned}
$$

With $w = V(t) - \chi$ we get

$$|\text{I}| \leq \sum_K \left(\|g\|_{L_2(K)} \|V(t) - \chi\|_{L_2(K)} \right.$$
$$\left. + \tfrac{1}{2} \|\epsilon [\partial U_j / \partial n]\|_{L_2(\partial K \backslash \partial \Omega)} \|V(t) - \chi\|_{L_2(\partial K)} \right).$$

We choose $\chi = \Pi V(t)$, where $\Pi : V \to V_h$ is an interpolation operator such that

$$\left\| D^l (w - \Pi w) \right\|_{L_2(K)} \leq C h_K^{m-l} \|D^m w\|_{L_2(S_K)}, \quad l = 0, 1, \ m = 1, 2, \quad (4.20)$$

where S_K is the union of all simplices adjacent to K; see [11]. Using also the (scaled) trace inequality

$$\|w\|_{L_2(\partial K)} \leq C \left(h_K^{-1/2} \|w\|_{L_2(K)} + h_K^{1/2} \|Dw\|_{L_2(K)} \right), \quad (4.21)$$

we obtain

$$\|V(t) - \Pi V(t)\|_{L_2(\partial K)} \leq C h_K^{m-1/2} \|D^m V(t)\|_{L_2(S_K)}, \quad m = 1, 2.$$

Hence

$$|\text{I}| \leq C \sum_K h_K^m \left(\|g\|_{L_2(K)} \right.$$
$$\left. + h_K^{-1/2} \|\epsilon [\partial U_j / \partial n]\|_{L_2(\partial K \backslash \partial \Omega)} \right) \|D^m V(t)\|_{L_2(S_K)}$$
$$\leq C \|h^m R_x\|_{L_2} \|D^m V(t)\|_{L_2} \leq C \|h^m R_x\|_{L_2} \left\| \int_{\tilde{I}_j} v \, d\sigma \right\|_m,$$

where the constant depends only on the constants in (4.20) and (4.21).

Note that the quantities R_t and R_x, in the previous theorem are computable a posteriori from the numerical solution U.

We now show that the local residuals tend to zero with h and k by using local a priori estimates from §2. Essentially, the idea is to compare U with an auxiliary local solution $\tilde{u}(t) = S(t - T_n, U(T_n))$ (we use the notation $T_n = n\tau$ from §3). In order to present the idea with a minimum of technical details we make a very strong assumption about the mesh. For a more general result we refer to [4].

Theorem 4.1. *Assume that the family of spatial meshes is quasi-uniform and that the the time step is constant with $h^2 k^{-1} \leq C$, where now $h = \max_K h_K$, $k = k_j$. Let $U \in \tilde{\mathcal{V}}_h$ be a solution of (4.16) with $\|U(t)\|_1 \leq R$ for $t \in [0, T]$. Let τ be as in Theorems 2.3 and 2.6 and assume that $t_{j-1} \geq \tau$. Then, for $l = 0, 1$, we have*

$$\|k R_t\|_{L_\infty(I_j, H^l)} + \left\| h^{2-l} R_x \right\|_{L_\infty(I_j, H)} \leq C(R) \left(h^{2-l} + k \right).$$

Proof. Since $t_{j-1} \geq \tau$ we have $I_j \subset [T_{n+1}, T_{n+2}]$ for some $n \geq 0$ (we use the notation $T_n = n\tau$ from §3). Let $\tilde{u}(t) = S(t - T_n, U(T_n))$ be the local exact solution starting with $\tilde{u}(T_n) = U(T_n)$. Theorem 2.6 gives for the error $\tilde{e} = U - \tilde{u}$ the bound

$$\|\tilde{e}\|_{L_\infty((T_{n+1}, T_{n+2}), H^l)} \leq C(R)(h^{2-l} + k).$$

Hence

$$\|kR_t\|_{L_\infty(I_j, H^l)} = \left\|[U]_{t_{j-1}}\right\|_l = \left\|[\tilde{e}]_{t_{j-1}}\right\|_l \leq 2\|\tilde{e}\|_{L_\infty((T_{n+1}, T_{n+2}), H^l)}$$

$$\leq C(R)(h^{2-l} + k).$$

Similarly, for the first term in R_x,

$$\left\|h^{2-l}k_j^{-1}[U]_{t_{j-1}}\right\|_{L_\infty(I_j, H)} \leq C(R)h^{2-l}k^{-1}(h^{2-l} + k) \leq C(R)(h^{2-l} + k).$$

Also, by Lemma 1.1,

$$\left\|h^{2-l}f(U)\right\|_{L_\infty(I_j, H)} \leq C(R)h^{2-l}.$$

Finally, for the last contribution to R_x, we use (4.21) and Theorem 1.2,

$$h^{2-l}\left(\sum_K h^{-1}\|\epsilon[\partial U_j/\partial n]\|_{L_2(\partial K)}^2\right)^{1/2} = h^{2-l}\left(\sum_K h^{-1}\|\epsilon[\partial\tilde{e}/\partial n]\|_{L_2(\partial K)}^2\right)^{1/2}$$

$$\leq Ch^{2-l}\left(\sum_K h^{-1}\|D\tilde{e}\|_{L_2(\partial K)}^2\right)^{1/2}$$

$$\leq Ch^{2-l}\left(\sum_K h^{-1}\left(h^{-1/2}\|D\tilde{e}\|_{L_2(K)} + h^{1/2}\left\|D^2\tilde{e}\right\|_{L_2(K)}\right)^2\right)^{1/2}$$

$$\leq Ch^{2-l}\left(h^{-1}\|D\tilde{e}\|_{L_\infty((T_{n+1}, T_{n+2}), H)} + \left\|D^2\tilde{e}\right\|_{L_\infty((T_{n+1}, T_{n+2}), H)}\right)$$

$$\leq C\left(h^{1-l}\|\tilde{e}\|_{L_\infty((T_{n+1}, T_{n+2}), H^1)} + h^{2-l}\left\|D^2\tilde{u}\right\|_{L_\infty((T_{n+1}, T_{n+2}), H)}\right)$$

$$\leq C(R)(h^{2-l} + k).$$

4.3. A Global Error Estimate

We now combine the results of §4.1 and §4.2 to obtain a global a posteriori error estimate. We estimate both the H and V norms of the error ($l = 0, 1$). To achieve this we use a duality argument:

$$\|e\|_l \leq C\left\|A^{l/2}e\right\| = C \sup_{\|\varphi\|=1} (A^{l/2}e, \varphi) = C \sup_{\|\varphi\|=1} (e, A^{l/2}\varphi).$$

We thus take $\psi = A^{l/2}\varphi$, $\|\varphi\| = 1$, in (4.13) and (4.10), so that $z(s) = K(s, t)A^{l/2}\varphi$. Using also Lemma 4.2 with $m = 2 - l$ and Lemma 4.1 we get the following.

Theorem 4.2. *Let U and u be solutions of (4.16) and (4.1), respectively. Assume that $\|U(t)\|_1 \le R$, $\|u(t)\|_1 \le R$, for $t \in [0,T]$. Then, for $t \in [0,T]$, $l = 0, 1$, with n such that $t \in [t_n, t_{n+1})$,*

$$\begin{aligned}
\left\| e(t^-) \right\|_l &\le CS_l^0(0,t) \left\| e(0^-) \right\|_l \\
&+ CS_l^1(0,t) \left\| kR_t \right\|_{L_\infty((0,t),H^l)} \\
&+ CS_l^2(0,t) \left\| h^{2-l}R_x \right\|_{L_\infty((0,t),H)} \\
&+ C(R)S_l^3(0,t) \|e\|_{L_\infty((0,t),H^1)} \|e\|_{L_\infty((0,t),H^1)},
\end{aligned}$$

where the so called stability factors are defined by

$$\begin{aligned}
S_l^0(0,t) &= \sup_{\|\varphi\|=1} \max_{\tau\in[0,t]} \left\| A^{-l/2} K(0,\tau) A^{l/2}\varphi \right\|, \\
S_l^1(0,t) &= \sup_{\|\varphi\|=1} \left(\int_0^{t_{n-1}} \left\| A^{-l/2} D_s K(s,t) A^{l/2}\varphi \right\| ds \right. \\
&\qquad\qquad \left. + \max_{s\in[t_{n-1},t]} \left\| A^{-l/2} K(s,t) A^{l/2}\varphi \right\| \right), \\
S_l^2(0,t) &= \sup_{\|\varphi\|=1} \left(\int_0^{t_{n-1}} \left\| A^{(2-l)/2} K(s,t) A^{l/2}\varphi \right\| ds \right. \\
&\qquad\qquad \left. + \left\| \int_{t_{n-1}}^t A^{(2-l)/2} K(s,t) A^{l/2}\varphi\, ds \right\| \right), \\
S_l^3(0,t) &= \sup_{\|\varphi\|=1} \int_0^t \left\| A^{(1-l)/2} K(s,t) A^{l/2}\varphi \right\| ds.
\end{aligned}$$

Note in this connection that

$$\left\| A^{-l/2} K(s,t) A^{l/2} \right\|_{\mathcal{L}(H)} = \left\| A^{l/2} L(t,s) A^{-l/2} \right\|_{\mathcal{L}(H)}.$$

The special treatment of the last interval (t_{n-1}, t) is necessitated by the expected singular behavior (cf. Lemma 3.1)

$$\left\| A^{(2-l)/2} K(s,t) A^{l/2}\varphi \right\| \sim \left\| A^{-l/2} D_s K(s,t) A^{l/2}\varphi \right\| \sim (t-s)^{-1}, \quad \text{as } s \to t.$$

The stability factors grow fast with t if the linearized problem is unstable.

Of course, the error estimate of the the previous theorem is not 'closed' because the square of the error occurs on the right hand side. However, this square remainder can be removed by means of the contraction mapping principle, if we compute with a sufficiently small tolerance. We demonstrate this in a slightly different context in the next section.

5. Shadowing—Second Approach

We finish these lectures by describing another approach to shadowing based on a posteriori error analysis. This is work in progress together with M. G. Larson, and we only sketch a simplified version.

The error estimate in Theorem 4.2 is based on the stability of the adjoint (4.10) of the linearized initial value problem (3.3). We now use the stability of the boundary value problem (3.7). Let $t \in (0,T)$, $\psi \in H$, and let z be a solution of

$$- z' + Az + B^*(s)z = 0, \quad s \in (0,T), \quad s \neq t; \quad [z]_t = -\psi. \tag{5.1}$$

Thus z has a jump at $s = t$. This equation does not determine z uniquely: we must provide appropriate boundary conditions at $s = 0, T$. But for the time being we simply assume that we have some solution operator, denoted $z(s) = G(t,s)^* \psi$, of (5.1).

Multiplying (from the left) by a test function $w \in \tilde{\mathcal{V}}$ and integrating by parts as in (4.8), taking into account that z is discontinuous at t, we get

$$
\begin{aligned}
0 &= \int_0^T (w, -z' + Az + B^*z) \, d\sigma \\
&= \int_{J_{0.T}} ((w', z) + a(w. z) + (Bw, z)) \, d\sigma \\
&\quad + \sum_{J_{0.T}^-} ([w]_{t_j}. z(t_j)) + (w(0^-), z(0)) - (w(T^-), z(T)) + (w(t^-), [z]_t).
\end{aligned}
$$

Thus z satisfies

$$\tilde{B}_{0,T}(w, z) = (w(t^-), \psi), \quad \forall w \in \tilde{\mathcal{V}}, \tag{5.2}$$

where now, instead of (4.8),

$$
\begin{aligned}
\tilde{B}_{0,T}(w, v) &= \int_{J_{0.T}} ((w', v) + a(w, v) + (Bw, v)) \, d\sigma \\
&\quad + \sum_{J_{0.T}^-} ([w]_{t_j}, v(t_j^+)) + (w(0^-), v(0^+)) - (w(T^-), v(T^+)).
\end{aligned}
\tag{5.3}
$$

Combining this definition with (4.7) yields, instead of (4.9),

$$
\begin{aligned}
e \in \tilde{\mathcal{V}}; \quad \tilde{B}_{0,T}(e, v) &= (e(0^-), v(0^+)) - (e(T^-), v(T^+)) \\
&\quad + R_{0,T}(U, v) + \int_0^T (\eta, v) \, d\sigma, \quad \forall v \in \tilde{\mathcal{V}}.
\end{aligned}
\tag{5.4}
$$

Together with (5.2) this gives, instead of (4.13),

$$(e(t^-), \psi) = (e(0^-), z(0)) - (e(T^-), z(T))$$

$$+ R_{0,T}(U, z) + \int_0^T (\eta, z) \, d\sigma, \quad \forall \psi \in H. \tag{5.5}$$

Note in this connection that $z(s)$ may have an additional jump at $s = t$ in which case $z \notin \tilde{\mathcal{V}}$. However, we may still use $v = z$ in (5.4) because $R_{0,T}(U, v)$ does not involve v'; see (4.4).

The error representation (5.5) is similar to (3.13). In fact, if the linearized problem has an exponential dichotomy, then we choose $z(s) = G(t, s)^* \psi$, where $G(t, s)$ is the Green's operator in (3.14).

With U given we want to solve equation (5.5) for e. We then obtain a shadow solution $u = U - e$. In order to achieve this we set the boundary data $(e(0^-), z(0)) = (e(T^-), z(T)) = 0$. With U fixed we consider $\eta = \eta(U, u) = \eta(U, U - e)$ as a function of e and define the nonlinear operator

$$N_{0,T}(e, v) = \int_0^T (\eta(U, U - e), v) \, d\sigma.$$

We then write (5.5) as a fixed point equation

$$(e(t^-), \psi) = R_{0,T}(U, z) + N_{0,T}(e, z), \quad \forall \psi \in H. \tag{5.6}$$

As before we take $\psi = A^{1/2}\varphi$, $\|\varphi\| = 1$, in (5.6), so that $z(s) = G(t, s)^* A^{1/2}\varphi$. Define the ball

$$\mathcal{B}_\rho = \{e \in L_\infty((0, T), V) : \|e\|_{L_\infty((0,T),H^1)} \le \rho\}.$$

Similarly to Lemma 4.1 it follows that $N_{0,T}$ has a small Lipschitz constant in $\mathcal{B}_\rho$: if $\psi = A^{1/2}\varphi$, $\|\varphi\| = 1$, then

$$|N_{0,T}(e_1, z) - N_{0,T}(e_2, z)| \le S_1^3 C(R)\rho \, \|e_1 - e_2\|_{L_\infty((0,T),H^1)}, \quad e_1, e_2 \in \mathcal{B}_\rho. \tag{5.7}$$

By Lemma 4.2 we get an error bound similar to that in Theorem 4.2. It is of the form

$$|R_{0,T}(U, z)| \le C S_1^1 \, \|k R_t\|_{L_\infty((0,T),H^1)} + C S_1^2 \, \|h R_x\|_{L_\infty((0,T),H)} =: \mathcal{R}(U). \tag{5.8}$$

Here the stability factors $S_1^i = S_1^i(0, T)$, $i = 1, 2, 3$, are defined as in Theorem 4.2, but using $G(t, s)^*$ instead of $K(s, t) = L(t, s)^*$ and giving special treatment to an interval near $s = t$ instead of the endpoint $s = T$.

We can now apply the contraction mapping principle to the equation (5.6) in the ball $\mathcal{B}_\rho$ and obtain the following.

Theorem 5.1 (Numerical shadowing theorem 2). *Assume that*

$$\|U\|_{L_\infty((0,T),H^1)} \le R.$$

Let ρ be so small that

$$S_1^3 C(R)\rho \leq \tfrac{1}{2}. \tag{5.9}$$

Let h and k be so small that

$$\mathcal{R}(U) \leq \tfrac{1}{2}\rho. \tag{5.10}$$

Then U has a shadow u with

$$\|U - u\|_{L_\infty((0,T),H^1)} \leq \rho. \tag{5.11}$$

Proof. The assumption (5.9) together with (5.7) guarantees that the right side of (5.6) defines a contraction on $\mathcal{B}_\rho$. The assumption (5.10) together with (5.8) guarantees that $\mathcal{B}_\rho$ is mapped into itself. Hence there is a unique solution $e \in \mathcal{B}_\rho$ of (5.6). The shadow solution is given by $u = U - e$.

As in Theorem 4.1 it follows from local a priori estimates that the quantity $\mathcal{R}(U)$ tends to zero with h and k, so that (5.10) can be achieved if the stability factors occurring in its definition are not too large. This would be the case if the linearized problem has an exponential dichotomy on the interval $(0, T)$.

Note that, in addition to the existence of u, we also get an error bound (5.11) in the V norm. By an additional application of the contraction property, now with respect to H, we get an error bound also in H, cf. the proof of Theorem 3.2.

The radius ρ plays the role of a computational tolerance; see (5.10) and (5.8). Thus, we have shadowing if we compute with a sufficiently small tolerance.

The fixed point iteration that we employ in the proof is a quasi-Newton iteration for the equation

$$u' + Au - f(u) = 0,$$

namely

$$e'_{n+1} + Ae_{n+1} - f'(U(t))e_{n+1} = -(u'_n + Au_n - f(u_n)), \quad u_n = U - e_n.$$

However, the derivatives that occur here must be understood in a weak sense. That is why we argue by duality as in (5.6).

Bibliography

1. W.-J. Beyn, *On the numerical approximation of phase portraits near stationary points*, SIAM J. Numer. Anal. **24** (1987), 1095–1113.
2. P. G. Ciarlet, *The Finite Element Method for Elliptic Problems*, North-Holland, Amsterdam, 1978.

3. C. M. Elliott and S. Larsson, *Error estimates with smooth and nonsmooth data for a finite element method for the Cahn-Hilliard equation*, Math. Comp. **58** (1992), 603–630, S33–S36.

4. D. Estep, M. G. Larson, and R. D. Williams, *Estimating the error of numerical solutions of systems of nonlinear reaction-diffusion equations*, Mem. Amer. Math. Soc., to appear.

5. L. C. Evans, *Partial Differential Equations*, Graduate Studies in Mathematics, vol. 19, American Mathematical Society, Providence, RI, 1998.

6. D. Henry, *Geometric Theory of Semilinear Parabolic Equations*, Lecture Notes in Mathematics, vol. 840, Springer-Verlag, 1981.

7. S. Larsson, *Nonsmooth data error estimates with applications to the study of the long-time behavior of finite element solutions of semilinear parabolic problems*, preprint 1992–36, Department of Mathematics, Chalmers University of Technology, 1992.

8. S. Larsson and S. Yu. Pilyugin, *Numerical shadowing near the global attractor for a semilinear parabolic equation*, preprint 1998–21, Department of Mathematics, Chalmers University of Technology and Göteborg University, 1998.

9. S. Larsson and J.-M. Sanz-Serna, *A shadowing result with applications to finite element approximation of reaction-diffusion equations*, Math. Comp. **68** (1999), 55–72.

10. _____, *The behavior of finite element solutions of semilinear parabolic problems near stationary points*, SIAM J. Numer. Anal. **31** (1994), 1000–1018.

11. L. R. Scott and S. Zhang, *Finite element interpolation of nonsmooth functions satisfying boundary conditions*, Math. Comp. **54** (1990), 483–493.

12. V. Thomée, *Galerkin Finite Element Methods for Parabolic Problems*, Springer Series in Computational Mathematics, vol. 25, Springer-Verlag, Berlin, 1997.

Integration Schemes for Molecular Dynamics and Related Applications

Robert D. Skeel

Department of Computer Science and Beckman Institute, University of Illinois, Urbana, Illinois 61801, USA

Summary. A variety of modern practical techniques are presented for the derivation of integration schemes that are useful for molecular dynamics and a variety of related applications. In particular, the emphasis is on Hamiltonian systems, including those with constraints, and to a lesser extent stochastic differential equations. Among the techniques discussed are operator splitting, multiple time stepping, and accuracy enhancement through "post-processing." Attention is also given to analytical tools for selecting among different integration schemes, for example, small-time-step analysis of the backward error, linear analysis, and small-energy analysis.

1. Introduction

This is not a comprehensive survey but more of a sampling of issues in the construction of numerical integrators. The emphasis is on Hamiltonian systems, including those with constraints, and to a lesser extent stochastic differential equations. This material is presented here in the belief that techniques developed for a particular application are invariably useful in other domains and can often be abstracted and applied to generic problems.

Three aspects of molecular dynamics deserve special mention: (i) the chaotic nature of the differential equations, (ii) the large number of degrees of freedom, and (iii) the long and multiple time scales. These are discussed in the next three paragraphs.

The topic of this article might be considered to be numerical integrators for chaotic ordinary differential equations (ODEs). Chaos is a feature not only of molecular dynamics but also of galaxy simulations and climate modeling. For longer integrations, accurate trajectories are not expected: rather, one wants statistically correct behavior and sampling of phase space. More specifically, initial velocities are randomly generated, and it is the evolution of the probabilities and their steady-state values that must be accurate. These probabilities are derived from the probability density function of the phase space variables, which satisfy an underlying linear homogeneous hyperbolic system of partial differential equations (in very many independent variables). Parts of this formulation are slightly speculative and nonrigorous. They depend on the hypothesis that the dynamical system is "mixing" (although for some purposes it suffices merely to be ergodic).

* This work was supported in part by NSF Grant DMS-9600088, NSF Grant BIR-9318159, and NIH Grant P41RR05969.

It is routine to simulate biomolecular systems with tens of thousands of atoms but desirable to simulate millions of atoms. Parallelism is an effective though expensive way to cope with large system size. Less expensive is to use Langevin dynamics, in which most degrees of freedom are omitted, and the missing forces are replaced by averaged and stochastic forces. Typically, the neglected degrees of freedom are those due to solvent molecules.

It is routine to simulate MD for hundreds of thousands of time steps but processes of biological interest are often on time scales that are yet millions of time longer. The highest frequency components of the dynamics are of small amplitude and generally of no intrinsic interest. Hence, it is reasonable to regard the equations to be "stiff-oscillatory" and longer time steps to be the major computational challenge. Only modest progress has been made in extending the length of the time step. A factor of two can be obtained by imposing constraints that remove the restraining effects of bond-length vibrations. A factor of four is possible with the use of multiple time steps to evaluate the more distant forces less often. For the Langevin equations, which contains friction forces, greater gains can been achieved. The friction inhibits instability and permits multiple time stepping to achieve its full potential. Also, the infinite friction limit sometimes provides an adequate approximation, reducing the second order Langevin equations to the first order stochastic ODEs of Brownian dynamics.

A secondary theme of this article concerns the subtleties of the numerical integration of special second order ODEs $(\mathrm{d}^2/\mathrm{d}t^2)q = F(q)$. Both the stability and the accuracy issues are special.

It is common for special second order ODEs to have "neutral" linear stability thereby requiring a more delicate, nonlinear stability analysis, necessitating excursions into dynamical systems theory. Hamiltonian systems of ODEs are a special case, for which it is desirable that the integrator be symplectic. The most successful technique for constructing effective symplectic integrators and preserving other qualities of the original problem is the use of operator splitting. Also, for problems as sensitive as these, the effects of finite precision must be considered. Happily, there are very simple ways to maintain the quality of the numerical solution.

Because the first derivative $(\mathrm{d}/\mathrm{d}t)q$ is not needed for calculating the "force" $F(q)$, it is a somewhat arbitrary decision as to how to define the numerical approximation to the first derivative. This leads to many different forms of the same method and to uncertainty regarding the accuracy of the method. The most illuminating way of understanding accuracy, especially for Hamiltonian systems, is by means of the "backward" error, usually as an asymptotic expansion in powers of the time step Δt. Also of value, particularly for special second order ODEs, is the concept of "effective" accuracy, in which the analyst does not impose an *a priori* interpretation on the meaning of values generated by a numerical integrator. Indeed, it is very practical to obtain the full effective accuracy of a numerical trajectory by "post-processing."

2. Newtonian Dynamics

The Hamiltonian systems that commonly model classical molecular dynamics are chaotic. Underlying these unstable systems of ODEs are *stable* partial differential equations that describe the evolution of the probability density function for ensembles of trajectories produced by the system of ODEs. Calculating ensemble averages is the goal of MD simulations.

In a classical Newtonian mechanics model, each atom is a particle with position $\mathbf{r}_i$ and mass m_i. These positions are determined by the second order ODEs

$$m_i \frac{\mathrm{d}^2}{\mathrm{d}t^2} \mathbf{r}_i = -\nabla_i U(\mathbf{r}_1, \mathbf{r}_2, \ldots, \mathbf{r}_N) \tag{2.1}$$

for $i = 1, 2, \ldots, N$ where the potential energy function

$$U(\mathbf{r}_1, \mathbf{r}_2, \ldots, \mathbf{r}_N) = \text{bonded interactions} + \sum_{j>i}^{N} U_{ij}^{\text{nonbonded}}.$$

The first of the two terms consists of 2-, 3-, and 4-body forces. The second term consists of electrostatic interactions and the van der Waals attraction and hard-core repulsion. The electron distribution is accounted for by assigning partial charges to atoms. A third term would be present if "hydrogen bonds" are treated explicitly. These force fields are derived from (i) *ab initio* quantum mechanics calculations, (ii) spectroscopic measurements of small molecules that are similar to parts of macromolecules, and (iii) fitting to measured constitutive properties such as diffusion coefficients and dielectric constants. Well known force fields include CHARMM, AMBER, GROMOS, and OPLS. Books on these and other aspects of molecular dynamics include [1, 33, 43, 65]. An article on MD integrators is [68].

An important matter is the treatment of boundaries. A simple treatment is to surround the system with a thick layer of solvent molecules in a vacuum. To prevent the occasional solvent molecule from evaporating, artificial restraining forces are often imposed. Alternatively, an implicit solvent model is used yielding a system of Langevin equations (discussed in section 8). However, the most popular method is to impose periodicity on the system.

2.1. Properties

Let $q = (\mathbf{r}_1^{\mathrm{T}}, \mathbf{r}_2^{\mathrm{T}}, \ldots, \mathbf{r}_N^{\mathrm{T}})^{\mathrm{T}}$ and define a collective momenta vector $p = (m_1 \mathbf{v}_1^{\mathrm{T}}, m_2 \mathbf{v}_2^{\mathrm{T}}, \ldots, m_N \mathbf{v}_N^{\mathrm{T}})^{\mathrm{T}}$ where $\mathbf{v}_i = (\mathrm{d}/\mathrm{d}t)\mathbf{r}_i$ are velocities. Then the system (2.1) can be written

$$\frac{d}{dt}q = M^{-1}p, \quad \frac{d}{dt}p = F(q)$$

where M is a diagonal matrix of masses (each mass replicated three times) and the collective force vector $F(q) = -U_q(q)$. These equations are a special case of a Hamiltonian system of differential equations

$$\frac{d}{dt}q = H_p(q,p), \quad \frac{d}{dt}p = -H_q(q,p), \tag{2.2}$$

with Hamiltonian $H(q,p) = K(p) + U(q)$ where the kinetic energy $K(p) = \frac{1}{2}p^T M^{-1}p$.

Propagating the solution $(q(t), p(t))$ of a Hamiltonian system (2.2) for a fixed time increment Δt defines a transformation, the flow. Let $\phi_{\Delta t H}$ denote the Δt-flow for the system with Hamiltonian H:

$$\phi_{\Delta t H}(q(t), p(t)) = (q(t + \Delta t), p(t + \Delta t)).$$

It is valid to use the product $\Delta t H$ as the subscripted argument (without a comma) because the Δt-flow for Hamiltonian H is simply the 1-flow for Hamiltonian $\Delta t H$.

Hamiltonian systems have special properties, one of which is that the flow is a *symplectic* map. A transformation $\bar{y} = X(y)$ from phase space to itself where $y = (q,p)$ is *symplectic* if its Jacobian matrix X_y is symplectic:

$$X_y^T J X_y = J$$

where

$$J = \begin{bmatrix} 0 & I \\ -I & 0 \end{bmatrix}.$$

(Note that a symplectic matrix is very easy to invert.) An important consequence of the symplectic property is that volume in phase space remains invariant under such a transformation.

Because $H(q, -p) = H(q,p)$, the Hamiltonian system possesses the property of time reversibility, which means that we can trace back the trajectory simply by reversing the momenta and integrating the same differential equation.

Linear and angular momentum are conserved if the potential is invariant under translation and rotation:

$$U(\{Q\mathbf{r}_j + \mathbf{d}\}) = U(\{\mathbf{r}_j\}).$$

The energy $H(q,p)$ is always conserved.

Letting $\phi = \phi_{\Delta t H}$, the properties given in the last three paragraphs can all be expressed in the form $\mathcal{F}(\phi)(y) = \mathcal{F}(\mathrm{id})(y)$ for some operator $\mathcal{F}$:

1. symplecticness:
$$\phi_y(y)^T J \phi_y(y) = J.$$

2. reversibility:
$$R\phi(R\phi(y)) = y$$
where $R = \text{diag}(I, -I)$.

3. conservation laws:

linear momentum	$a_i^T \phi(y) = a_i^T y, \quad i = 1, 2, 3,$
angular momentum	$\phi(y)^T A_i \phi(y) = y^T A_i y, \quad i = 1, 2, 3$
energy	$H(\phi(y)) = H(y).$

Which of these are preserved by numerical integrators?

2.2. The Liouville Equation

The chaotic nature of atomic trajectories is pronounced: a roundoff error introduced into the computations will after a short time overwhelm the theoretically correct trajectory, even with the use of double precision. Over a 50-picosecond duration growth is by a factor 10^{15} [1].

Appropriate accuracy criteria depend on the purpose of the computation:

1. exploration of phase space [65, p. 434]. MD is one of many ways to search for conformations of a biomolecules having minimal or near minimal free energy.

2. sampling of phase space. The goal is to sample ensembles of molecular systems from some given probability distribution and to compute averages of certain quantities or to find structures having high probabilities. Choosing suitably random initial conditions is difficult for position variables, so the sampling is preceded by an equilibration phase that randomizes the initial conditions. For the sampling phase, Monte Carlo methods are popular for small molecules but they suffer from high rejection rates in the case of macromolecules, where MD and hybrid methods are more often used. The use of one long MD simulation for sampling relies on the ergodic hypothesis that the time average is equal to the ensemble average.

3. accurate short-time trajectories. The calculation of time correlation functions requires a collection of trajectories with initial conditions drawn from an appropriate ensemble. One long MD trajectory can be used for this purpose.

4. statistically accurate long-time behavior [65, p. 354]. The intent is to simulate the dynamics of ensembles of molecular systems, with initial conditions chosen at random from an appropriate distribution. The simulation should give the correct probabilities of making a transition from one structure to another after some designated time interval. It is typical practice for biomolecules to simulate only a small number of instances. However, this may not be a serious matter because, unlike general macromolecules (such as polymers from which materials are constructed), biomolecules are designed to function in a fairly consistent and

reproducible manner on longer time scales even though they experience thermal fluctuations on short time scales.

Let us formalize a probabilistic accuracy criterion for constant energy dynamics. Assume that the initial conditions are drawn from a distribution with probability density (function) $\rho_0(y)$. Imagine generating an "ensemble" of trajectories from these initial conditions. Let $\rho(y,t)$ be the probability density that results. The quantities of interest are statistical and generally involve integrals of the form

$$\int g(y)\rho(y,t)\mathrm{d}y.$$

(Structure determination can also be formulated in this way.) It is quantities such as these that are required to be accurate. Because of the approximate nature of the force fields, it is necessary that these quantities be insensitive in some sense to small changes in the energy function. (Otherwise, molecular dynamics as a methodology is on shaky ground.)

For the ODE system $(\mathrm{d}/\mathrm{d}t)y = f(y)$, the probability density $\rho(y,t)$ for an ensemble of trajectories obeys the Liouville equation

$$\rho_t + \nabla^{\mathrm{T}}(\rho f(y)) = 0$$

where ∇^{T} denotes the divergence operator. This is a homogeneous linear hyperbolic partial differential equation. The solutions of $(\mathrm{d}/\mathrm{d}t)y = f(y)$ constitute the characteristics of this partial differential equation. If $\nabla^{\mathrm{T}}f(y) = 0$, as it is for Hamiltonian systems, where $f(y) = JH_y(y)$, this becomes

$$\rho_t + f(y)^{\mathrm{T}}\rho_y = 0.$$

For the special Hamiltonian $H(q,p) = \frac{1}{2}p^{\mathrm{T}}M^{-1}p + U(q)$, we get

$$\rho_t + (M^{-1}p)^{\mathrm{T}}\rho_q - U_q(q)^{\mathrm{T}}\rho_p = 0.$$

Assume that initial conditions are restricted so that center of mass, linear momentum, and angular momentum are always zero and that the only nonzero conserved quantity is energy. Does $\rho(y,t)$ have a limit as $t \to \infty$? No, this would contradict reversibility. But it is conceivable that there is a limit in the weak sense:

$$\lim_{t\to\infty}\int g(y)\rho(y,t)\mathrm{d}y = \int g(y)\bar\rho(y)\mathrm{d}y$$

for any $g \in C^\infty$. If there is such a limit, the dynamical system is said to be *mixing* [64, 43]. The meaning of this weak limit is that the difference $\rho(y,t) - \bar\rho(y)$ becomes more and more highly fluctuating as a function of y as $t \to \infty$. A yet weaker property of a dynamical system is *ergodicity*:

$$\lim_{t\to\infty}\frac{1}{t}\int_0^t \rho(y,\tau)\mathrm{d}\tau = \bar\rho(y).$$

In either case the equilibrium density must satisfy

$$(\bar{\rho}_y(y))^{\mathrm{T}} J H_y(y) = 0,$$

which implies that $\bar{\rho}(y)$ is a conserved quantity, which implies $\bar{\rho}(y) = \text{function}(H(y))$. Hence $\bar{\rho}(y)$ is essentially uniform in a very thin energy shell $E \leq H(y) \leq E + \epsilon$. Letting $\epsilon \to 0$, we get

$$\bar{\rho}(y) \propto \frac{1}{\|H_y(y)\|_2} \qquad \text{for } H(y) = E.$$

If the density almost always tends to some limit density $\bar{\rho}(y)$ that is bounded away from zero, this suggests that almost every trajectory passes arbitrarily close to every point on its energy surface (if we wait long enough). This ergodic property of trajectories justifies the use of a time average to calculate an ensemble average. For ensembles other than the microcanonical ensemble (constant energy), the Hamiltonian needs to be extended in some way. Examples are given in section 7.

For further discussion of mathematical issues and for references to the literature, the preprints [103, 29] are recommended.

3. The Leapfrog Method

The equations of motion are typically integrated by the leapfrog/Störmer-[115]/Verlet method, which is reported [121] to date as far back as 1793 [21]. The method used by Verlet [126] was expressed in the form

$$M \frac{q^{n+1} - 2q^n + q^{n-1}}{\Delta t^2} = F(q^n), \quad v^n = \frac{q^{n+1} - q^{n-1}}{2\Delta t} \qquad (3.1)$$

where v^n is the collective velocity vector and $q^1 = q^0 + \Delta t v^0 + \frac{\Delta t^2}{2} M^{-1} F(q^0)$. However, there are a surprising variety of other ways to write the leapfrog method. These and numerous other interesting but hardly well known facts about the leapfrog method will emerge as we use this scheme to illustrate a variety of techniques for analysis and for algorithm construction.

Analytical error analysis of numerical methods must necessarily make simplifications—if we knew precisely what the error was, we would also know the solution and have little need for the numerical method. Three types of simplifying assumptions are popular, each giving a different mode of error analysis:

1. error bounds. By making repeated worst case assumptions one can derive rigorous upper bounds on the magnitude of the error. These are generally much greater than the worst possible error. Nonetheless, they do convey qualitative information and are useful in comparing alternative methods. However, good error bounds are difficult to obtain. Overall, this approach is the least rewarding.

2. exact error analysis for simple test problems. Usually these are linear constant coefficient problems. Nonetheless, it is not difficult to relate the conclusions to real problems, and these conclusions generally have good predictive value.

3. asymptotic expansions in some small parameter. Typically this small parameter is Δt, but another possibility is to analyze small deviations from equilibrium. Conclusions from these analyses also have good predictive value.

In sections 3.2–3.4 we pursue the second and third of these for the leapfrog method.

Discretization error for leapfrog can be rather dramatically reduced by the use of post-processing discussed in section 3.5. Another kind of error, roundoff error, is treated in section 3.6.

3.1. Derivation

The second order difference equation formulation (3.1) given above is known to accelerate by a factor of n the accumulation of roundoff error. Here we derive a form of the method less affected by roundoff and more befitting of the name "leapfrog."

The exact calculation of the flow $\phi_{\Delta t(K+U)}$ is, of course, generally intractable. However, the special cases $\phi_{\Delta t K}$ and $\phi_{\Delta t U}$ have a very simple closed form. For example, $\phi_{(\Delta t/2)U}(q^n, p^n)$ can be obtained by solving an initial value problem for

$$
\frac{\mathrm{d}}{\mathrm{d}t}q = 0, \quad \frac{\mathrm{d}}{\mathrm{d}t}p = -U_q(q),
$$

yielding

$$
\phi_{(\Delta t/2)U}(q^n, p^n) = \left(\begin{array}{c} q^n \\ p^n - \frac{1}{2}\Delta t U_q(q^n) \end{array} \right).
$$

This is the motivation for operator splitting. In particular, the leapfrog method is the result of a Strang splitting:

$$
\begin{aligned}
\phi^{LF}_{\Delta t(K+U)} &= \phi_{(\Delta t/2)U} \circ \phi_{\Delta t K} \circ \phi_{(\Delta t/2)U} \\
&= \phi_{\Delta t(K+U)} + O(\Delta t^3).
\end{aligned}
$$

Over many steps these errors accumulate to yield trajectory errors that are, in fact, $O(\Delta t^2)$. That this composition of propagators should be a reasonable approximation, let alone of second order accuracy, may not be obvious. The justification is due to the exponential nature of solution propagators, a subject that is discussed further in appendix A. The inaccuracy is due to the fact that the rule for a product of exponentials is not exact for operators that do not commute.

The leapfrog method can be expressed as follows. We begin the step with values q^n, p^n, and $F^n = F(q^n)$ obtained from the previous step. Then we compute

$\phi_{(\Delta t/2)U}$ (half-kick):

$$p^{n+1/2} = p^n + \frac{\Delta t}{2} F^n,$$

$\phi_{\Delta t K}$ (drift):

$$q^{n+1} = q^n + \Delta t M^{-1} p^{n+1/2},$$

$\phi_{(\Delta t/2)U}$ (half-kick):

$$F^{n+1} = F(q^{n+1}),$$
$$p^{n+1} = p^{n+1/2} + \frac{\Delta t}{2} F^{n+1}.$$

(If we combine the last half kick of one step with the first half kick of the next step, we get a form of the method that uses p values only at midpoints of intervals. This form of the method is associated with the name "leapfrog." Alternatively, we can define q values at midpoints of intervals using $q^{n+1/2} = q^n + \frac{1}{2}\Delta t M^{-1} p^{n+1/2}$, $q^{n+1} = q^{n+1/2} + \frac{1}{2}\Delta t M^{-1} p^{n+1/2}$.) These equations exactly model the following situation: an impulse of $\frac{1}{2}\Delta t F^n$ at time $t^n +$ followed by an impulse of $\frac{1}{2}\Delta t F^{n+1}$ at time $t^{n+1}-$. Note that because the term $\frac{1}{2}\Delta t F^{n+1}$ is treated as an impulse just before time t^{n+1}, it has no effect on the value of q^{n+1} and the method is explicit. In fact, as noted by Wisdom [129], the leapfrog method exactly satisfies the equations of motion

$$M \frac{d^2}{dt^2} q = \sum_n \Delta t\, \delta(t - n\Delta t) F(q).$$

There is an alternative form of the leapfrog method based on the splitting

$$\varphi_{\Delta t(K+U)} = \phi_{(\Delta t/2)K} \circ \phi_{\Delta t U} \circ \phi_{(\Delta t/2)K} + O(\Delta t^3).$$

We can interpret this as half a drift followed by a kick followed by half a drift. We call this the *midpoint* form of the leapfrog method because the force is evaluated at the middle. The first form we call the *endpoint* form. We prefer the endpoint form because it has a more efficient generalization to multiple time stepping. The two forms are equivalent in the sense that there exists a coordinate transformation that relates numerical trajectories produced by each of them [113].

The flow $\varphi_{\Delta t H}^{LF}$ defined by the leapfrog method is symplectic. This is a consequence of two important facts. One is that Hamiltonian flows are symplectic. The other is that the composition of two symplectic transformations is also symplectic (as is the inverse of a symplectic transformation). Also, the leapfrog method preserves the conservation laws for linear and angular momentum and the property of time reversibility. One conservation law that is not maintained is conservation of energy. More is said about the significance of these properties in section 4.

3.2. Small-Δt Analysis

At least three types of error expansions in powers of Δt are possible:

1. "global" error expansions. For example, time-reversible methods like the leapfrog method have an expansion

$$y^n = y(n\Delta t) + \Delta t^2 e_1(n\Delta t) + \Delta t^4 e_2(n\Delta t) + \cdots$$

   containing only even powers of Δt. The identity of the coefficients $e_1(t)$, $e_2(t)$, ... in such an expansion is determined by substituting the expansion into the difference equation, expanding about $t = n\Delta t$, and equating like powers of Δt^2. One can by this type of analysis determine those more important terms of the error (the secular terms) that grow in time, e.g., [69]. This type of analysis has been fruitful in studying discretizations of periodic orbits. For such problems, symplectic integrators have been shown to yield global errors that increase only linearly as time t grows rather than quadratically as they do for non-symplectic integrators [15].

2. local error expansions. The local error for the leapfrog method is

$$\phi^{LF}_{\Delta t H}(y(t)) - y(t + \Delta t).$$

   Because the first derivative v^n is not used, as q^n is, in an evaluation of a function, it hardly matters how v^n is defined. This arbitrariness is inherited by the definition of the local error whose value depends on how we choose to define v^n. However, the accuracy of the numerical approximation to q^n is not affected except by the manner in which the initial value $v(0)$ is incorporated into the solution. The arbitrariness can be avoided by treating the method as a second order difference approximation to a second order differential equation.

3. backward error expansions. The numerical solution can be expressed formally as the exact solution of an ODE with a perturbed right hand side expressed as an expansion in powers of Δt. This may be the most enlightening analysis. It was first developed in the context of computational fluid dynamics under the name "method of modified equations" [127]. It is interesting that a truncated asymptotic backward analysis yields a more accurate approximation to the numerical solution than does a truncated asymptotic expansion of the global error [48].

Here we explore only the third of these, namely, backward error analysis. To present the concept with the greatest clarity, we apply this technique first to the forward Euler approximation

$$\frac{y^{n+1} - y^n}{\Delta t} = f(y^n)$$

to the ODE system $(d/dt)y = f(y)$. The idea is to assume that $y^n = \tilde{y}(n\Delta t)$ where $\tilde{y}(t)$ satisfies a slightly different ODE system $(d/dt)\tilde{y} = \tilde{f}(\tilde{y})$ with $\tilde{f}$

assumed to possess an asymptotic expansion in powers of Δt. We begin with an expansion about some chosen point in time, $t = n\Delta t$:

$$\frac{d}{dt}\tilde{y} + \frac{\Delta t}{2}\frac{d^2}{dt^2}\tilde{y} + \frac{\Delta t^2}{6}\frac{d^3}{dt^3}\tilde{y} + \cdots = f(\tilde{y}).$$

Then we use the ODE $(d/dt)\tilde{y} = \tilde{f}(\tilde{y})$ and its derivatives $(d^2/dt^2)\tilde{y} = \tilde{f}_y(\tilde{y})\tilde{f}(\tilde{y})$, $(d^3/dt^3)\tilde{y} = \tilde{f}_{yy}(\tilde{y})\tilde{f}(\tilde{y})\tilde{f}(\tilde{y}) + \tilde{f}_y(\tilde{y})\tilde{f}_y(\tilde{y})\tilde{f}(\tilde{y})$, $\ldots$ to get an equation involving only functions of $\tilde{y}(t)$ and not of its derivatives. (In general we have $(d^k/dt^k)\tilde{y}(t) = (\tilde{f}^T\nabla)^k\mathrm{id}(\tilde{y}(t))$. See appendix A.) Moving all but the first term to the right-hand side and suppressing the dependence on $y = \tilde{y}(t)$, we get

$$\tilde{f} = f - \frac{\Delta t}{2}\tilde{f}_y\tilde{f} - \frac{\Delta t^2}{6}(\tilde{f}_{yy}\tilde{f}\tilde{f} + \tilde{f}_y\tilde{f}_y\tilde{f}) - \cdots.$$

We can develop an expansion for $\tilde{f}$ by successive substitution into the right-hand side. Initially take

$$\tilde{f} = f + O(\Delta t).$$

Then

$$\tilde{f} = f - \frac{\Delta t}{2}f_y f + O(\Delta t^2),$$

and then

$$\tilde{f} = f - \frac{\Delta t}{2}f_y f + \frac{\Delta t^2}{12}(f_{yy}ff + 4f_y f_y f) + O(\Delta t^3).$$

etc.

We now do a backward analysis for the leapfrog method. By eliminating the intermediate stage in the leapfrog method, we get

$$\frac{p^{n+1} - p^n}{\Delta t} = \frac{1}{2}(F^n + F^{n-1}), \tag{3.2}$$

$$\frac{q^{n+1} - q^n}{\Delta t} = \frac{1}{2}M^{-1}(p^{n+1} + p^n) - \frac{\Delta t}{4}M^{-1}(F^{n-1} - F^n). \tag{3.3}$$

Expanding (3.3) and (3.2) in a Taylor series about the midpoint $t = (n+\frac{1}{2})\Delta t$, we get

$$\frac{d}{dt}\tilde{q} + \frac{\Delta t^2}{24}\frac{d^3}{dt^3}\tilde{q} = M^{-1}\tilde{p} + \frac{\Delta t^2}{8}M^{-1}\frac{d^2}{dt^2}\tilde{p} + \frac{\Delta t^2}{4}M^{-1}\frac{d}{dt}U_q(\tilde{q}) + O(\Delta t^4),$$

$$\frac{d}{dt}\tilde{p} + \frac{\Delta t^2}{24}\frac{d^3}{dt^3}\tilde{p} = -U_q(\tilde{q}) - \frac{\Delta t^2}{8}\frac{d^2}{dt^2}U_q(\tilde{q}) + O(\Delta t^4)$$

where $\tilde{q} = \tilde{q}((n + 1/2)\Delta t)$, $\tilde{p} = \tilde{p}((n + 1/2)\Delta t)$. Hence

$$\frac{d}{dt}\tilde{q} = M^{-1}\tilde{p} + \Delta t^2 \left(\frac{1}{8}M^{-1}\frac{d^2}{dt^2}\tilde{p} + \frac{1}{4}M^{-1}U_{qq}(\tilde{q})\frac{d}{dt}\tilde{q} - \frac{1}{24}\frac{d^3}{dt^3}\tilde{q}\right)$$
$$+ O(\Delta t^4),$$

$$\frac{d}{dt}\tilde{p} = -U_q(\tilde{q}) + \Delta t^2 \left(-\frac{1}{8}U_{qqq}(\tilde{q})(\frac{d}{dt}\tilde{q}, \frac{d}{dt}\tilde{q}) - \frac{1}{8}U_{qq}(\tilde{q})\frac{d^2}{dt^2}\tilde{q} - \frac{1}{24}\frac{d^3}{dt^3}\tilde{p}\right)$$
$$+ O(\Delta t^4),$$

where $U_{qq}(\tilde{q})$ is the Hessian of the potential energy function and a double summation involving $U_{qqq}(\tilde{q})$ is intended by the notation $(\frac{d}{dt}\tilde{q}, \frac{d}{dt}\tilde{q})$. By one application of successive substitution we get

$$\frac{d}{dt}\tilde{q} = M^{-1}\tilde{p} + \Delta t^2 \left(\frac{1}{6}M^{-1}U_{qq}(\tilde{q})M^{-1}\tilde{p}\right) + O(\Delta t^4),$$

$$\frac{d}{dt}\tilde{p} = -U_q(\tilde{q}) + \Delta t^2 \left(-\frac{1}{12}U_{qqq}(\tilde{q})(M^{-1}\tilde{p}, M^{-1}\tilde{p}) + \frac{1}{12}U_{qq}(\tilde{q})M^{-1}U_q(\tilde{q})\right)$$
$$+ O(\Delta t^4),$$

whence follows

$$\tilde{H}(q, p) = \frac{1}{2}p^{\mathrm{T}}M^{-1}p + U(q)$$
$$+ \Delta t^2 \left(\frac{1}{12}(M^{-1}p)^{\mathrm{T}}U_{qq}(q)M^{-1}p - \frac{1}{24}U_q(q)^{\mathrm{T}}M^{-1}U_q(q)\right) \quad (3.4)$$
$$+ O(\Delta t^4).$$

The forgoing calculation can be done more efficiently using the formalism of Lie operators and the BCH formula, as described in appendix A.

3.3. Linear Analysis

Normal mode analysis examines the dynamics in the vicinity of a potential energy minimum, for which the gradient satisfies

$$U_q(q^*) = 0$$

and the Hessian satisfies

$$U_{qq}(q^*) \geq 0,$$

meaning that it is semi-positive definite. The dynamics in a neighborhood of $q = q^*$ is described by

$$\frac{d}{dt}q = M^{-1}p, \quad \frac{d}{dt}p = -U_{qq}(q^*)(q - q^*)$$

if the $O(||q - q^*||^2)$ terms are neglected, which is the case for normal mode analysis. The resulting linear problem can be decoupled by a similarity transformation and with a translation and symplectic scaling of variables the system reduces to a set of harmonic oscillators

$$\frac{\mathrm{d}}{\mathrm{d}t}q_i = p_i, \quad \frac{\mathrm{d}}{\mathrm{d}t}p_i = -w_i^2 q_i \tag{3.5}$$

where the w_i^2 are the eigenvalues of the mass-weighted Hessian

$$M^{-1/2}U_{qq}(q^*)M^{-1/2}.$$

Dropping the subscripts and writing in vector form as

$$\frac{\mathrm{d}}{\mathrm{d}t}\begin{bmatrix} q \\ p \end{bmatrix} = \begin{bmatrix} 0 & 1 \\ -w^2 & 0 \end{bmatrix}\begin{bmatrix} q \\ p \end{bmatrix},$$

we see that the solution is

$$\begin{bmatrix} q(t) \\ p(t) \end{bmatrix} = \exp\left(t\begin{bmatrix} 0 & 1 \\ -w^2 & 0 \end{bmatrix}\right)\begin{bmatrix} q(0) \\ p(0) \end{bmatrix}$$

$$= \begin{bmatrix} \cos wt & w^{-1}\sin wt \\ -w\sin wt & \cos wt \end{bmatrix}\begin{bmatrix} q(0) \\ p(0) \end{bmatrix}.$$

If $w > 0$ is assumed, the variables can be scaled to yield a pure rotation by $-wt$ in phase space:

$$\begin{bmatrix} wq(t) \\ p(t) \end{bmatrix} = \begin{bmatrix} \cos wt & \sin wt \\ -\sin wt & \cos wt \end{bmatrix}\begin{bmatrix} wq(0) \\ p(0) \end{bmatrix}.$$

For leapfrog and other standard integrators applied to the original linear problem, precisely the same transformations decouple the difference equation to yield a leapfrog scheme applied to each of the harmonic oscillators (3.5). If $w\Delta t \geq 2$, the leapfrog method is unstable; if $w\Delta t < 2$, it is stable. In such a case, the Δt-flow is

$$\begin{bmatrix} wq^{n+1} \\ p^{n+1} \end{bmatrix} = D\begin{bmatrix} \cos\tilde{w}\Delta t & \sin\tilde{w}\Delta t \\ -\sin\tilde{w}\Delta t & \cos\tilde{w}\Delta t \end{bmatrix}D^{-1}\begin{bmatrix} wq^n \\ p^n \end{bmatrix} \tag{3.6}$$

where

$$\tilde{w} = \frac{2}{\Delta t}\arcsin\frac{w\Delta t}{2} = w(1 + O((w\Delta t)^2))$$

and

$$D = \mathrm{diag}\left(1, \sqrt{1 - w^2\Delta t^2/4}\right) = I + O((w\Delta t)^2).$$

Hence, we note that the leapfrog scheme imparts a "blue shift," to angular frequency w, which is 1.5% for Δt equal to $\frac{1}{10}$ of a period.

There now follows a derivation from [113] of the modified Hamiltonian for the harmonic oscillator. The leapfrog solution (3.6) has

$$\begin{bmatrix} w\tilde{q}(t) \\ \tilde{p}(t) \end{bmatrix} = D\begin{bmatrix} \cos\tilde{w}t & \sin\tilde{w}t \\ -\sin\tilde{w}t & \cos\tilde{w}t \end{bmatrix}D^{-1}\begin{bmatrix} wq^0 \\ p^0 \end{bmatrix}.$$

Differentiating with respect to t gives

$$\frac{d}{dt}\left[\begin{array}{c} \omega\tilde{q}(t) \\ \tilde{p}(t) \end{array}\right] = \tilde{\omega}DJ\left[\begin{array}{cc} \cos\tilde{\omega}t & \sin\tilde{\omega}t \\ -\sin\tilde{\omega}t & \cos\tilde{\omega}t \end{array}\right]D^{-1}\left[\begin{array}{c} \omega q^0 \\ p^0 \end{array}\right],$$

and eliminating initial values from this and the previous equation gives the ODE system

$$\frac{d}{dt}\left[\begin{array}{c} \tilde{q}(t) \\ \tilde{p}(t) \end{array}\right] = \tilde{\omega}J\mathrm{diag}\left(\omega\left(1-\tfrac{1}{4}\omega^2\Delta t^2\right)^{1/2}, \frac{1}{\omega\left(1-\tfrac{1}{4}\omega^2\Delta t^2\right)^{1/2}}\right)\left[\begin{array}{c} \tilde{q}(t) \\ \tilde{p}(t) \end{array}\right].$$

This is a Hamiltonian system arising from

$$\tilde{H}(q,p) = \frac{1}{\omega\Delta t}\arcsin\left[\frac{\omega\Delta t}{2}\left(\left(1-\tfrac{1}{4}\omega^2\Delta t^2\right)^{-1/2}p^2 + \left(1-\tfrac{1}{4}\omega^2\Delta t^2\right)^{1/2}\omega^2q^2\right)\right],$$

valid for $\omega\Delta t < 2$. The "unmodified" Hamiltonian is $\frac{1}{2}(p^2 + \omega^2q^2)$.

3.4. Small-Energy Analysis

Here we consider how nonlinearity affects behavior in the neighborhood of a stable equilibrium, for which

$$p^* = 0, \quad U_q(q^*) = 0, \quad \text{and} \quad U_{qq}(q^*) > 0,$$

meaning that $U_{qq}(q^*)$ is positive definite. Without neglecting any higher powers of $\|q - q^*\|$, it is possible to make exact mathematical statements about the stability of a symplectic map in a "sufficiently small" neighborhood of a fixed point y^*. Unstable behavior in this situation would dictate the use of a smaller time step or a different integrator. Stability for an integrator means bounded growth in the error uniformly for all time for initial values in some neighborhood of its fixed point y^*. More precisely, let $y^{n+1} = \tilde{\phi}_{\Delta tH}(y^n)$ be a numerical integrator with fixed point $y^* = \tilde{\phi}_{\Delta tH}(y^*)$. The integrator is stable at y^* if for any $\varepsilon > 0$ there exists a $\delta > 0$ such that

$$\|y^0 - y^*\| < \delta \quad \Rightarrow \quad \|y^n - y^*\| < \varepsilon \quad \text{for all } n.$$

Expanding the numerical flow $\tilde{\phi}_{\Delta tH}(y^n)$ about $y^n = y^*$, we can write

$$y^{n+1} - y^* = Q(y^n - y^*) + O(\|y^n - y^*\|^2)$$

where Q is the Jacobian matrix of the map at y^*.

Consider the scalar case. Let us assume that Δt is less than the linear stability threshold $2/\omega$. Then, as can be seen from eq. (3.6), Q has eigenvalues $\exp(\pm i\tilde{\omega}\Delta t)$. Stability at equilibrium of the nonlinear map can be determined using the Moser twist theorem [4, p. 411]. If $\exp(i\tilde{\omega}\Delta t)$ is not a root of unity or equivalently if the ratio of $\tilde{\omega}\Delta t$ to 2π is irrational, the map is stable, unless all the coefficients in a certain expansion happen to vanish. If $\exp(i\tilde{\omega}\Delta t)$ is an nth root of unity, we have nth order *resonance* and the possibility of instability [4, pp. 390–398]. Third order resonances are normally unstable in the sense that stability is possible only if a certain coefficient in some expansion happens to vanish. Fifth and higher order resonances are normally stable in the sense that instability is possible only if a certain coefficient happens to vanish. Both stability and instability are normal for fourth order resonances. The specifics depend on both the integrator and the function $U(q)$. For this reason both third and fourth order resonances should be regarded as generally unacceptable. In the case of the leapfrog method, $\tilde{\omega}\Delta t = \frac{2}{3}\pi$ (3:1 resonance) corresponds to $\omega\Delta t = \sqrt{3}$ and $\tilde{\omega}\Delta t = \frac{1}{2}\pi$ (4:1 resonance) corresponds to $\omega\Delta t = \sqrt{2}$. For rotations $\tilde{\omega}\Delta t$ near an unstable resonance, the stable neighborhood of y^* is very small. These positive results on stability were first elucidated in [100]. The article [80] explains that stable resonances induce oscillations with an amplitude that depends on the order of the resonance. Empirical evidence of instability due to third order resonance was first observed for the implicit midpoint method [73], which is unconditionally stable for linear problems.

The situation in higher dimensions is less conclusive. Assume the eigenvalues $\lambda_1, \overline{\lambda_1}, \lambda_2, \overline{\lambda_2}, \ldots, \lambda_d, \overline{\lambda_d}$ of the Jacobian matrix Q of the map are all of unit modulus. If there are integers $k_1, k_2, \ldots, k_d$ such that $\lambda_1^{k_1}\lambda_2^{k_2}\cdots\lambda_d^{k_d} = 1$, we have a resonance of order $n = |k_1| + |k_2| + \cdots + |k_d|$. In this more general case it is possible to have instability even if unstable resonances are avoided. However, this instability will be of a less virulent type known as *Arnol'd diffusion*. Instability in higher dimensions has been observed for the implicit midpoint rule applied to models in structural mechanics [107, 39]. There are a large number of different frequencies present in a vibrating structure, so third order resonance is a plausible explanation for instability. In MD, resonance-induced instability has been demonstrated for the implicit midpoint rule applied to a biomolecule [102]. The same article proposes other implicit methods that avoid third and even fourth order resonances. Resonances can, in fact, be observed during normal MD simulations using the leapfrog method. In Fig. 3.1 a plot is shown of energy as a function of time; it was produced during the course of testing the correctness of a simulation program, NAMD 1.5 [84], on a protein–water system. The graph is cycling through 10 different values, which is evidence of 10th order resonance. This is not surprising since a time step of 1 fs is about one tenth the period of the fastest normal mode in the simulation. Because typical dynamical simulations involve a spectrum of different frequencies, it seems advisable to select the time step Δt for the leapfrog method so that the rotation $\tilde{\omega}\Delta t$ per step is less than $\pi/2$. Hence, *the stability condition for the leapfrog method is effectively*

$$\omega\Delta t < \sqrt{2}$$

rather than $\omega\Delta t < 2$. This, in fact, is consistent with an observed stability limit of 2 fs or less for MD simulations.

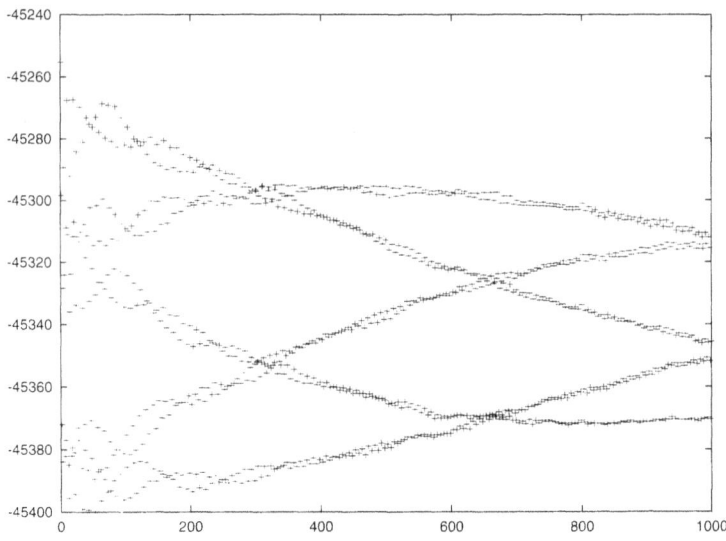

Fig. 3.1. Energy as a function of time for a leapfrog simulation of MD.

3.5. Effective Accuracy and Post-Processing

It has been observed first in [14] and later in the context of Hamiltonian systems in [96, 119, 130, 79, 69] that the "intrinsic" accuracy of an integrator may be greater than what it seems—a slightly different interpretation of the numerical solution (q^n, p^n) may make the method look more accurate, e.g., as assessed by local error or backward error analysis. More specifically, the standard interpretation is that $y^n \approx y(n\Delta t)$ and a more favorable interpretation may be that

$$y^n \approx \hat{y}(n\Delta t) = X(y(n\Delta t))$$

where X is a coordinate transformation in phase space only slightly different from the identity mapping id. In other words we associate with a method a transformation X^{-1} that obtains more desirable values $X^{-1}(y^n)$ from the raw values y^n used by the integration scheme. Clearly, there are limits to what can be achieved, and, in particular, there is no transformation that removes the growing secular part of the error. In the case of the leapfrog method applied to the harmonic oscillator discussed previously, the error $D - I$ that causes the energy to oscillate is completely removable; whereas the error $\tilde{\omega} - \omega$ that causes a growing phase error is not at all removable.

A simple yet dramatic example of this phenomenon is given by the method [105],[37, p. 24]

$$p^{n+1} = p^n - \Delta t U_q(q^n), \quad q^{n+1} = q^n + \Delta t M^{-1} p^{n+1} \tag{3.7}$$

and its twin

$$q^{n+1} = q^n + \Delta t M^{-1} p^n, \quad p^{n+1} = p^n - \Delta t U_q(q^{n+1}), \tag{3.8}$$

either of which might be called the *symplectic Euler method*. As noted by Suzuki [119], either of (3.7) or (3.8) is equivalent to the leapfrog method. In particular, perform the change of coordinates

$$
\begin{aligned}
q &= \bar{q}, \\
p &= \bar{p} + \frac{\Delta t}{2} U_q(\bar{q}),
\end{aligned}
$$

and the first-order accurate method (3.7) becomes the second-order accurate leapfrog method for the new coordinates $\bar{q}^n, \bar{p}^n$. Note that $X = \mathrm{id} + O(\Delta t)$ and that X is symplectic.

In general, a transformation X such that a method better propagates $X(y(t))$ than it does $y(t)$ can be used to create a better numerical trajectory as follows:

$$
\begin{aligned}
y^n &= X(\bar{y}^n), \\
y^{n+1} &= \tilde{\phi}_{\Delta t H}(y^n), \\
\bar{y}^{n+1} &= X^{-1}(y^{n+1}).
\end{aligned}
$$

Thus we create a more accurate integrator

$$X^{-1} \circ \tilde{\phi}_{\Delta t H} \circ X.$$

If X is symplectic, there exists a modified Hamiltonian that can be used to assess the accuracy of the improved integrator. In practice it not necessary to pre- and post-process the numerical solution at every step. Rather the preprocessing $y^n = X(\bar{y}^n)$ need be done only at the beginning, for $n = 0$, and the post-processing $\bar{y}^{n+1} = X^{-1}(y^{n+1})$ only when output is desired. Moreover, because the transformation X is not part of the propagator, it can be approximated without any dire effects, and this approximation need not be symplectic. In the case of MD, the initial positions are not known precisely and the initial velocities are chosen at random, so preprocessing can be omitted. Similarly for the post-processing, if the accuracy of the computed quantities is unaffected by the high frequency error that is present in an unprocessed trajectory. Even in this extreme case where we forgo all pre- and post-processing, the concept is still a valuable theoretical or experimental tool [113] for comparing the intrinsic accuracy of two integrators.

It is natural to ask whether the order of accuracy of the leapfrog method itself might be increased. This is not, however, generally possible [69]. All that can be accomplished by a symplectic transformation X is to reduce the size of the coefficient of Δt^2 in an expansion of the error. For example, consider $\bar{y} = X^{-1}(y)$ defined by

$$
\begin{aligned}
\bar{q} &= q + \beta \Delta t^2 M^{-1} U_q(q), \\
\bar{p} &= \left(I + \beta \Delta t^2 M^{-1} U_{qq}(q)\right)^{-1} p.
\end{aligned}
$$

The accuracy of this processed leapfrog method can be assessed in terms of the modified Hamiltonian because the transformation X is symplectic. For leapfrog with the given processing, the modified Hamiltonian given by Eq. (3.4) becomes

$$\tilde{H}^{\mathrm{proc}}(\bar{q},\bar{p}) \quad = \quad H(\bar{q},\bar{p}) + \Delta t^2 \left(\left(\beta + \frac{1}{2} \right) \bar{p}^{\mathrm{T}} M^{-1} U_{qq}(\bar{q}) M^{-1} \bar{p} \right.$$
$$\left. - \left(\frac{1}{24} + \beta \right) U_q(\bar{q})^{\mathrm{T}} M^{-1} U_q(\bar{q}) \right). \tag{3.9}$$

The best choice of β is not obvious. However, the choice $\beta = -\frac{1}{16}$, which equalizes the coefficients. leads to energy conservation errors of only $O(\Delta t^4)$ for quadratic Hamiltonians. This is likely a near optimal choice for MD. where the dynamics is harmonic to a significant degree. A more complicated symplectic transformation which exactly conserves energy in the case of a quadratic Hamiltonian is given in [102]. A highly efficient approximation to symplectic post-processing transformations based on differencing is given in [70, 113]. If both the time integrator and the processing are symplectic. then the fluctuations in total energy of the processed solution represent a sampling of the perturbation to the Hamiltonian due to the finite time step Δt of the processed integrator. Also, with the removal of most of the non-growing part of the systematic error (which is rich in high frequency components). the energy of the processed trajectory becomes more revealing as an error indicator and hence as a detector of programming bugs and perhaps collision events. Indeed, this may be reason enough to actually perform the processing. or at least the post-processing.

3.6. Finite-Precision Effects

In reality, numerical methods are implemented in finite-precision arithmetic. and it is to be expected that rapid exponential growth of even tiny roundoff errors would overwhelm numerical trajectories well before the end of the simulation. Apparently these effects are not catastrophic; otherwise no one would be doing MD.

The best way to begin thinking about the matter is to realize that the effect of finite precision is to project the computed trajectory onto a finite lattice in phase space. Assuming that the trajectory is not terminated by overflow, it will eventually start repeating itself and settle into a limit cycle.

More can be said if we make certain idealized assumptions about the computation and if we assume that the lattice is uniform, which is the case for fixed point arithmetic but not for floating point. Building on the work of others, it is shown in [104] that the effect of computing the trajectory of a symplectic integrator on a uniform lattice is equivalent to applying that symplectic method to a slightly different Hamiltonian system with Hamiltonian $H_\varepsilon = H + O(\varepsilon^2)$ where ε is the machine epsilon. (The fact that the numerical flow is defined only for lattice points provides latitude in defining H_ε on the continuum.) Note that what is being said is that the finite precision trajectory is the exact *leapfrog* trajectory for a slightly different Hamiltonian, not the exact analytical trajectory for a different Hamiltonian. Nonetheless, this does mean that the finite precision map is symplectic.

In a subsequent paper, it is shown [109] that the idealized assumptions about the computation are unnecessary and that a *lattice leapfrog method* can be implemented in floating-point arithmetic at very little extra cost. Also shown is how the analysis extends to include errors other than those due to finite precision: the method of proof allows for errors in the force $F(q)$ that could be due to

- a fast electrostatic solver such as the fast multipole method,
- Lennard–Jones cutoffs, and
- errors due to unconverged iterations such as those needed to implement implicit integration schemes.

Being symplectic, however, is no guarantee of stability, and errors much larger than ε can produce instability [9].

Example 1. The Kepler problem in polar coordinates is given by

$$H(q,p) = \frac{p^2}{2m} - \frac{GMm}{q} + \frac{l^2}{2mq^2}.$$

One can choose units in q and t so that $H(q,p) = \frac{1}{2}p^2 - q^{-1} + \frac{1}{2}q^{-2}$. The potential energy function has a minimum at $q = 1$ and is negative for $\frac{1}{2} < q < +\infty$. Pieces of orbits for $\varepsilon = 2^{-13}, 2^{-14}, 2^{-15}, 2^{-16}$ are shown in Figs. 3.2–3.5. The time step is $\Delta t = 0.2$. We observe that increasing precision does not reduce the thickness of the orbit.

4. Other Methods

Methods other than leapfrog have also been used extensively for MD. These other methods have mostly been of higher order and have all been non-symplectic. Notable examples are the "other" Gear methods (the higher order Störmer methods in Nordsieck form) [37, p. 154]. However, the leapfrog method has emerged as the favorite. Does the symplectic property of the leapfrog method explain its success in comparison with (non-symplectic) higher order methods? A case for symplectic integration is made at the end of this section.

Experience with numerical methods in general suggests that fourth-order methods are likely to be optimally efficient for low-accuracy solutions. For typical accuracies in MD there is some limited experimental evidence in favor of higher order symplectic integrators [89, 57].

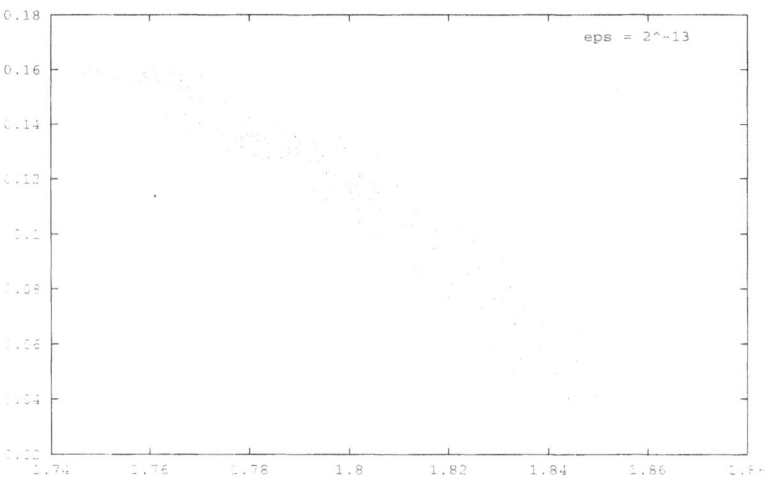

Fig. 3.2. Piece of infinite orbit for polar-coordinate Kepler problem with $\varepsilon = 2^{-13}$.

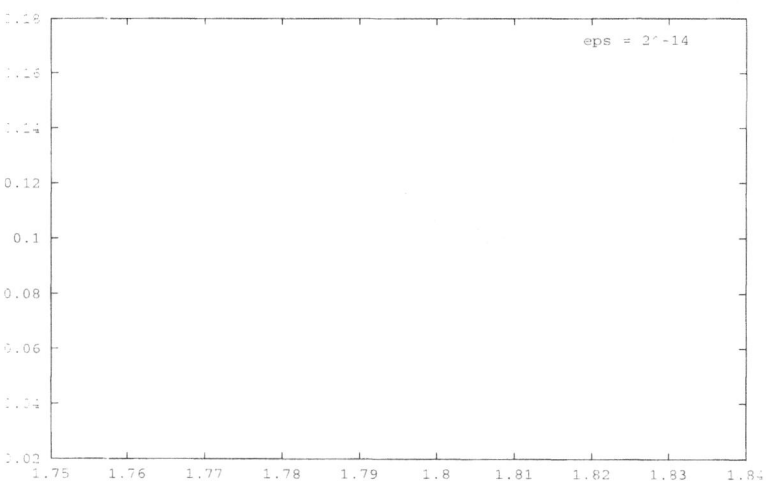

Fig. 3.3. Piece of infinite orbit for polar-coordinate Kepler problem with $\varepsilon = 2^{-14}$.

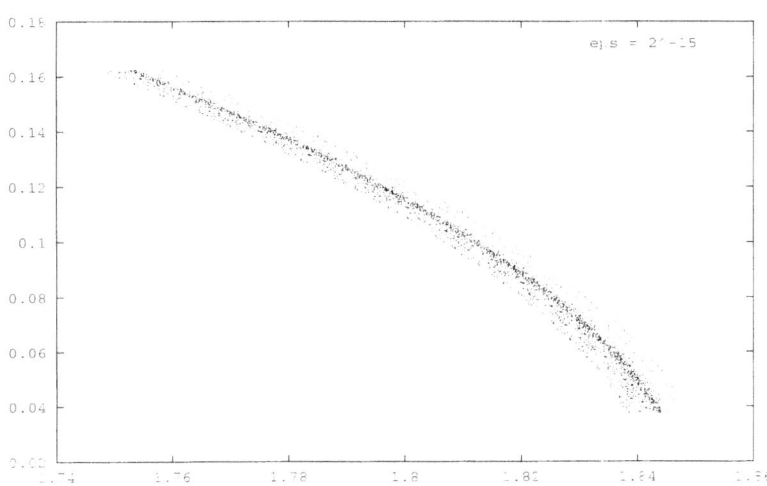

Fig. 3.4. Piece of infinite orbit for polar-coordinate Kepler problem with $\varepsilon = 2^{-15}$.

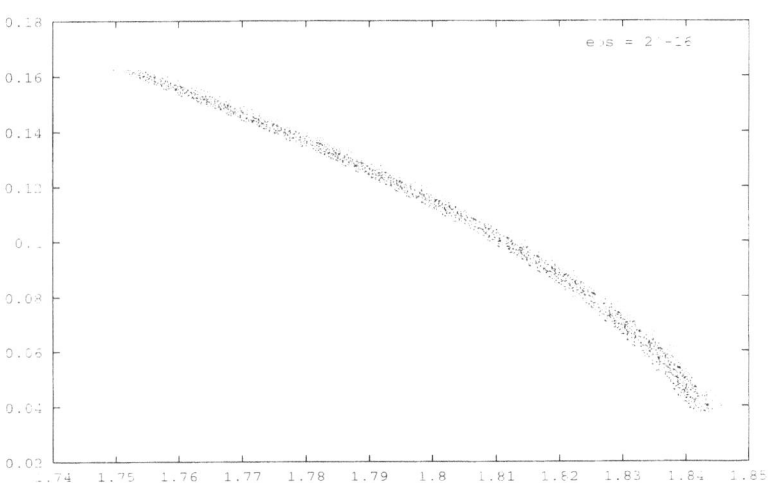

Fig. 3.5. Piece of infinite orbit for polar-coordinate Kepler problem with $\varepsilon = 2^{-16}$.

Our brief look at other methods brings us into contact with a number of interesting results. One is the remarkable construction, by Yoshida, of symplectic methods of arbitrary order obtained by taking a basic low order symplectic method and executing a sequence of substeps of specially chosen sizes. Another is the Ge–Marsden result on the incompatibility between energy conservation and the symplectic property. A third is the result of Chawla on the optimal stability of the leapfrog among all explicit methods of standard type. Much more about numerical methods for Hamiltonian systems is to be found in the book by Sanz-Serna and Calvo [98].

4.1. A Family of Methods

We consider, as an example, a method more elaborate than leapfrog and potentially of higher order. Here is one step, given in algorithmic style:

$$q = q + (\frac{1}{2} - a)\Delta t M^{-1}p;$$
$$p = p + (\frac{1}{2} - \frac{1}{2}a - \frac{1}{2}b)\Delta t F(q);$$
$$q = q + a\Delta t M^{-1}p;$$
$$p = p + (a + b)\Delta t F(q);$$
$$q = q + a\Delta t M^{-1}p;$$
$$p = p + (\frac{1}{2} - \frac{1}{2}a - \frac{1}{2}b)\Delta t F(q);$$
$$q = q + (\frac{1}{2} - a)\Delta t M^{-1}p;$$

Here a and b are method parameters that can be chosen to optimize accuracy and stability. This particular parameterization is chosen for convenience in obtaining methods of order four and the method of Rowlands [96]. Note that the total increment for the positions q is exactly $\Delta t M^{-1}p$ and the total for the momenta is $\Delta t F(q)$. The choice $a = 0$ or $a + b = 1$ gives the midpoint form of the leapfrog method. This method is symplectic, as is any method based on splitting the Hamiltonian.

The modified Hamiltonian can be calculated using the BCH formula given in appendix A; the result is

$$\tilde{H}(q,p) \doteq H(q,p) + \Delta t^2 \left(c_1 p^T M^{-1} U_{qq} M^{-1} p + c_2 U_q^T M^{-1} U_q\right)$$

where $c_1 = -\frac{1}{24} + \frac{1}{2}a^2 - \frac{1}{2}a^3 - \frac{1}{2}a^2 b$ and $c_2 = \frac{1}{12} - \frac{1}{4}a + \frac{1}{4}a^3 + \frac{1}{2}a^2 b + \frac{1}{4}ab^2$. With the post-processing transformation

$$\bar{q} = q + \beta \Delta t^2 M^{-1} U_q(q),$$
$$\bar{p} = \left(I + \beta \Delta t^2 M^{-1} U_{qq}(q)\right)^{-1} p,$$

the modified Hamiltonian for the processed method is

$$\tilde{H}^{\text{proc}}(\bar{q}, \bar{p}) = H(\bar{q}, \bar{p}) + \Delta t^2 \left((c_1 - \beta)\bar{p}^{\mathrm{T}} M^{-1} U_{qq}(\bar{q}) C M^{-1} \bar{p} \right.$$
$$\left. + (c_2 + \beta) U_q(\bar{q})^{\mathrm{T}} M^{-1} U_q(\bar{q}) \right).$$

To attain effective order 4 requires $\beta = c_1 = -c_2$, which imposes the following condition on the two method parameters:

$$6ab^2 = 6a(1-a)^2 - 1.$$

This can be solved for b if either $a < 0$ or $a \geq \Omega$ where $\Omega = \frac{1}{3}(2 + 2^{1/3} + 2^{-1/3})$. Some interesting special choices are the following:

1. The choice $a = \frac{1}{2} - \frac{1}{2}\Omega$ and $b = \frac{1}{2} - \frac{3}{2}\Omega$ yields a method of *conventional* order 4 found first by Forest and Ruth [31, 16]. This scheme is simply the composition of three steps of the midpoint form of leapfrog with substeps of sizes $\Omega \Delta t$, $(1 - 2\Omega)\Delta t$, $\Omega \Delta t$. This scheme is, however, far from the best choice because of large error coefficients. There is one other scheme of conventional order 4, corresponding to choosing b to have the opposite sign. It has even larger error coefficients.
2. The choice $a = 1 - \Omega$ and $b = 2\Omega - 1$ yields a method with a maximal stability threshold for $\omega \Delta t$ for the harmonic oscillator $(\mathrm{d}^2/\mathrm{d}t^2)q = -\omega^2 q$ [71]. This can be regarded as the composition of a first order scheme followed by midpoint leapfrog followed by the "backward" form of the first order scheme with substeps of sizes $(\frac{1}{2} - \frac{1}{2}\Omega)\Delta t$, $\Omega \Delta t$, $(\frac{1}{2} - \frac{1}{2}\Omega)\Delta t$.
3. The limit $a \to 0$ and hence $b \to \infty$ yields a method found first by Rowlands [96, 72]:

$$q = q + \frac{1}{2}\Delta t M^{-1} p;$$
$$p = p + \Delta t (I - \frac{1}{12}\Delta t^2 U_{qq}(q) M^{-1}) F(q);$$
$$q = q + \frac{1}{2}\Delta t M^{-1} p.$$

It is not necessary to calculate the Hessian as such; rather one can form a Hessian–vector product in as little work as is needed to compute the force vector. In fact, the work is even less because expensive square roots and elementary function values can be re-used. (A more straightforward derivation of Rowlands' method is to choose $\beta = -\frac{1}{12}$ in (3.9) and then replace U in the leapfrog method by $U - \frac{1}{24}\Delta t^2 U_q^{\mathrm{T}}(q) M^{-1} U_q(q)$ in order to cancel the dominant error term in the nearby Hamiltonian.)

4.2. Quest for Accuracy and Stability

We call the method above, in the case $a \neq 0$ and $a + b \neq 1$, a 3-stage method because it uses three force evaluations. Of course, even more stages can be used for the purpose of improving error coefficients, order of accuracy, and/or intervals of stability.

Awareness of the existence of symplectic numerical integrators is reported [17] to date back to unpublished work in 1956 by De Vogelaere [20]. The first published paper on symplectic methods is that of Ruth [97] in 1983, which shows that the leapfrog method possesses this property. Ruth also derives new symplectic methods, including an explicit method of order 3 not requiring analytical derivatives. Other early papers are [28, 17].

Then began a search for higher order methods based on splittings. In a rather remarkable development, Yoshida [131] gave a systematic construction for schemes of order $2r$ having 3^{r-1} stages. The first step of this construction creates the Forest–Ruth method from the leapfrog method. These methods all have negative fractional time steps and hence large coefficients and not-too-surprisingly large error coefficients. Therefore, current searches for higher order methods also seek modest error coefficients. However, for methods of order > 2 it is not possible to avoid negative fractional time steps in either q or p [106, 118, 38].

In general it is advantageous to use symmetric methods. If we have a good non-symmetric method, it can be composed with its adjoint (its "backward" form). The asymptotic expansion of the error for the composed method will have the same coefficients as the original method for even powers of Δt but the odd powers will be absent. And very often, but not always, the adjoint of a method is as efficient to execute as the original method.

A measure of (linear) stability is the length of the *interval of periodicity*, which is the longest interval of values of $\omega\Delta t$ starting at zero for which the integrator's propagation matrix has only eigenvalues of unit modulus and is power-bounded when applied to the test equation with Hamiltonian $\frac{1}{2}p^2 + \frac{1}{2}\omega^2 q^2$. A better measure is the length of the *scaled* interval of periodicity, based on the *scaled time step*, which is the time step Δt divided by the number of force evaluations. (A force evaluation at the very beginning of a step counts only as one half an evaluation if the same value is used also at the end of the previous step.) In this sense the leapfrog method is optimal—Chawla [18] has shown that no "conventional" explicit method has a scaled interval of periodicity which extends beyond $\omega\Delta t/m = 2$, where m is the number of force evaluations per step. The optimally stable effectively 4th order 3-stage method given above as item 2 is periodic for $(\omega\Delta t)/3 \leq 1.89\cdots$. Rowlands' method has a stability barrier of $\sqrt{3}$, if it is regarded as a 2-stage method. A long stability interval seems to have the added benefit of yielding smaller method and error coefficients; for example, the optimally stable scheme just mentioned has an error coefficient that is smaller than that of the Forest–Ruth method by a factor of one hundred! Optimally stable schemes are extensively investigated in [81].

Multistep methods cannot be symplectic [46], but symmetric multistep methods [63] are reversible. These have been used by astronomers [91]. They exhibit a resonance due to parasitic principal roots of the difference equation.

4.3. The Case for Symplectic Integration

A minimum reasonable requirement for an integrator for Hamiltonian systems is that it have an interval of periodicity of positive length, i.e., that it be neutrally stable for sufficiently small time steps Δt, when applied to a stable linear problem. This alone may not suffice (see [90] for an example of quadratic error growth where a symplectic method would give only linear error growth). Two stronger properties (neither of which implies the other) that might be adequate are the Liouville (or volume-preserving) property and the property of reversibility. A mapping $X(y)$ preserves volume if its Jacobian $\det X_y = 1$, which is a consequence of the property $X_y^{\mathrm{T}} J X_y = J$ of a symplectic mapping. Some consequences of these properties, and their possible implications for molecular dynamics, are considered below.

Perhaps the strongest motivation for using symplectic methods is given by *backward error analysis*. The numerical trajectories produced by any integrator can be expressed formally as the exact solution of perturbed system of ODEs whose right hand side function $f(y)$ is expressible as a formal series in powers of Δt. It has been shown [114, 98] that this perturbed system is a Hamiltonian system if and only if the integrator is symplectic. This series is generally divergent regardless of how small Δt is, although for linear problems it converges for small enough Δt. If the series is truncated at an appropriate place, the solution of the corresponding Hamiltonian system will differ only by an amount $O(\Delta t^p)$ from the numerical solution of the symplectic integrator on a time interval of length $O(1/\Delta t)$ where p is the order of accuracy. See [114, section IV3], [101], [44], [7], [47], and [92]. Unfortunately $1/\Delta t$ is much shorter than integration intervals in MD. Nonetheless, there is no reason to believe that the backward error analysis cannot be extended to much longer times.

As suggested from the discussion in section 2.2 of the goals of MD simulations and the Liouville equation, what we want from the numerical integrator are probability densities $\rho^n(y)$ that reasonably well approximate the theoretical densities $\rho(y, n\Delta t)$ even for very large $n\Delta t$. Hence, it would seem that the foremost concern is to have correct limiting behavior as $t \to \infty$. The numerical method ought to have a unique equilibrium density not too different from the correct one and it ought to maintain either the ergodic or mixing property so that convergence to equilibrium occurs. For the sake of discussion assume that the numerical integrator approximates the solution of a modified ODE system $(\mathrm{d}/\mathrm{d}t)y = \tilde{f}(y)$ on an infinite time interval, and also assume that there is an equilibrium probability density function $\tilde{\rho}(y)$ for ensembles of trajectories. Hence $\nabla^{\mathrm{T}}(\tilde{\rho}\tilde{f}) = 0$. Recall from section 2.2 that the true equilibrium probability density $\bar{\rho}(y)$ is a conserved quantity for the true dynamics. The density $\tilde{\rho}(y)$ will be conserved for the modified dynamics only if $\nabla^{\mathrm{T}}\tilde{f} = 0$. This apparently requires the integrator to have the Liouville property. Assume now that this is the case. There remains the concern that there is more than one equilibrium density for the modified ODE system or that the ergodic or mixing property is absent, preventing convergence to equilibrium. If this modified ODE system is Hamiltonian, then it is unlikely that there are additional integrals of motion because non-integrability is generic for Hamiltonian systems.

Concerning ergodicity, a comparison [17] with ordinary Runge–Kutta methods shows the symplectic method to have superior ergodic behavior. However, these Runge–Kutta methods do not satisfy the bare minimum linear stability requirement given at the beginning of this subsection, so this may be a case of the solution being attracted to a lower dimensional manifold by damping.

All known symplectic methods also conserve angular momentum [99, 132]. (Linear momentum is preserved by any standard method.) These seem like good properties to maintain if we want to have good infinite time behavior.

Stability is essential and KAM theory [4, p. 411] suggests superior stability properties for symplectic methods as a consequence of the preservation of invariant tori around an elliptic equilibrium. Also, stability for at least a time interval $O(1/\Delta t)$ is guaranteed by the conservation of energy for a nearby Hamiltonian.

Calculations requiring equilibrium MD can be performed rigorously with hybrid Monte Carlo methods which consist of a sequence of short MD trajectories, a small fraction of which are rejected. It has been shown that the property of detailed balance needed by this method is satisfied if the integrator is both reversible and volume-preserving [82].

A result of Ge and Marsden [35] shows that if a symplectic scheme conserves energy exactly for a Hamiltonian system having no integral of motion other than energy, then it computes the exact trajectory except for a reparameterization of time. This result suggests that small errors in energy conservation for a symplectic integrator correspond to a trajectory that in some sense also has small errors, and hence that the conservation of energy might be a useful error indicator *for symplectic methods* (whereas for non-symplectic methods it could be deceptively small). If the energy is left free to vary, it can also indicate the presence of programming bugs. For example a jump in energy is likely to be caused by a bug. Also, a continual downward drift in energy is almost certainly a bug, if a volume-preserving integrator is being used. (There is less phase space volume at lower energy values.)

The use of a trajectory's history to enhance the accuracy of integration tends to be incompatible with being symplectic [88, 46]. As a consequence, the storage requirements of symplectic methods tend to be minimal.

Symplecticness does not imply reversibility. However, as suggested in the preceding subsection, symmetric methods, which are also reversible, offer more accuracy for the same computational cost. A case for time reversibility is given in [13].

5. Multiple Time Steps

Many of the terms that govern the dynamics of molecules correspond to distant Coulomb interactions which vary much more slowly than the short-range forces. Appropriate time steps vary from the 0.5 to 1 fs value needed to resolve the 9 fs period of the fastest normal modes associated with methyl groups to a value of perhaps 16 fs for the most distant electrostatics [32]. For bonded interactions a fixed time step is appropriate but for non-bonded interactions, the appropriate time step is distance dependent and thus varies with time. This need—each interaction having an appropriate individual time step and non-bonded interactions having variable time steps—is met by multiple time stepping (MTS) methods, to a significant extent.

Multiple time stepping means the use of different time steps for different parts of the right-hand-side function $f(y)$ of a system of ODEs $(\mathrm{d}/\mathrm{d}t)y = f(y)$. In the case of an additive partitioning of $f(y)$, this is formally equivalent to a multirate method [36]. The use of MTS dates back to at least 1967 [50], in astronomy, and to 1978 in MD [116]. The program GROMOS [125] has long used a *twin-ranged method* in which longer-range forces are evaluated less frequently, for example, every ten steps. For the sake of discussion suppose that we separate the forces into two parts at time t^n,

$$F = F^{\mathrm{fast}} + F^{\mathrm{slow}},$$

and that the slow force is to be evaluated only once for the next two time steps. Commonly this is done by re-using $F^{\mathrm{slow},n}$ at time t^{n+1}. Such a method is only first order accurate, and is not symplectic. In an effort to construct methods that retain the good behavior of the leapfrog/Verlet, a property termed "Verlet-equivalence" was proposed [40, 42]. This condition requires that the method reduce to the Verlet method in the case where all forces are zero except those associated with just (any) one of the time steps. The simplest such method, called Verlet-I, is based on delivering the slow force as widely spaced impulses:

$$M\frac{\mathrm{d}^2}{\mathrm{d}t^2}q = \sum_{n'} \frac{\Delta t}{m}\, \delta\left(t - n'\frac{\Delta t}{m}\right) F^{\mathrm{fast}}(q) + \sum_{n} \Delta t\, \delta(t - n\Delta t) F^{\mathrm{slow}}(q).$$

Due to the possibility of resonance artifacts, this method was rejected in favor of a more elaborate scheme, Verlet-II, which distributes the slow force throughout the interval between slow force evaluations. Both methods retain the second order accuracy of Verlet. The latter method was tested using as many as 8 distance classes with sampling frequencies that grow by factors of two. A third method Verlet-X [8] is as accurate as Verlet-II [41, Tabelle 4.1, DC-1c] but is more economical in storage and is self-starting. The impulsive Verlet-I method was proposed independently and tested under the name r-RESPA [122], an equivalence first noted in [41, p. 28]. The Verlet-I method was observed to be symplectic in [8] and shown to be good at conserving energy unlike Verlet-II or Verlet-X. However, the possibility of resonance was confirmed on a simple nonlinear test problem. Nonetheless, the Verlet-I/r-RESPA/impulse MTS method gives acceptable results for time steps as great as 4 fs [56, 128] and has become quite popular.

The original technique for effecting variable time steps for non-bonded interactions was to move them from one time step class to another. A refinement of this technique that makes the transition smoother is used in [51, Verlet-IId]. An alternative method is the use in [122] of switching functions for partitioning. A similar technique based on artificial splittings of potentials is proposed in [110] as a method for doing symplectic variable time steps.

Methods based on natural or artificial splittings of Hamiltonians are quite application-specific. At the same time, conventional techniques for varying the time step lead to a loss of the symplectic property [111] and stability [108]. Recently, general techniques have been proposed for varying the time step and maintaining reversibility [55] and even the symplectic property [45, 92]. These, of course. permit only a single time step to be used for all variables and interactions.

5.1. The Verlet-I/r-RESPA/Impulse MTS Method

To obtain a symplectic scheme, we partition the potential energy $U = U^{\text{fast}} + U^{\text{slow}}$ as a basis for partitioning the force vector. Then the integration for $\Delta t H$ is partitioned as

$$(\Delta t/2)U^{\text{slow}},$$
$$\Delta t(U^{\text{fast}} + K) \text{ numerically—} m \text{ steps of leapfrog.}$$
$$(\Delta t/2)U^{\text{slow}}.$$

To avoid complicated subscripting, we express the details algorithmically:

```
void step() {
    p = p + ½ΔtF^slow(q);
    for (k = 1; k ≤ m; k = k + 1) {
        p = p + ½(Δt/m)F^fast(q);
        q = q + (Δt/m)M⁻¹p;
        p = p + ½(Δt/m)F^fast(q);
    }
    p = p + ½ΔtF^slow(q);
}
```

Partitioning the potential energy among different interactions is adequate for bonded interactions of differing speeds but not for *nonbonded* interactions because their speed can vary greatly during the course of the simulation depending on the interparticle distance r. However, we can employ the machinery of MTS to vary the stepsize for each nonbonded interaction $U_{ij}(q)$ if we introduce an artificial splitting $U_{ij}(q) = U_{ij}^{\text{fast}}(q) + U_{ij}^{\text{slow}}(q)$ such that

$U_{ij}^{\text{fast}}(q)$ vanishes for $\|\mathbf{r}_j - \mathbf{r}_i\|$ beyond some cutoff r_{cut} and $U_{ij}^{\text{slow}}(q)$ is slow for all q and never requires a small time step. The effect of this is to permit a large time step whenever $\|\mathbf{r}_j - \mathbf{r}_i\|$ exceeds the cutoff. An example of how to choose the splitting is given in the subsection that follows.

For a larger number of different time steps use a splitting

$$U = U_0 + U_1 + \cdots + U_\nu$$

where the term U_k is used with time step $2^{-k}\Delta t$. This defines a splitting of the force vector into terms $F_k(q) = -\nabla_q U_k(q)$. As an example, if $\nu = 2$, we can define the method from the following progression of splittings of the Hamiltonian $\Delta t H$:

$\frac{1}{2}\Delta t U_0$

$+\quad \frac{1}{4}\Delta t U_1$

$+\quad \frac{1}{8}\Delta t U_2$

$+\quad \frac{1}{4}\Delta t K$

$+\quad \frac{1}{8}\Delta t U_2$

$+\quad \frac{1}{8}\Delta t U_2$

$+\quad \frac{1}{4}\Delta t K$

$+\quad \frac{1}{8}\Delta t U_2$

$+\quad \frac{1}{4}\Delta t U_1$

$+\quad \frac{1}{4}\Delta t U_1$

$+\quad \frac{1}{8}\Delta t U_2$

$+\quad \frac{1}{4}\Delta t K$

$+\quad \frac{1}{8}\Delta t U_2$

$+\quad \frac{1}{8}\Delta t U_2$

$+\quad \frac{1}{4}\Delta t K$

$+\quad \frac{1}{8}\Delta t U_2$

$+\quad \frac{1}{4}\Delta t U_1$

$+\quad \frac{1}{2}\Delta t U_0$

The potential energy terms can be merged into groups to produce the following set of *partial* force vectors:

$$F_0(q) + \tfrac{1}{2}F_1(q) + \tfrac{1}{4}F_2(q),$$
$$\tfrac{1}{4}F_2(q),$$
$$\tfrac{1}{2}F_1(q) + \tfrac{1}{4}F_2(q),$$
$$\tfrac{1}{4}F_2(q),$$
$$F_0(q) + \tfrac{1}{2}F_1(q) + \tfrac{1}{4}F_2(q).$$

An algorithm for general ν, which loops through 2^ν substeps in each step, is given below. Before the very first step a *partial* force vector F is computed by

$$F = F_0(q) + 2^{-1}F_1(q) + \cdots + 2^{-\nu}F_\nu(q);$$

Then each step proceeds as follows:

```
void step() {
    for (i = 1; i ≤ 2ᵛ; i = i + 1) {
        p = p + ½ΔtF;
        q = q + 2⁻ᵛΔtM⁻¹p;
        l = the smallest integer such that 2ᵛ⁻ˡ divides i;
        F = 2⁻ˡFₗ(q) + 2⁻ˡ⁻¹Fₗ₊₁(q) + ⋯ + 2⁻ᵛFᵥ(q);
        p = p + ½ΔtF;
    }
}
```

If the algorithm is implemented as described above, it will not be much faster than a single time step algorithm for three reasons.

1. At time steps when slower forces are being evaluated, there are more interactions to be evaluated, e.g., $F_0(q) + \tfrac{1}{2}F_1(q) + \tfrac{1}{4}F_2(q)$ instead of just $F(q)$.
2. Even though the evaluations of most interactions are avoided at most time steps because they are zero, there remains the cost of determining whether or not to evaluate an interaction. This test can be done by computing the square of a distance and comparing this to the square of the cutoff, thus avoiding a square root calculation. However, this alone will not provide dramatic savings in computing time.
3. If the shortest possible time step $2^{-\nu}\Delta t$ is chosen smaller than what is needed for bonded forces in order to resolve rare close encounters between a non-bonded pair of atoms, it would inefficient to descend to this level except when necessary.

Remedies for these inefficiencies are given in [49], two of which are described in section 5.3. It is helpful first to discuss in greater detail the method of partitioning the non-bonded interactions.

5.2. Partitioning of Interactions

There are two issues: (i) how to assign appropriate time steps to each interaction, and (ii) how to split a non-bonded interaction. The choice of parameters in the splitting technique depends on the assignment of time steps.

A rule for the assignment of time steps to differing interactions in a multiple time stepping scheme is given in [69]. It is based on the spectral radius of the mass-weighted Hessian of an interaction potential $U(q)$:

$$\Delta t \propto \left(\max_q \rho(M^{-1/2} U_{qq}(q) M^{-1/2}) \right)^{-1/2}.$$

This rule is derived for bonded interactions from a truncation error analysis. It is a reasonable rule of thumb for non-bonded interactions as well.

The question of how to split a non-bonded interaction is addressed first for electrostatics. The potential energy term is given by

$$U_{ij}(q) = \frac{1}{2} C_{ij} \phi(\|\mathbf{r}_j - \mathbf{r}_i\|^2)$$

where

$$\phi(s) = 2s^{-1/2}.$$

Note that ϕ is defined in such a way that its argument $\|\mathbf{r}_j - \mathbf{r}_i\|^2$ is a smooth (in fact, quadratic) function of q. The appropriate time step depends on the value of s. Let L denote the following linear operator

$$L\gamma(s) = \begin{cases} \gamma(r_{\text{cut}}^2) + (s - r_{\text{cut}}^2)\gamma'(r_{\text{cut}}^2), & s \le r_{\text{cut}}^2, \\ \gamma(s), & s \ge r_{\text{cut}}^2, \end{cases}$$

in which a function $\gamma(s)$ is replaced by its linear Taylor interpolant for $s \le r_{\text{cut}}^2$. Define the splitting

$$\phi^{\text{slow}}(s) = L\phi(s), \quad \phi^{\text{fast}}(s) = (1 - L)\phi(s).$$

Then $U_{ij}(q) = \frac{1}{2} C_{ij} \phi^{\text{slow}}(\|\mathbf{r}_j - \mathbf{r}_i\|^2) + \frac{1}{2} C_{ij} \phi^{\text{fast}}(\|\mathbf{r}_j - \mathbf{r}_i\|^2)$ with

$$\frac{1}{2} \phi^{\text{slow}}(r^2) = \begin{cases} \frac{3}{2} r_{\text{cut}}^{-1} - \frac{1}{2} r_{\text{cut}}^{-3} r^2, & r \le r_{\text{cut}}, \\ r^{-1}, & r \ge r_{\text{cut}}, \end{cases}$$

$$\frac{1}{2} \phi^{\text{fast}}(r^2) = \begin{cases} r^{-1} - \frac{3}{2} r_{\text{cut}}^{-1} + \frac{1}{2} r_{\text{cut}}^{-3} r^2, & r \le r_{\text{cut}}, \\ 0, & r \ge r_{\text{cut}}. \end{cases}$$

Each piece has a continuous first derivative. In such splittings it is desirable to avoid cancellation to keep the temporal discretization error small [117].

The technique applies also to the Lennard–Jones potential, given by

$$U_{ij}^{\text{LJ}}(q) = \frac{1}{2} \varepsilon_{ij} \phi^{\text{LJ}}(\|\mathbf{r}_j - \mathbf{r}_i\|^2 / \sigma_{ij}^2) - \varepsilon_{ij}$$

where

$$\phi^{\mathrm{LJ}}(s) = 2(2s^{-3} - 1)^2.$$

If r_{cut} is chosen where $\phi^{\mathrm{LJ}}(r^2/\sigma_{ij}^2)$ takes its minimum value, the result is a splitting of the Lennard–Jones potential into a repulsive part and an attractive part, known as the Weeks–Chandler–Andersen (WCA) breakup .

We can do multiple time stepping for an electrostatic potential energy term by defining a multiple splitting of the function $\phi(s) = 2s^{-1/2}$. We partition the range of values of s by means of a sequence of cutoffs $r_1 > r_2 > \cdots > r_k > \cdots > 0$ such that $2^{-k}\Delta t$ is the appropriate time step when $r_{k-1}^2 \leq s < r_k^2$. Here it is convenient to define $r_0 = +\infty$. Then we construct an *artificial* splitting

$$\phi = \phi_0 + \phi_1 + \cdots + \phi_k + \cdots \tag{5.1}$$

such that

1. $2^{-k}\Delta t$ is an appropriate time step for ϕ_k and
2. $\phi_k(s) = 0$ for $s \geq r_k^2$.

This means that if $r_{k+1}^2 \leq s < r_k^2$, then $2^{-k}\Delta t$ is the smallest time step needed for any nonzero term in the splitting. To obtain the splitting, we can repeatedly apply the two-time-step idea

$$
\begin{aligned}
\phi(s) &= \phi_0(s) + \psi_1(s) \\
&= \phi_0(s) + \phi_1(s) + \psi_2(s) \\
&= \phi_0(s) + \phi_1(s) + \phi_2(s) + \psi_3(s) \\
&\;\;\vdots
\end{aligned}
$$

This can be done by the following recursive construction:

$$\psi_0(s) = \phi(s), \tag{5.2}$$

and, for $k = 1, 2, \ldots, \nu$,

$$\phi_{k-1}(s) = L_k \psi_{k-1}(s), \quad \psi_k(s) = (1 - L_k)\psi_{k-1}(s) \tag{5.3}$$

where L_k denotes the linear operator

$$L_k \gamma(s) = \begin{cases} \gamma(r_k^2) + (s - r_k^2)\gamma'(r_k^2), & s \leq r_k^2, \\ \gamma(s), & s \geq r_k^2. \end{cases}$$

In particular,

$$\frac{1}{2} L_k \phi(r^2) = \begin{cases} \frac{3}{2} r_k^{-1} - \frac{1}{2} r_k^{-3} r^2, & r \leq r_k, \\ r^{-1}, & r \geq r_k. \end{cases}$$

The recurrence (5.2), (5.3) can be solved by using the fact that $L_k L_{k-1} = L_{k-1}$ (linear Taylor interpolation reproduces a linear polynomial) to get

$$\phi_{k-1}(s) = (L_k - L_{k-1})\phi(s), \quad \psi_k(s) = (1 - L_k)\phi(s).$$

The second of these holds also for $k = 0$ if we define $L_0 = 0$.

5.3. Efficient Implementation

Remedies are given here for two of the three inefficiencies listed at the end of section 5.1. This is also a good place to point out that the force F_0 normally consists only of long-range electrostatics and that this is most efficiently done by some fast tree algorithm or grid-based solver.

We consider first how to avoid the extra work created by the different parts of a split force. Consider the calculation of the lth level partial forces

$$F = 2^{-l} F_l(q) + 2^{-l-1} F_{l+1}(q) + \cdots + 2^{-\nu} F_\nu(q);$$

With U_k defined according to eq. (5.1) the contribution to the partial force vector of the electrostatic force on particle i due to particle j is the gradient of

$$\frac{1}{2} C_{ij} \chi_l(\|\mathbf{r}_j - \mathbf{r}_i\|^2),$$

where

$$\chi_l(s) = 2^{-l} \phi_l(s) + 2^{-l-1} \phi_{l+1}(s) + \cdots + 2^{-\nu+1} \phi_{\nu-1}(s) + 2^{-\nu} \psi_\nu(s).$$

This simplifies:

$$
\begin{aligned}
\chi_l(s) &= (2^{-l}(L_{l+1} - L_l) + 2^{-l-1}(L_{l+2} - L_{l+1}) + \cdots \\
&\qquad \cdots + 2^{-\nu+1}(L_\nu - L_{\nu-1}) + 2^{-\nu}(1 - L_\nu))\phi(s) \\
&= (-2^{-l} L_l + 2^{-l-1} L_{l+1} + 2^{-l-2} L_{l-2} + \cdots + 2^{-\nu} L_\nu + 2^{-\nu})\phi(s).
\end{aligned}
$$

Let $k(s)$ be defined by

$$r^2_{k(s)-1} \le s < r^2_{k(s)}$$

unless $s < r^2_\nu$, in which case $k(s)$ is defined to be ν. Hence, $L_k \phi(s) = \phi(s)$ for $k \ge k(s) + 1$. There are two cases. If $k(s) < l$, then *by construction*

$$\chi_l(s) = 0.$$

If $k(s) \ge l$,

$$
\begin{aligned}
\chi_l(s) &= (-2^{-l} L_l + 2^{-l-1} L_{l+1} + 2^{-l-2} L_{l+2} + \cdots \\
&\qquad \cdots + 2^{-k(s)} L_{k(s)})\phi(s) + 2^{-k(s)} \phi(s) \\
&= a_{l,k(s)} s + b_{l,k(s)} + 2^{-k(s)} \phi(s)
\end{aligned}
$$

where $a_{l,k(s)}$ and $b_{l,k(s)}$ are constants depending on $\phi(s)$ and the cutoff radii. The values $b_{l,k(s)}$ are not needed for force calculations and the values $a_{l,k(s)}$ and $2^{-k(s)}$ can be pre-computed and stored in tables.

The question of avoiding unnecessary distance calculations is now considered. As noted already, the test $\|\mathbf{r}_j - \mathbf{r}_i\|^2 < r^2_{\text{cut}}$ is considerably cheaper to perform than is the computation of a force with potential energy $C_{ij}\|\mathbf{r}_j - \mathbf{r}_i\|^{-1}$, because of the additional square root calculation. Nonetheless, it is highly desirable to avoid doing this test in the vast majority of cases. The cost of doing each collection of fast force evaluations should be only linear in the number of particles N (although cubic in the cutoff r_k). A linear time algorithm can be attained by using "cell lists" [26]. The computational box is divided into cubic or nearly cubic cells, and periodically, e.g., at the beginning of every (full) step, a list is constructed for each cell of the atoms that it contains. (For efficiency the looping is done over atoms.) Also, we choose a *safety margin* δ to be a distance that is unlikely to be traversed by any atom within the chosen period. This is monitored by checking that each particle stays within a distance δ of its container during the period. In this way distance testing is needed only for those atoms whose containers are within a distance $r_k + 2\delta$ of each other. Efficiency may be further enhanced by using the cell lists to construct a neighbor list, which would contain a list of all atom pairs (i, j) for which $r_{ij} \leq r_{\text{cut}} + 2\delta$.

5.4. Mollified Impulse MTS Methods

Stability analysis of the Verlet-I/r-RESPA/impulse MTS method [34] shows that for linear problems it is numerically unstable not only for time steps near the period of the fastest normal mode but also for time steps just less than half the period. Also an "order reduction" effect is observed; i.e., the second order accuracy in terms of the *long* time steps is not independent of the fast force. The stability limitations are observed in practice for MD simulations with time steps just less than the half period of the fastest normal mode, which is about 5 fs [56, 9]. The resonance observable in various graphs in [59] is almost certainly the same phenomenon.

The "five femtosecond time step barrier" of the impulse method is overcome by the *mollified* impulse method (MOLLY) [34], which replaces $U^{\text{slow}}(q)$ by $U^{\text{slow}}(\mathcal{A}(q))$ where $\mathcal{A}(q)$ is a time averaging of vibrational motion due to the fastest forces, e.g., bond stretching and angle bending. This modification has the effect of replacing $F^{\text{slow}}(q)$ by $\mathcal{A}_q(q)^{\text{T}} F^{\text{slow}}(\mathcal{A}(q))$. The main benefit is to filter out those components of the impulse that would excite the fastest forces. Stability can be shown for a linear problem with two degrees of freedom for time steps less than a period, as can uniform second order accuracy [34]. Various averaging functions $\mathcal{A}(q)$ with B-spline weights are tested for MD in [112].

More effective than averaging is a choice for $\mathcal{A}(q)$ based on projection [58]. If $U^{\text{fastest}}(q)$ consists only of bonded terms, we can write down a set of equations

$$g^k(q) = 0, \quad k = 1, 2, \ldots, \mu,$$

that represent equilibrium, $U_q^{\text{fastest}}(q) = 0$, for the fastest interactions. These are, in fact, distance constraints, and we can write

$$g^k(q) = \|\mathbf{r}_{j(k)} - \mathbf{r}_{i(k)}\|^2 - l_k^2.$$

Projection onto these constraints can be accomplished by defining

$$\mathcal{A}(q) = q + M^{-1} \sum_{l=1}^{\mu} \lambda_l g_q^l(q)$$

where the $\{\lambda_l\}$ are chosen to satisfy $g(\mathcal{A}(q)) = 0$. The solution of these quadratic systems of equations is a topic of the section that follows. Experiments indicate an increase of about 50% in the longest time step for this "equilibrium" version of MOLLY over the impulse method for a given level of energy conservation.

6. Constrained Dynamics

Even with multiple time stepping the time step is restricted by the small-amplitude high-frequency components of the motion that are due to bond-length stretching. To enable larger time steps, it is common to make some or all of the bond lengths rigid. Sometimes angles are also constrained. This is accomplished by appending constraint equations to the equations for Newton's law of motion and by including one Lagrange multiplier in the force term for each constraint equation. This results in a system of differential equations with quadratic constraints. As in the previous section we express these constraints as

$$g(q) = 0$$

where

$$g^k(q) = \|\mathbf{r}_{j(k)} - \mathbf{r}_{i(k)}\|^2 - l_k^2, \quad k = 1, 2, \dots, \mu.$$

Constraints on angles can be imposed as distance constraints if bond lengths are also being constrained. The set of constraints is used to determine Lagrange multipliers in the equations of motion:

$$\frac{d}{dt}q = M^{-1}p, \quad \frac{d}{dt}p = F(q) + g_q(q)^T \lambda.$$

Constraints on positions clearly reduce the number of degrees of freedom in the system. Differentiating the position constraints gives velocity constraints:

$$g_q(q)M^{-1}p = 0.$$

(Differentiating the velocity constraints gives a linearly implicit formula for the Lagrange multipliers, so this is an index 3 differential-algebraic equation [11].) Hence phase space is of dimension $6N - 2\mu$. Ideally, the initial conditions satisfy the constraints. The differential system is Hamiltonian in an appropriate set of $6N - 2\mu$ coordinates [66].

The most important question concerns the inaccuracy introduced by imposing rigidity. Computer simulations indicate that the application of constraints to bond lengths has only a slight effect on dynamics, but that constraining all bond lengths and angles—*torsion dynamics*—has a significant effect. The argument justifying the use of constraints is based on averaging. In the case of the canonical (NVT) ensemble this averaging introduces a correction to the potential, first obtained by Fixman [30]. This correction is small for bond length constraints but not for angle constraints. The Fixman potential for angle constraints is generally difficult to compute, but a practical approximation has been found [95, 94], having the form of a correction to the dihedral potential. Rigid angles also introduce excessive rigidity in the structure, which is often compensated for by softening the Lennard–Jones potential. A method termed *soft-SHAKE* [95] avoids this artificial rigidity by formulating the constrained form of the ODEs as a singular perturbation expansion and taking the expansion a little further.

Because of a small mass, interactions involving hydrogen are faster than similar interactions involving heavy atoms, and it makes sense to select only such interactions for constraints. In particular, it is common to do simulations with constraints on only the lengths of covalent bonds to hydrogens. Also, independently of whether or not other constraints are imposed, it is common to use fully rigid water models.

6.1. Discretization

A popular generalization of the leapfrog method to constrained dynamics is the SHAKE method [123], which replaces the drift step of leapfrog by a "targeted" drift. The Lagrange multipliers at one time level n are used to satisfy the constraint equations at time level $n+1$. An enhancement known as RATTLE [3] also forces velocities to satisfy their constraints. However, these adjustments to the velocities are effectively undone by the next SHAKE step, and the dynamics is unaffected. Hence, the RATTLE step is just cosmetic. It has been shown that RATTLE, and therefore SHAKE, is reversible and symplectic [67].

Given below is the algorithm for SHAKE with the RATTLE option. The step begins with values q^n, p^n, F^n obtained from the previous step. Then we compute

half-kick:
$$p^{n+\epsilon} = p^n + \frac{1}{2}\Delta t F^n,$$

SHAKE:
$$p^{n+1-\epsilon} = p^{n+\epsilon} + \Delta t g_q(q^n)^\mathrm{T}\lambda^n,$$
$$q^{n+1} = q^n + \Delta t M^{-1}p^{n+1-\epsilon},$$
$$\text{where } \lambda^n \text{ satisfies } g(q^{n+1}) = 0,$$

half-kick:

$$F^{n+1} = F(q^{n+1}),$$

$$\bar{p}^{n+1} = p^{n+1-\epsilon} + \frac{1}{2}\Delta t F^{n+1},$$

*RATTLE:

$$p^{n+1} = \bar{p}^{n+1} + \Delta t g_q(q^{n+1})^{\mathrm{T}}\lambda_{\mathrm{p}}^{n+1}$$

where $\lambda_{\mathrm{p}}^{n+1}$ satisfies $g_q(q^{n+1})M^{-1}p^{n+1} = 0$.

The SHAKE step involves solving nonlinear equations and this raises the question of existence and uniqueness of the solution. Even with one constraint the quadratic will have two solutions and in extreme cases none. More generally the kth constraint equation is dominated by the kth unknown Lagrange multiplier and is quadratic in this unknown. Hence, there are typically about 2^μ solutions. It is suggested that the physically correct solution is one obtained by analytical continuation as the atoms drift from q^n to q^{n+1}. If analytical continuation is terminated by bifurcation, then the time step is simply too large.

Application of constraints to a small molecule frequently results in a completely rigid molecule, in which case it is feasible to use a more direct description of rigid body dynamics. If $\mathbf{r}_1, \mathbf{r}_2, \ldots, \mathbf{r}_m$ are the atomic positions of a rigid molecule, they can be expressed as

$$\mathbf{r}_i = \mathbf{c} + Q\mathbf{d}_i$$

where $\mathbf{c}$ is a varying center of mass, Q is a varying orthogonal rotation matrix, and the $\mathbf{d}_i$ are constant positions of the atoms in body-fixed coordinates. The rotation matrix Q embraces three degrees of freedom, but in order to avoid singularities it has been common to use quaternions [25], which involves using four variables with one constraint. The equations of motion for quaternions can be expressed as a constrained Hamiltonian system. However, positions and momenta are coupled in the expression for the kinetic energy, and this forces the use of implicit symplectic methods.

A better foundation for discretization uses as variables all nine elements of the rotation matrix Q (see [77, 93]) subject to the six constraints $g(Q) = Q^{\mathrm{T}}Q - I = 0$ (usually enforced by the incorporation of Lagrange multipliers). Again, this choice of variables admits a constrained Hamiltonian formulation, which can be solved using the symplectic SHAKE algorithm, as discussed in [77]. Recently, a variation of this method was evaluated for molecular applications in [62], with an efficient treatment of the nonlinear equations for the Lagrange multipliers, resulting in improved conservation of energy for simulations of a dipolar soft-sphere fluid compared to a more traditional quaternion-based scheme. Moreover, explicit symplectic integration is possible. The idea is to split the rotational Hamiltonian into kinetic and potential terms, and integrate the kinetic term (describing a free rigid body) by reduction to the Euler equations for the free rigid body [78, 93].

This method was recently refined and applied to treat rigid body molecular problems in [23], with manifest improvements in observed long term stability and accuracy.

6.2. Solution of the Nonlinear Equations

There remains the question of solving the quadratic system of constraint equations. If we define $\bar{q} = q^n + \Delta t M^{-1} p^{n+\epsilon}$ and $d^l = \Delta t^2 M^{-1} g_q^l(q^n)$, then the constraint equations simplify somewhat to

$$q^{n+1} = \bar{q} + \sum_l \lambda_l d^l$$

where the $\{\lambda_l\}$ solve

$$\{g^k(q^{n+1}) = 0\}.$$

Nonlinear Gauss–Seidel applied to these equations for the $\{\lambda_l\}$ is

$\lambda_k = 0,\ k = 1, 2, \ldots, m;$
while (not converged)
    for $(k = 1;\ k \le m;\ k = k + 1)$
        solve $g^k(\bar{q} + \sum_l \lambda_l d^l) == 0$ for $\lambda_k;$

Each equation is quadratic in its associated unknown. A further simplification is to replace the quadratic solve by a single Newton–Raphson iteration:

$$\lambda_k = \lambda_k - \frac{g^k(\bar{q} + \sum_l \lambda_l d^l)}{g_q^k(\bar{q} + \sum_l \lambda_l d^l)^T d^k};$$

yielding Newton-Gauss–Seidel. This algorithm is the originally proposed and popular way of solving the SHAKE constraint equations. The number of iterations can be halved by using over-relaxation [5] (Newton-SOR). The same study finds that the use of full Newton methods (with matrices) is no faster than Newton-SOR in CPU time.

7. Constant-Temperature and Constant-Pressure Ensembles

For equilibrium MD calculations it is preferable to sample from ensembles more similar to physiological conditions of constant temperature and pressure (NpT). Methods for doing this are given in [83, 76, 75, 120, 74, 27].

As an illustration of the ideas, we consider two ensembles for which the methods are simpler: a constant-temperature ensemble (NVT) and a constant-pressure ensemble (isobaric–isoenthalpic). The methods considered are based on extending the ODE system with an additional degree of freedom [1, 65]. The extended system is designed to yield time averages that are the same as those for the desired ensemble. Both extended systems have formulations that are Hamiltonian but non-separable. Nonetheless, both can be integrated by symplectic extensions of leapfrog that are effectively explicit.

7.1. Constant-Temperature Ensembles

The instantaneous temperature of an unconstrained system at time t is given by

$$T(t) = \frac{2}{dk_{\mathrm{B}}} \sum_i \frac{1}{2} m_i |\mathbf{v}_i|^2$$

where k_{B} is Boltzmann's constant and d is the number of degrees of freedom ($3N - 3$ if the total momentum is fixed). The temperature T is the time average of $T(t)$.

The extended system method was introduced by Nosé [86] with additional work by Hoover [54]. The idea is to model the dynamics of a system in contact with a thermal reservoir, so an additional degree of freedom is included, which represents the reservoir, and the extended system is simulated. Heat flows dynamically between the reservoir and the system, so the reservoir has a "thermal inertia" associated with it analogous to the mass of a particle.

Hoover's equations of motion for the system are

$$\frac{\mathrm{d}}{\mathrm{dt}} q = M^{-1} p, \quad \frac{\mathrm{d}}{\mathrm{dt}} p = F(q) - \xi p,$$

where the friction coefficient ξ is an extra degree of freedom, which satisfies

$$\frac{\mathrm{d}}{\mathrm{dt}} \xi = \frac{1}{Q} \left(p^{\mathrm{T}} M^{-1} p - dk_{\mathrm{B}} T \right)$$

where Q is a thermal inertia parameter. It can be seen that ξ changes so that it restores the kinetic energy to its desired value.

Hoover's equations are not a Hamiltonian system, but they can be formulated as such, which is how Nosé originally expressed them. Introduce an integrating factor s, $(\mathrm{d}/\mathrm{dt})s = \xi s$, for the momentum equation and a fictitious momentum

$$\bar{p} = sp,$$

and the system becomes

$$\frac{\mathrm{d}}{\mathrm{dt}} q = s^{-1} M^{-1} \bar{p}, \qquad \frac{\mathrm{d}}{\mathrm{dt}} \bar{p} = s F(q), \tag{7.1}$$

$$\frac{\mathrm{d}}{\mathrm{dt}} s = \xi s, \qquad \frac{\mathrm{d}}{\mathrm{dt}} \xi = \frac{1}{Q} \left(s^{-2} \bar{p}^{\mathrm{T}} M^{-1} \bar{p} - dk_{\mathrm{B}} T \right). \tag{7.2}$$

Define $p_{\mathrm{s}} = Q\xi$ and we have nearly a Hamiltonian system,

$$\frac{1}{s} \frac{\mathrm{d}}{\mathrm{dt}} q = s^{-2} M^{-1} \bar{p}, \qquad \frac{1}{s} \frac{\mathrm{d}}{\mathrm{dt}} \bar{p} = F(q), \tag{7.3}$$

$$\frac{1}{s} \frac{\mathrm{d}}{\mathrm{dt}} s = \frac{1}{Q} p_{\mathrm{s}}, \qquad \frac{1}{s} \frac{\mathrm{d}}{\mathrm{dt}} p_{\mathrm{s}} = s^{-3} \bar{p}^{\mathrm{T}} M^{-1} \bar{p} - dk_{\mathrm{B}} T s^{-1}, \tag{7.4}$$

for Hamiltonian

$$\frac{1}{2}s^{-2}\bar{p}^{\mathrm{T}}M^{-1}\bar{p} + U(q) + \frac{1}{2Q}p_{\mathrm{s}}^2 + dk_{\mathrm{B}}T\ln s.$$

All that remains is to reparameterize time according to

$$\frac{\mathrm{d}}{\mathrm{d}\tau}t = \frac{1}{s},$$

where τ is fictitious time and the system becomes Hamiltonian:

$$\frac{\mathrm{d}}{\mathrm{d}\tau}q = s^{-2}M^{-1}\bar{p}, \qquad \frac{\mathrm{d}}{\mathrm{d}\tau}\bar{p} = F(q), \tag{7.5}$$

$$\frac{\mathrm{d}}{\mathrm{d}\tau}s = \frac{1}{Q}p_{\mathrm{s}}, \qquad \frac{\mathrm{d}}{\mathrm{d}\tau}p_{\mathrm{s}} = s^{-3}\bar{p}^{\mathrm{T}}M^{-1}\bar{p} - dk_{\mathrm{B}}Ts^{-1}. \tag{7.6}$$

The extended Hamiltonian is the sum of the usual kinetic and potential energy plus an extra kinetic energy term $p_{\mathrm{s}}^2/(2Q)$ and an extra potential energy term $dk_{\mathrm{B}}T\ln s$.

Although the system is non-separable, it has enough special structure to permit the use of explicit symplectic integrators. For example, the leapfrog method for general Hamiltonians $H(q,p)$,

$$
\begin{aligned}
p^{n+1/2} &= p^n - \Delta t H_q(q^n, p^{n+1/2}), \\
q^{n+1/2} &= q^n + \Delta t H_p(q^n, p^{n+1/2}), \\
q^{n+1} &= q^{n+1/2} + \Delta t H_p(q^{n+1}, p^{n+1/2}), \\
p^{n+1} &= p^{n+1/2} - \Delta t H_q(q^{n+1}, p^{n+1/2}),
\end{aligned}
$$

is explicit if in the first equation $\bar{p}$ is updated before p_{s} and in the third equation s is updated before q. Alternatively, one can advance by analytical integrations for Hamiltonian $(\Delta t/2)(K + \frac{1}{2}Q^{-1}p_{\mathrm{s}}^2)$ followed by $\Delta t(U + dk_{\mathrm{B}}T\ln s)$ followed by $\frac{1}{2}\Delta t(K + \frac{1}{2}Q^{-1}p_{\mathrm{s}}^2)$.

Canonical ensemble (NVT) averages can be computed from time averages using real momentum and real time, under the assumption that the extended system is ergodic. More precisely, it can be shown that

$$\lim_{t\to\infty} \frac{\int_0^t g(q(\tau), s(\tau)^{-1}\bar{p}(\tau))s(\tau)^{-1}\,\mathrm{d}\tau}{\int_0^t s(\tau)^{-1}\,\mathrm{d}\tau}$$
$$= \frac{\int\int g(q,p)\exp(-H(q,p)/(k_{\mathrm{B}}T))\mathrm{d}q\mathrm{d}p}{\int\int \exp(-H(q,p)/(k_{\mathrm{B}}T))\mathrm{d}q\mathrm{d}p}.$$

Numerical integration of the extended system yields data at unequally spaced points in real time. This is awkward for computing time correlation functions. However, there is a real time Hamiltonian formulation of Nosé–Hoover dynamics, termed the Nosé–Poincaré method [10]. It is based on a Poincaré transformation of the Nosé extended Hamiltonian with thermostat and yields canonical ensemble time averages. An efficient symplectic integrator is provided. The paper includes substantial numerical experiments as well as many details such as constraints, rigid bodies, and "Nosé chains."

7.2. Constant-Pressure Ensembles

The instantaneous pressure is

$$P(t) = \frac{1}{3V} \left(p^T M^{-1} p - q^T F(q) \right).$$

An extended system method of Andersen [2] includes an external variable V, the volume of the simulation box, as an additional degree of freedom. This mimics the action of a piston on the real system. The kinetic and potential energy for the piston are given by

$$K_V = \frac{1}{2} Q^{-1} p_V^2, \quad U_V = PV$$

where p_V is conjugate momentum for V and P is the specified pressure. Introduce fictitious positions and momenta

$$\bar{q} = V^{-1/3} q, \quad \bar{p} = V^{1/3} p$$

and the kinetic and potential energies of the particles become

$$K = \frac{1}{2} V^{-2/3} \bar{p}^T M^{-1} \bar{p}, \quad U = U(V^{1/3} \bar{q}).$$

The Hamiltonian

$$H_V = K + K_V + U + U_V$$

gives rise to the equations of motion

$$\frac{d}{dt} \bar{q} = V^{-2/3} M^{-1} \bar{p}, \qquad \frac{d}{dt} \bar{p} = V^{1/3} F(V^{1/3} \bar{q}),$$

$$\frac{d}{dt} V = Q^{-1} p_V, \qquad \frac{d}{dt} p_V = P(t) - P$$

where the pressure function $P(t)$ is calculated using

$$P(t) = \frac{1}{3V} \left(V^{-2/3} \bar{p}^T M^{-1} \bar{p} - V^{1/3} \bar{q}^T F(V^{1/3} \bar{q}) \right).$$

The trajectories sample the isobaric–isenthalpic ensemble. Although the system is non-separable, it has some special structure and can be treated explicitly by the generalized leapfrog method.

8. Stochastic Dynamics

Langevin dynamics is an approximation to MD that greatly reduces the number of atoms by omitting explicit solvent atoms and replacing solvent forces by homogenized deterministic and stochastic forces. Brownian dynamics is a further approximation, in which particle masses are set to zero. It speeds up simulations by making longer time steps possible.

Stochastic ordinary differential equations are briefly introduced in appendix B. A good reference on numerical methods for stochastic ODEs is the book [60]. Other references are [53, 61].

8.1. Langevin Dynamics

Simulations of a biomolecule in water spend the bulk of the time doing calculations with water molecules. Hence, it is popular to use an *implicit solvent*, in which the effect of the solvent is replaced by averaged and random forces. By considering N large spherical particles and a number of fluid particles, a second order system of stochastic ODEs known as the Langevin equation can be derived [22].

The Langevin equation is

$$dq = vdt, \quad Mdv = F(q)dt - k_BTD(q)^{-1}vdt + \sqrt{2k_BT}D_{1/2}(q)^{-T}dW(t) \tag{8.1}$$

where k_B is Boltzmann's constant, T is temperature, D is a diffusion tensor, $W(t)$ is a collection of independent standard Wiener processes, and $D_{1/2}$ is a matrix satisfying $D_{1/2}D_{1/2}^T = D$, e.g., a Choleski factorization [24]. In this equation k_BTD^{-1} is the friction tensor. The diffusion tensor D could be non-diagonal; if diagonal, $D_{ii} = (k_BT)/(6\pi\eta a_i)$ where η is the solvent viscosity and a_i the effective radius of the ith particle. Other possible diffusion tensors are given in [24]. Electrostatic effects of the solvent are represented by a potential of mean force incorporated into the force vector F and van der Waals/steric effects are represented by added friction and noise terms. The potential of mean force is obtained from the numerical solution of the Poisson–Boltzmann equation for continuum electrostatics [52].

There are numerous generalizations of the leapfrog method for the Langevin equation. A popular choice is the BBK integrator [12]. An integrator based on the idea of splitting, and very similar to a discretization proposed in [124], is given here. For convenience we consider the common special case of the Langevin equation (8.1) where the friction tensor k_BTD^{-1} equals γM for some scalar $\gamma \geq 0$:

$$dq = vdt. \quad dv = M^{-1}F(q)dt - \gamma vdt + \sqrt{2\gamma k_BT}M^{-1/2}dW(t).$$

The fractional steps of the integration consist of half a kick, with just the drift term $M^{-1}F(q)dt$ active, followed by a "fluctuation" with all terms active except the drift term, followed by another half a kick. The method is

half-kick:
$$v^{n+\epsilon} = v^n + \frac{1}{2}\Delta tM^{-1}F^n,$$

fluctuate:
$$v^{n+1-\epsilon} = e^{-\gamma\Delta t}v^{n+\epsilon} + \sqrt{2\gamma k_BT}M^{-1/2}R_1^{n+1},$$
$$q^{n+1} = q^n + \Delta t\frac{1-e^{-\gamma\Delta t}}{\gamma\Delta t}v^{n+\epsilon} + \sqrt{2\gamma k_BT}\frac{1}{\gamma}M^{-1/2}R_2^{n+1},$$

half-kick:

$$F^{n+1} = F(q^{n+1}),$$

$$v^{n+1} = v^{n+1-\epsilon} + \frac{1}{2}\Delta t M^{-1}F^{n+1}.$$

where

$$R_1^{n+1} = \int_{t^n}^{t^{n+1}} e^{-\gamma(t^{n+1}-t)}dW(t), \quad R_2^{n+1} = \int_{t^n}^{t^{n+1}} (1 - e^{-\gamma(t^{n+1}-t)})dW(t).$$

The pair R_1^{n+1} and R_2^{n+1} are joint Gaussian random variables of zero mean and known covariance matrix C and can be generated as described in appendix B. The elements of the covariance matrix C are the two variances

$$c_{11} = \int_{t^n}^{t^{n+1}} \left(e^{-\gamma(t^{n+1}-t)} \right)^2 dt, \quad c_{22} = \int_{t^n}^{t^{n+1}} \left(1 - e^{-\gamma(t^{n+1}-t)} \right)^2 dt$$

and the covariance

$$c_{12} = c_{21} = \int_{t^n}^{t^{n+1}} e^{-\gamma(t^{n+1}-t)} \left(1 - e^{-\gamma(t^{n+1}-t)} \right) dt.$$

We choose $[R_1^{n+1}, R_2^{n+1}]^T = C_{1/2}[Z_1^{n+1}, Z_2^{n+1}]^T$ where Z_1^{n+1}, Z_2^{n+1} are independent Gaussian random numbers of mean 0 and variance 1 and where $C_{1/2}$ is any matrix satisfying $C_{1/2}C_{1/2}^T = C$.

Because of damping, Langevin dynamics is not so sensitive to instabilities, and multiple time stepping is especially advantageous, for example, using the integration method LN [6].

8.2. Brownian Dynamics

The stiffness in the Langevin equations due to the friction term suggests the possibility of a high friction limit approximation. This is attained by setting the masses M to zero in eq. (8.1) and the resulting equations are those of Brownian dynamics:

$$dq = \frac{1}{k_B T}D(q)F(q)dt + \sqrt{2}D_{1/2}(q)dW(t).$$

This approximation is derived in [24, sec. III]. The spatial scale of the trajectory grows as $\sqrt{t}$ but approximation error from neglecting the inertial term remains bounded as t increases, so Brownian dynamics gives a reasonable approximation over long distances.

The customary choice of numerical integrator is the Euler(–Maruyama) method [60, p. 305], introduced in this context by Ermak and McCammon [24]. This method advances coordinates for a time step Δt by the simple recipe[1]

---

[1] This can be obtained as the limit as $M \to 0$ of the method given for Langevin dynamics.

$$q^{n+1} - q^n = \frac{1}{k_{\mathrm{B}}T} D^n F^n \Delta t + \sqrt{2} D^n_{1/2} \sqrt{\Delta t} Z^{n+1}$$

where Z^{n+1} is a collection of independent random numbers from a Gaussian distribution with mean 0 and variance 1. The time step Δt is chosen so that the force on the particle changes by only a few percent during any step (10% or less according to [19]). The Euler method has *weak* order of accuracy 1 meaning that expected values computed from the trajectories have an error whose magnitude has an expected value of $O(\Delta t)$. There are other stochastic integrators with greater approximation power. The book [60] gives a variety of integrators of higher weak (and strong) order. For MD, weak order is the appropriate concept [60, sec. 17.2].

We conclude with an example illustrating the statistical context of molecular simulations: a number of reactions in the cell have rates that are limited by diffusion, and their rate constant can be calculated by computing large numbers of trajectories using Brownian dynamics. Assuming a dilute solution, only a pair of reactants need to be considered in the calculation. A practical way to get a rate constant from trajectories is the NAM technique, given in [85]. The simulation domain is restricted to intermolecular distances r less than some value r_{cut}, which is chosen large enough so that to a good approximation the probability of reaction beyond r_{cut} depends on only r. Simulations are initiated with the centers of the two reactants at a distance $b < r_{\mathrm{cut}}$ apart where b is sufficiently large that for $r \geq b$ the drift term $F(r)$ depends to a good approximation on only the distance r of separation. Initial values for rotational and internal degrees of freedom are chosen at random from their equilibrium distribution. Each trajectory is simulated until either reaction occurs or separation exceeds r_{cut}. Thousands of trajectories are computed to yield a probability of reaction β. The rate constant can then be calculated from β and a couple of one-dimensional integrals. The time it takes for a particle to reach the end of its trajectory can be arbitrarily long, so it is necessary to put a time limit on the duration of a trajectory and record as "unknown" the value of one which exceeds the time limit.

Acknowledgments

I am grateful to undergraduate student Tony Surma for Figs. 3.2–3.5; to graduate students David Hardy, Jesús Izaguirre, Hui Lu, James Phillips, and Gang Zou for contributing content and feedback that improved the manuscript; and to Ben Leimkuhler for also helping to improve the manuscript.

A. Lie Series and the BCH Formula

The main purpose of this appendix is to obtain a formula for the composition of Hamiltonian flows based on the Baker–Campbell–Hausdorff (BCH) formula for Lie operators. The details are given below. The conclusion is the following formula:

$$\phi_{(\Delta t/2)N} \circ \phi_{\Delta t H} \circ \phi_{(\Delta t/2)N} = \phi_{\Delta t H^+}$$

where

$$
\begin{aligned}
H^+ &= H + N + \frac{\Delta t^2}{12}\{H,H,N\} - \frac{\Delta t^2}{24}\{N,N,H\} \\
&+ \frac{\Delta t^4}{5760}\{N,N,N,N,H\} - \frac{\Delta t^4}{720}\{H,H,H,H,N\} \\
&+ \frac{\Delta t^4}{360}\{N,H,H,H,N\} + \frac{\Delta t^4}{360}\{H,N,N,N,N\} \\
&- \frac{\Delta t^4}{480}\{N,N,H,H,N\} + \frac{\Delta t^4}{120}\{H,H,N,N,H\} \\
&+ O(\Delta t^6).
\end{aligned}
$$

The Poisson bracket of two scalar fields is defined by

$$\{H,N\} = H_y^{\mathrm{T}} J N_y.$$

The first step is to express the flow defined by a general ODE system as the exponential of a Lie operator. Let $y(t)$ be a trajectory for an ODE system $(\mathrm{d}/\mathrm{d}t)y = f(y)$. Let $g(y)$ be a scalar function of y. Normally for small enough Δt one can write

$$g(y(t + \Delta t)) = \sum_{k=0}^{\infty} \frac{\Delta t^k}{k!} \frac{\mathrm{d}^k}{\mathrm{d}t^k} g(y(t)). \tag{A.1}$$

The idea is to use the ODE to generate the first and higher derivatives. Differentiating along a trajectory gives

$$\frac{\mathrm{d}}{\mathrm{d}t} g(y(t)) = \frac{\mathrm{d}}{\mathrm{d}t} y(t)^{\mathrm{T}} g_y(y(t)) = ((f^{\mathrm{T}}\nabla)g)(y(t)) \tag{A.2}$$

where the Lie operator $f^{\mathrm{T}}\nabla$ is defined by $(f^{\mathrm{T}}\nabla)g(y) = f(y)^{\mathrm{T}} g_y(y)$. It is straightforward to show by induction that

$$\frac{\mathrm{d}^k}{\mathrm{d}t^k} g(y(t)) = ((f^{\mathrm{T}}\nabla)^k g)(y(t))$$

where $(f^{\mathrm{T}}\nabla)^k$ is defined recursively by $(f^{\mathrm{T}}\nabla)^k g = (f^{\mathrm{T}}\nabla)((f^{\mathrm{T}}\nabla)^{k-1}g)$. Substituting the above into (A.1), we get

$$g(y(t + \Delta t)) = (\exp(\Delta t f^{\mathrm{T}}\nabla)g)(y(t)). \tag{A.3}$$

We generalize from $g(y)$ to a vector field by defining $f^{\mathrm{T}}\nabla$ component-wise. For the vector field $\mathrm{id}(y)$, eq. (A.3) gives the formula for the flow:

$$y(t + \Delta t) = (\exp(\Delta t f^{\mathrm{T}} \nabla) \mathrm{id})(y(t)).$$

Therefore, eq. (A.3) can be written

$$g \circ (\exp(\Delta t f^{\mathrm{T}} \nabla) \mathrm{id}) = \exp(\Delta t f^{\mathrm{T}} \nabla) g. \qquad (\mathrm{A.4})$$

Next, we explain how the BCH formula can be used to compose flows defined by general ODE systems. Two applications of the identity (A.4) yield

$$
\begin{aligned}
g \circ \exp(\Delta t f_2^{\mathrm{T}} \nabla) \mathrm{id} \circ \exp(\Delta t f_1^{\mathrm{T}} \nabla) \mathrm{id} &= \exp(\Delta t f_1^{\mathrm{T}} \nabla)(g \circ \exp(\Delta t f_2^{\mathrm{T}} \nabla) \mathrm{id}) \\
&= \exp(\Delta t f_1^{\mathrm{T}} \nabla) \exp(\Delta t f_2^{\mathrm{T}} \nabla) g,
\end{aligned}
$$

which gives us a way to write a composition of flows as a product of exponentials:

$$
\begin{aligned}
\exp(\Delta t f_k^{\mathrm{T}} \nabla) \mathrm{id} \circ \cdots \circ \exp(\Delta t f_2^{\mathrm{T}} \nabla) \mathrm{id} \circ \exp(\Delta t f_1^{\mathrm{T}} \nabla) \mathrm{id} \\
= \exp(\Delta t f_1^{\mathrm{T}} \nabla) \exp(\Delta t f_2^{\mathrm{T}} \nabla) \cdots \exp(\Delta t f_k^{\mathrm{T}} \nabla) \mathrm{id}.
\end{aligned}
$$

A product of exponentials can be combined using the Baker–Campbell–Hausdorff formula. Of most use to us is the special case of a symmetric product [98, p. 161]:

$$\exp(\tfrac{1}{2} B) \exp(A) \exp(\tfrac{1}{2} B) = \exp(C)$$

where

$$
\begin{aligned}
C &= A + B + \frac{1}{12}[A, A, B] - \frac{1}{24}[B, B, A] + \frac{7}{5760}[B, B, B, B, A] \\
&\quad - \frac{1}{720}[A, A, A, A, B] + \frac{1}{360}[B, A, A, A, B] + \frac{1}{360}[A, B, B, B, A] \\
&\quad - \frac{1}{480}[B, B, A, A, B] + \frac{1}{120}[A, A, B, B, A] + \cdots.
\end{aligned}
$$

The commutator of two operators is defined by $[A, B] = AB - BA$, and the iterated commutator is defined recursively by

$$[A_1, A_2, \ldots, A_k] = [A_1, [A_2, \ldots, A_k]].$$

Finally, we specialize the BCH formula to the case of a Hamiltonian flow $f(y) = J H_y(y)$ by noting the following homomorphism from scalar fields with the negative Poisson bracket to Lie operators with the commutator:

$$[(J H_y)^{\mathrm{T}} \nabla, (J N_y)^{\mathrm{T}} \nabla] = (J(-\{H, N\})_y)^{\mathrm{T}} \nabla.$$

B. Stochastic Processes

Some about stochastic processes which is helpful for understanding the derivation of simple numerical integrators is given here. Books on stochastic processes include [87, 60].

2.1. Wiener Processes

A *continuous time stochastic process* is a family of random variables $y(t)$ indexed for some continuous range of parameter values $0 \le t \le T$. For any outcome, $y(t)$ will be a function of t called a *realization* or *sample path* of the process. Many useful stochastic processes can be constructed from *Wiener processes*, and we restrict our attention to such processes.

A standard Wiener process $W(t)$, $t \ge 0$, can be defined [60, p. 28] as follows:

1. $W(0) = 0$ with probability 1 (w. p. 1),
2. $W(t + \Delta t) - W(t)$, $\Delta t > 0$, is independent of $W(\tau)$ for $\tau \le t$ and is Gaussian with mean zero and variance Δt.

Hence, the increment $Y = W(t + \Delta t) - W(t)$ has probability density function

$$\frac{1}{\sqrt{2\pi \Delta t}} \exp\left(-\frac{y^2}{2\Delta t} \right).$$

To make this more concrete, three methods for generating a Wiener process are given below. In all cases Z^n designates a sequence of independent standard Gaussian random variables (with mean 0 and variance 1).

1. $W(0) = 0$, $W(t^n) = W(t^{n-1}) + \sqrt{t^n - t^{n-1}} Z^n$.
2. In the case where $W(t)$ is being evaluated at a sequence of arbitrary points not known in advance and not necessarily increasing (such as might occur if it were being plotted by an adaptive procedure), then use

$$W(t^n) = \frac{b^n - t^n}{b^n - a^n} W(a^n) + \frac{t^n - a^n}{b^n - a^n} W(b^n) + \sqrt{\frac{(t^n - a^n)(b^n - t^n)}{b^n - a^n}} Z^n$$

where a^n is the point from the set $\{0, t^1, \ldots, t^{n-1}\}$ immediately to the left of t^n and b^n is the point from the same set immediately to the right of t^n.

3. The Karhunen-Loève expansion, good for a finite interval $[0, T]$, is

$$\sum_{n=0}^{\infty} Z^n \frac{2\sqrt{2T}}{(2n+1)\pi} \sin\left(\frac{(2n+1)\pi t}{2T} \right).$$

These constructions indicate that a realization of $W(t)$ is continuous but *not differentiable* at any given point t w. p. 1.

2.2. The Ito Integral

The construction of useful stochastic processes from the Wiener process requires integrals of the form

$$I(f) = \int_0^T f(t) \mathrm{d}W(t)$$

For a differentiable deterministic function $f(t)$ we can use the Riemann–Stieltjes integral. which is the limit as $\Delta t \to 0$ of

$$\sum_{n=1}^{N} f(\tau^n)(W(t^n) - W(t^{n-1}))$$

where $0 = t^0 < t^1 < \cdots < t^N = T$, $\Delta t = \max_n(t^n - t^{n-1})$, and each τ^n is an arbitrary value in the closed interval from t^{n-1} to t^n. We have that

- $I(f)$ is a Gaussian random variable,
- $E[I(f)] = 0$,
- $E[I(f)^2] = \int_0^T f(t)^2 \mathrm{d}t$,
- $E[I(f)I(g)] = \int_0^T f(t)g(t)\mathrm{d}t$.

We can generate $\int_0^T f(t)\mathrm{d}W(t)$ as $(\int_0^T f(t)^2 \mathrm{d}t)^{1/2} Z$. If other integrals of the form $\int_0^T g(t)\mathrm{d}W(t)$ are also to be generated, the covariance of these random variables has to be taken into account. Jointly Gaussian random variables $Y_1, Y_2, \ldots, Y_k$ with mean vector μ and covariance matrix C have joint p.d.f.

$$\rho(y) = (2\pi \det C)^{-k/2} \exp(-\frac{1}{2}(y - \mu)^{\mathrm{T}} C(y - \mu)).$$

The following formula generates appropriately random values:

$$Y = C_{1/2} Z + \mu$$

where Z is a vector of independent standard Gaussian random variables and $C_{1/2}$ is any matrix satisfying $C_{1/2} C_{1/2}^{\mathrm{T}} = C$, for example, a Choleski factor.

The Riemann–Stieltjes integral is inadequate for a non-differentiable stochastic integrand such as $W(t)$, because the result of the limit depends on the choice of the τ^n. The choice $\tau^n = t^{n-1}$, where the integrand is evaluated at the very beginning of each subinterval, is known as the Ito integral. It is computationally convenient [87, p. 37], [60, p. 228] because it is a *martingale* and yields an integral with expected value zero, although it may not be the most physically reasonable choice [60, p. 228]. For the integrand $W(t)$ the Ito integral is

$$\int_0^T W(t)\mathrm{d}W(t) = \lim_{\Delta t \to 0} \sum_{n=1}^N W(t^{n-1})(W(t^n) - W(t^{n-1}))$$

$$= \lim_{\Delta t \to 0} \sum_{n=1}^N \left(\frac{1}{2}W(t^n)^2 - \frac{1}{2}W(t^{n-1})^2 - \frac{1}{2}(W(t^n) - W(t^{n-1}))^2 \right)$$

$$= \frac{1}{2}W(T)^2 - \lim_{\Delta t \to 0} \frac{1}{2} \sum_{n=1}^N (t^n - t^{n-1})(Z^n)^2,$$

where the Z^n are independent Gaussian random variables, for which we have $E((Z^n)^2) = 1$ and $\mathrm{Var}((Z^n)^2) = 2$. Hence,

$$E\left(\frac{1}{2} \sum_{n=1}^N (t^n - t^{n-1})(Z^n)^2 \right) = \frac{1}{2}T,$$

$$\mathrm{Var}\left(\frac{1}{2} \sum_{n=1}^N (t^n - t^{n-1})(Z^n)^2 \right) = \sum_{n=1}^N (t^n - t^{n-1})^2 \le \Delta tT.$$

Therefore,

$$\int_0^T W(t)\mathrm{d}W(t) = \frac{1}{2}W(T)^2 - \frac{1}{2}T \quad \text{w.p.1.} \tag{B.1}$$

2.3. Stochastic Differential Equations

It is by means of stochastic differential equations that most interesting stochastic processes are generated. The general stochastic ODE has the form

$$\mathrm{d}y(t) = f(y(t), t)\mathrm{d}t + B(y(t), t)\mathrm{d}W(t), \tag{B.2}$$

a drift term plus a diffusion term. Because $W(t)$ is not differentiable, this should be understood as an integral equation. Perhaps the easiest way to understand its meaning is as the limit as $\Delta t \to 0$ of the Euler–Maruyama discretization given in section 8.2.

2.4. The Fokker–Planck Equation

Let $\rho(y, t)$ be the p.d.f. of $y(t)$:

$$\Pr(y(t) \in \Omega) = \int_\Omega \rho(y, t)\mathrm{d}y.$$

It satisfies the following generalization of the Liouville equation:

$$\rho_t + \nabla^{\mathrm{T}}(\rho f - \frac{1}{2}(\nabla^{\mathrm{T}}(\rho BB^{\mathrm{T}}))^{\mathrm{T}}) = 0$$

where $f = f(y, t)$ and $B = B(y, t)$.

2.5. The Ito Formula

The *Ito formula* [60, pp. 92, 97], [87, pp. 44, 49] provides for the Ito integral the needed generalizations of the fundamental theorem of calculus and of (an integral formulation of) the chain rule for differentiation. Following is a Lie operator version of the Ito formula [60, pp. 163, 177], which generalizes eq. (A.2) to a stochastic process $y(t)$ defined by the stochastic ODE (B.2):

$$g(y(T), T) = g(y(0), 0) + \int_0^T (L^0 g)(y(t), t) dt + \int_0^T ((Lg)(y(t), t))^T dW(t)$$

$$(B.3)$$

where $g(y, t)$ is any smooth scalar function,

$$L^0 g(y, t) = g_t(y, t) + g_y(y, t)^T f(y, t) + \frac{1}{2} \text{tr} \left(B(y, t)^T g_{yy}(y, t) B(y, t) \right),$$

$$(Lg(y, t))^T = g_y(y, t)^T B(y, t),$$

and tr denotes the trace. A more heuristic formulation of the Ito formula, easier to remember, is given by

$$dg = g_t dt + g_y^T dy + \frac{1}{2} (dy)^T g_{yy} (dy)$$

where

$$dy = f dt + B dW(t)$$

and

$$dt dt = 0, \quad dW(t) dt = 0, \quad dW(t)(dW(t))^T = dt I.$$

The Ito formula for the special case of a Wiener process $y(t) = W(t)$ ($f = 0$, $B = I$) is

$$g(W(T), T) = g(0, 0) + \int_0^T \left(g_t(W(t), t) + \frac{1}{2} \text{tr}(g_{yy}(W(t), t)) \right) dt$$

$$+ \int_0^T g_y(W(t), t)^T dW(t).$$

As examples, $g(y) = \frac{1}{2} y^2$ gives (B.1), and $g(y) = ty$ gives

$$TW(T) = \int_0^T W(t) dt + \int_0^T t dW(t).$$

2.6. Weak Ito–Taylor Expansions

The Lie series expansion of a solution to an ODE given in appendix A generalizes to the stochastic ODE (B.2) through the repeated use of the Ito formula (B.3). This avoids the direct differentiation of a stochastic process $y(t)$ with respect to time. For simplicity we consider the case of a scalar stochastic ODE. Application of the Ito formula (B.3) to $g(y, t) = y$ yields

$$y(T) = y(0) + \int_0^T (L^0 \mathrm{id})(y(t), t) \mathrm{d}t + \int_0^T (L \mathrm{id})(y(t), t)^{\mathrm{T}} \mathrm{d}W(t)$$

where

$$L^0 \mathrm{id} = f, \quad L \mathrm{id} = B.$$

Substituting the Ito formula for $g(y, t) = (L^0 \mathrm{id})(y, t)$ and for $g(y, t) = (L \mathrm{id})(y, t)$ into this equation gives

$$
\begin{aligned}
y(T) = {}& y(0) + \int_0^T \left((L^0 \mathrm{id})(y(0), 0) + \int_0^t (L^0 L^0 \mathrm{id})(y(\tau), \tau) \mathrm{d}\tau \right. \\
& \left. + \int_0^t (L L^0 \mathrm{id})(y(\tau), \tau) \mathrm{d}W(\tau) \right) \mathrm{d}t \\
& + \int_0^T \left((L \mathrm{id})(y(0), 0) + \int_0^t (L^0 L \mathrm{id})(y(\tau), \tau) \mathrm{d}\tau \right. \\
& \left. + \int_0^t (L L \mathrm{id})(y(\tau), \tau) \mathrm{d}W(\tau) \right) \mathrm{d}W(t)
\end{aligned}
$$

where

$$L^0 L^0 \mathrm{id} = f_t + f f_y + \frac{1}{2} B^2 f_{yy}, \quad L L^0 \mathrm{id} = B f_y,$$

$$L^0 L \mathrm{id} = B_t + f B_y + \frac{1}{2} B^2 B_{yy}, \quad L L \mathrm{id} = B B_y.$$

If we carry the expansion one step further and discard the triple integral remainer terms, we get a truncated Ito-Taylor expansion of weak order 2 [60, p. 211]:

$$
\begin{aligned}
y(T) = {}& y(0) + (L^0 \mathrm{id})(y(0), 0) \int_0^T \mathrm{d}t + (L \mathrm{id})(y(0), 0) \int_0^T \mathrm{d}W(t) \\
& + (L^0 L^0 \mathrm{id})(y(0), 0) \int_0^T \int_0^t \mathrm{d}\tau \mathrm{d}t + (L L^0 \mathrm{id})(y(0), 0) \int_0^T \int_0^t \mathrm{d}W(\tau) \mathrm{d}t \\
& + (L^0 L \mathrm{id})(y(0), 0) \int_0^T \int_0^t \mathrm{d}\tau \mathrm{d}W(t) + (L L \mathrm{id})(y(0), 0) \int_0^T \int_0^t \mathrm{d}W(\tau) \mathrm{d}W(t).
\end{aligned}
$$

Weak order of accuracy 2 means that expected values computed from the trajectories have an error of expected magnitude $O(\Delta t^2)$. To use this as a numerical method, note that

$$\int_0^T \mathrm{d}t = T,$$

$$\int_0^T \mathrm{d}W(t) =: R_1,$$

$$\int_0^T \int_0^t \mathrm{d}\tau \mathrm{d}t = \frac{1}{2}T^2,$$

$$\int_0^T \int_0^t \mathrm{d}W(\tau)\mathrm{d}t = TW(T) - \int_0^T t\mathrm{d}W(t) =: TR_1 - R_2,$$

$$\int_0^T \int_0^t \mathrm{d}\tau \mathrm{d}W(t) = \int_0^T t\mathrm{d}W(t) = R_2,$$

$$\int_0^T \int_0^t \mathrm{d}W(\tau)\mathrm{d}W(t) = \frac{1}{2}W(T)^2 - \frac{1}{2}T^2 = \frac{1}{2}R_1^2 - \frac{1}{2}T^2$$

where R_1 and R_2 can be computed as explained in sec. 8.1. It is possible to obtain both R_1 and R_2 with a single random number [60, p. XXXII] (when weak convergence is sufficient).

Bibliography

1. M. P. Allen and D. J. Tildesley. *Computer Simulation of Liquids.* Clarendon Press, Oxford, New York, 1987. Reprinted in paperback in 1989 with corrections.
2. H. C. Andersen. Molecular dynamics simulations at constant pressure and/or temperature. *J. Chem. Phys.*, 72:2384–2393, 1980.
3. H. C. Andersen. Rattle: A 'velocity' version of the shake algorithm for molecular dynamics calculations. *J. Comput. Phys.*, 52:24–34, 1983.
4. V. I. Arnol'd. *Mathematical Methods of Classical Mechanics.* Springer-Verlag, New York, second edition, 1989.
5. E. Barth, K. Kuczera, B. Leimkuhler, and R. D. Skeel. Algorithms for constrained molecular dynamics. *J. Comput. Chem.*, 16(10):1192–1209, Oct. 1995.
6. E. Barth, M. Mandziuk, and T. Schlick. A separating framework for increasing the timestep in molecular dynamics. In W. F. van Gunsteren, P. K. Weiner, and A. J. Wilkinson, editors, *Computer Simulation of Biomolecular Systems: Theoretical and Experimental Applications*, volume 3, chapter 4, pages 97–121. ESCOM, Leiden, The Netherlands, 1996.
7. G. Benettin and A. Giorgilli. On the Hamiltonian interpolation of near to the identity symplectic mappings with application to symplectic integration algorithms. *J. Statist. Phys.*, 74:1117–1143, 1994.
8. J. J. Biesiadecki and R. D. Skeel. Dangers of multiple-time-step methods. *J. Comput. Phys.*, 109(2):318–328, Dec. 1993.
9. T. Bishop, R. D. Skeel, and K. Schulten. Difficulties with multiple timestepping and the fast multipole algorithm in molecular dynamics. *J. Comput. Chem.*, 18(14):1785–1791, Nov. 15, 1997.
10. S. D. Bond, B. J. Leimkuhler, and B. B. Laird. The Nosé–Poincaré method for constant temperature molecular dynamics. Manuscript, 1998.

11. K. Brenan, S. Campbell, and L. Petzold. *Numerical Solution of Initial-Value Problems in Differential-Algebraic Equations*. North-Holland, 1989.

12. A. Brünger, C. B. Brooks, and M. Karplus. Stochastic boundary conditions for molecular dynamics simulations of ST2 water. *Chem. Phys. Lett.*, 105:495–500, 1982.

13. O. Buneman. Time-reversible difference procedures. *J. Comput. Phys.*, 1:517–535, 1967.

14. J. C. Butcher. The effective order of Runge–Kutta methods. In A. Dold, Z. Heidelberg, and B. Eckmann, editors, *Conference on the Numerical Solution of Differential Equations, Lecture Notes in Mathematics*, volume 109, pages 133–139. Springer-Verlag, New York, 1969.

15. M. Calvo and J. Sanz-Serna. The development of variable-step symplectic integrators, with applications to the two-body problem. *SIAM J. Sci. Statist. Comput.*, 14:936–952, 1993.

16. J. Candy and W. Rozmus. A symplectic integration algorithm for separable Hamiltonian functions. *J. Comput. Phys.*, 92:230–256, 1991.

17. P. J. Channell and J. C. Scovel. Symplectic integration of Hamiltonian systems. *Nonlinearity*, 3:231–259, 1990.

18. M. M. Chawla. On the order and attainable intervals of periodicity of explicit Nyström methods for $y'' = f(t, y)$. *SIAM J. Numer. Anal.*, 22:127–131, Feb. 1985.

19. M. E. Davis, J. D. Madura, B. A. Luty, and J. A. McCammon. Electrostatics and diffusion of molecules in solution: Simulations with the University of Houston Brownian dynamics program. *Computer Phys. Comms.*, 62:187–197, 1991.

20. R. De Vogelaére. Methods of integration which preserve the contact transformation property of Hamiltonian equations. Technical Report 4, Department of Mathematics, University of Notre Dame, 1956.

21. J. Delambre. *Mem. Acad. Turin*, 5:143, 1790–1793.

22. J. M. Deutsch and I. Oppenheim. *J. Chem. Phys.*, 54:3547. 1971.

23. A. Dullweber, B. Leimkuhler, and R. McLachlan. A symplectic splitting method for rigid-body molecular dynamics. *J. Chem. Phys.*, 107:5840, 1997.

24. D. L. Ermak and J. A. McCammon. Brownian dynamics with hydrodynamic interactions. *J. Chem. Phys.*, 69(4):1352–1360, Aug. 15, 1978.

25. D. J. Evans. On the representation of orientation space. *Mol. Phys.*, 34(2):317–325, 1977.

26. D. J. Evans and G. P. Moriss. Non-Newtonian molecular dynamics. *Comput. Phys. Rep.*, 1:297–344, 1984.

27. S. E. Feller, Y. Zhang, R. W. Pastor, and B. R. Brooks. Constant pressure molecular dynamics simulation: The Langevin piston method. *J. Chem. Phys.*, 103(11):4613–4621, 1995.

28. K. Feng. On difference schemes and symplectic geometry. In K. Feng, editor, *Proc. 1984 Beijing Symposium on Differential Geometry and Differential Equations—Computation of Differential Equations*, pages 42–58, Science Press, Beijing, 1985.

29. A. Fischer, F. Cordes, and C. Schütte. Hybrid Monte Carlo with adaptive temperature in a mixed-canonical ensemble: Efficient conformational analysis of rna. Technical Report SC 97-67, Konrad-Zuse-Zentrum für Informationstechnik Berlin, Dec. 1997. Available via http://www.zib.de/bib/pub/pw/.

30. M. Fixman. Simulation of polymer dynamics. I. General theory. *J. Chem. Phys.*, 69:1527–1537, 1978.

31. E. Forest and R. D. Ruth. Fourth-order symplectic integration. *Physica D*, 43:105–117, 1990.

32. T. Forester and W. Smith. On multiple time-step algorithms and the Ewald sum. *Mol. Sim.*, 13(3):195–204, 1994.
33. D. Frenkel and B. Smit. *Understanding Molecular Simulation: From Algorithms to Applications.* Academic Press, 1996.
34. B. García-Archilla, J. M. Sanz-Serna, and R. D. Skeel. Long-time-step methods for oscillatory differential equations. *SIAM J. Sci. Comput.* To appear. [Also Tech. Rept. 1996/7, Dep. Math. Aplic. Comput., Univ. Valladolid, Valladolid, Spain].
35. Z. Ge and J. E. Marsden. Lie–Poisson Hamilton–Jacobi theory and Lie–Poisson integrators. *Phys. Lett. A*, 133(3):134–139, 1988.
36. C. Gear and D. Wells. Multirate linear multistep methods. *BIT*, 24:484–502, 1984.
37. C. W. Gear. *Numerical Initial Value Problems in Ordinary Differential Equations.* Prentice-Hall, Englewood Cliffs, N. J., 1971.
38. D. Goldman and T. J. Kaper. Nth-order operator splitting schemes and non-reversible systems. *SIAM J. Numer. Anal.*, 33:349–367, 1996.
39. O. Gonzalez and J. Simo. On the stability of symplectic and energy–momentum algorithms for nonlinear Hamiltonian systems with symmetry. *Computer Methods in Applied Mechanics and Engineering*, 134:197–222, 1996.
40. H. Grubmüller. Dynamiksimulation sehr großer Makromoleküle auf einem Parallelrechner. Master's thesis, Physik-Dept. der Tech. Univ. München, Munich, 1989.
41. H. Grubmüller. *Molekular dynamik von Proteinen auf langen Zeitskalen.* PhD thesis, Physik-Dept. der Tech. Univ. München, Munich, 1994.
42. H. Grubmüller, H. Heller, A. Windemuth, and K. Schulten. Generalized Verlet algorithm for efficient molecular dynamics simulations with long-range interactions. *Molecular Simulation*, 6:121–142, 1991.
43. J. M. Haile. *Molecular Dynamics Simulation.* John Wiley and Sons, 1992.
44. E. Hairer. Backward error analysis of numerical integrators and symplectic methods. *Annals of Numer. Math.*, 1:107–132, 1994.
45. E. Hairer. Variable time step integration with symplectic methods. *Appl. Numer. Math.*, 25(2–3):219–227, Nov. 1997.
46. E. Hairer and P. Leone. Order barriers for symplectic multi-value methods. In D. Griffiths, D. Higham, and G. Watson, editors, *Proceedings of the 17th Dundee Biennial Conference, June 24–27, 1997*, volume 380 of *Pitman Research Notes in Mathematics*, pages 133–149, 1998.
47. E. Hairer and C. Lubich. The lifespan of backward error analysis for numerical integrators. *Numer. Math.*, 76:441–462, 1997.
48. E. Hairer and C. Lubich. Asymptotic expansions and backward analysis for numerical integrators. manuscript, 1998.
49. D. J. Hardy, D. I. Okunbor, and R. D. Skeel. Symplectic variable stepsize integration for N-body problems. *Appl. Numer. Math.*, 1998. To appear.
50. A. Hayli. Le problème des N corps dans un champ extérieur application a l'évolution dynamique des amas ouverts-I. *Bulletin Astronomique*, 2:67–89, 1967.
51. H. Heller, M. Schaefer, and K. Schulten. Molecular dynamics simulation of a bilayer of 200 lipids in the gel and in the liquid crystal-phases. *J. Phys. Chem.*, 97:8343–8360, 1993.
52. M. Holst, R. Kozack, F. Saied, and S. Subramaniam. Treatment of electrostatic effects in proteins: Multigrid-based Newton iterative method for solution of the full nonlinear Poisson–Boltzmann equation. *Proteins: Structure, Function, and Genetics*, 18(3):231–245, 1994.

53. J. Honerkamp. *Stochastic Dynamical Systems: Concepts, Numerical Methods, Data Analysis.* VCH, 1994.

54. W. G. Hoover. Canonical dynamics: Equilibrium phase-space distributions. *Phys. Rev. A*, 31:1695–1697, 1985.

55. W. Huang and B. Leimkuhler. The adaptive Verlet method. *SIAM J. Sci. Comput.*, 18:239–256, 1997.

56. D. D. Humphreys, R. A. Friesner, and B. J. Berne. A multiple-time-step molecular dynamics algorithm for macromolecules. *J. Phys. Chem.*, 98(27):6885–6892, July 7, 1994.

57. H. Ishida, Y. Nagai, and A. Kidera. Symplectic integrator for molecular dynamics of a protein in water. *Chem. Phys. Letts.*, 282(2):115–120, Jan. 9, 1998.

58. J. Izaguirre, S. Reich, and R. D. Skeel. Longer time steps for molecular dynamics. Submitted for publication.

59. D. Janežič and F. Merzel. An efficient symplectic integration algorithm for molecular dynamics simulations. *J. Chem. Inf. Comput. Sci.*, 35:321–326, 1995.

60. P. E. Kloeden and E. Platen. *Numerical Solution of Stochastic Differential Equations*, volume 23 of *Applications of Mathematics: Stochastic Modelling and Applied Probability*. Springer-Verlag, New York, 1992. Second corrected printing 1995.

61. P. E. Kloeden, E. Platen, and H. Schurz. *Numerical Solution of SDE Through Computer Experiments.* Springer-Verlag, 1994.

62. A. Kol, B. Laird, and B. Leimkuhler. A symplectic method for rigid-body molecular simulation. *J. Chem. Phys.*, 107(7), 1997.

63. J. D. Lambert and I. A. Watson. Symmetric multistep methods for periodic initial value problems. *J. Inst. Math. Applics.*, 18:189–202, 1976.

64. A. Lasota and M. C. Mackey. *Chaos, Fractals, and Noise: Stochastic Aspects of Dynamics.* Springer-Verlag, New York, second edition, 1994.

65. A. R. Leach. *Molecular Modelling: Principles and Applications.* Addison-Wesley Longman, Reading, Mass., July 1996.

66. B. Leimkuhler and S. Reich. The numerical solution of constrained Hamiltonian systems. *Math. Comput.*, 63:589–605, 1994.

67. B. Leimkuhler and R. D. Skeel. Symplectic numerical integrators in constrained Hamiltonian systems. *J. Comput. Phys.*, 112(1):117–125, May 1994.

68. B. J. Leimkuhler, S. Reich, and R. D. Skeel. Integration methods for molecular dynamics. In J. P. Mesirov, K. Schulten, and D. W. Sumners, editors, *Mathematical Approaches to Biomolecular Structure and Dynamics*, volume 82 of *IMA Volumes in Mathematics and its Applications*, pages 161–185. Springer-Verlag, 1996.

69. T. R. Littell, R. D. Skeel, and M. Zhang. Error analysis of symplectic multiple time stepping. *SIAM J. Numer. Anal.*, 34(5):1792–1807, Oct. 1997.

70. M. López-Marcos, J. M. Sanz-Serna, and R. D. Skeel. Cheap enhancement of symplectic integrators. In D. F. Griffiths and G. A. Watson, editors, *Numerical Analysis 1995*, pages 107–122, London, 1996. Longman Group.

71. M. López-Marcos, J. M. Sanz-Serna, and R. D. Skeel. Explicit symplectic integrators with maximal stability intervals. In D. F. Griffiths and G. A. Watson, editors, *Numerical Analysis, A. R. Mitchell 75th Birthday Volume*, pages 163–176, World Scientific, Singapore, June 1996.

72. M. López-Marcos, J. M. Sanz-Serna, and R. D. Skeel. Explicit symplectic integrators using Hessian–vector products. *SIAM J. Sci. Comput.*, 18:223–238, Jan. 1997.

73. M. Mandziuk and T. Schlick. Resonance in the dynamics of chemical systems simulated by the implicit midpoint scheme. *Chem. Phys. Letters*, 237:525–535, 1995.

74. G. J. Martyna. Remarks on 'constant-temperature molecular dynamics with momentum conservation'. *Phys. Rev. E*, 50(4):3234–3236, 1994.

75. G. J. Martyna, M. L. Klein, and M. Tuckerman. Nosé–Hoover chains: The canonical ensemble via continuous dynamics. *J. Chem. Phys.*, 97(5):2635–2643, 1992.

76. G. J. Martyna, D. J. Tobias, and M. L. Klein. Constant pressure molecular dynamics algorithms. *J. Chem. Phys*, 101(5):4177–4189, Sept. 1, 1994.

77. R. McLachlan and J. Scovel. Equivariant constrained symplectic integration. *Nonlinear Sci.*, 5:233–256, 1995.

78. R. I. McLachlan. Explicit Lie–Poisson integration and the Euler equations,. *Phys. Rev. Lett.*, 71:3043–3046, 1993.

79. R. I. McLachlan. More on symplectic correctors. In J. E. Marsden, G. W. Patrick, and W. F. Shadwick, editors, *Integration Algorithms and Classical Mechanics*, pp. 141–149, volume 10 of *Fields Institute Communications*. Fields Institute, American Mathematical Society, July 1996.

80. R. I. McLachlan and P. Atela. The accuracy of symplectic integrators. *Nonlinearity*, 5:541–562, March 1992.

81. R. I. McLachlan and S. K. Gray. Optimal stability polynomials for splitting methods, with application to the time-dependent Schrödinger equation. *Appl. Numer. Math.*, 25(2–3):275–286, Nov. 1997.

82. B. Mehlig, D. W. Heermann, and B. M. Forrest. Hybrid Monte Carlo method for condensed-matter systems. *Phys. Rev. B*, 45(2):679–685, Jan. 1, 1992.

83. S. Melchionna, G. Ciccotti, and B. L. Holian. Hoover NPT dynamics for systems varying in shape and size. *Mol. Phys.*, 78(3):533–544, 1993.

84. M. Nelson, W. Humphrey, A. Gursoy, A. Dalke, L. Kalé, R. D. Skeel, and K. Schulten. NAMD—a parallel, object-oriented molecular dynamics program. *Intl. J. Supercomput. Applics. High Performance Computing*, 10(4):251–268. Winter 1996.

85. S. H. Northrup, S. A. Allison, and J. A. McCammon. Brownian dynamics simulation of diffusion-influenced bimolecular reactions. *J. Chem. Phys.*, 80(4):1517–1524, 1984.

86. S. Nosé. A unified formulation of the constant temperature molecular dynamics methods. *J. Chem. Phys.*, 81:511, 1984.

87. B. Øksendal. *Stochastic Differential Equations: An Introduction with Applications*. Springer-Verlag, fifth edition, 1998.

88. D. Okunbor and R. D. Skeel. Explicit canonical methods for Hamiltonian systems. *Math. Comput.*, 59(200):439–455, Oct. 1992.

89. D. Okunbor and R. D. Skeel. Canonical numerical methods for molecular dynamics simulations. *J. Comput. Chem.*, 15(1):72–79, Jan. 1994.

90. A. Portillo and J. M. Sanz-Serna. Lack of dissipativity is not symplecticness. *BIT Numer. Math.*, 35(2):269–276, 1995.

91. G. D. Quinlan and S. Tremaine. Symmetric multistep methods for the numerical integration of planetary orbits. *Astron. J.*, 100:1694–1700, 1990.

92. S. Reich. Backward error analysis for numerical integrators. *SIAM J. Numer. Anal.*, to appear.

93. S. Reich. Momentum preserving symplectic integrators. *Physica D*, 76(4):375–383, Sept. 10, 1994.

94. S. Reich. A free energy approach to the torsion dynamics of macromolecules. Technical Report SC 95-17, Konrad-Zuse-Zentrum für Informationstechnik Berlin, June 1995.

95. S. Reich. Smoothed dynamics of highly oscillatory Hamiltonian systems. *Physica D*, 89(1 and 2):28–42, Dec. 21, 1995.
96. G. Rowlands. A numerical algorithm for Hamiltonian systems. *J. Comput. Phys.*, 97:235–239, Nov. 1991.
97. R. D. Ruth. A canonical integration technique. *IEEE Trans. Nucl. Sci.*, 30(4):2669–2671, 1983.
98. J. Sanz-Serna and M. Calvo. *Numerical Hamiltonian Problems.* Chapman and Hall, London, 1994.
99. J. M. Sanz-Serna. Runge-Kutta schemes for Hamiltonian systems. *BIT*, 28:877–883, 1988.
100. J. M. Sanz-Serna. Two topics in nonlinear stability. In W. Light, editor, *Advances in Numerical Analysis*, pages 147–174. Clarendon Press, Oxford, 1991.
101. J. M. Sanz-Serna. Symplectic integrators for Hamiltonian problem: an overview. *Acta Numerica*, 1:243–286, 1992.
102. T. Schlick, M. Mandziuk, R. D. Skeel, and K. Srinivas. Nonlinear resonance artifacts in molecular dynamics simulations. *J. Comput. Phys.*, 139:1–29, 1998.
103. C. Schütte, A. Fischer, W. Huisinga, and P. Deuflhard. A hybrid Monte Carlo method for essential molecular dynamics. Technical Report SC 98-04, Konrad-Zuse-Zentrum für Informationstechnik Berlin, Feb. 1998. Available via http://www.zib.de/bib/pub/pw/.
104. C. Scovel. On symplectic lattice maps. *Physics Letters A*, 159:396–400, Aug. 1991.
105. C. Scovel. Symplectic numerical integration of Hamiltonian systems. In T. Ratiu, editor, *The Geometry of Hamiltonian Systems: Proceedings of Workshop Held June 5-16, 1989*, pages 463–496. Mathematical Sciences Research Institute, Springer-Verlag, 1991.
106. Q. Sheng. *Solving Partial Differential Equations by Exponential Splitting.* PhD thesis, King's College, Cambridge, Oct. 1988.
107. J. C. Simo and N. Tarnow. A review of conserving algorithms for nonlinear dynamics. Technical Report SUDAM Report 92-7, Dept. Mechanical Engineering, Stanford Univ., Calif., 1992.
108. R. D. Skeel. Variable step size destabilizes the Störmer/leapfrog/Verlet method. *BIT*, 33:172–175, 1993.
109. R. D. Skeel. Symplectic integration with floating-point arithmetic and other approximations. *Appl. Numer. Math.*, 1998. To appear.
110. R. D. Skeel and J. J. Biesiadecki. Symplectic integration with variable stepsize. *Annals of Numer. Math.*, 1:191–198, 1994.
111. R. D. Skeel and C. W. Gear. Does variable step size ruin a symplectic integrator? *Physica D*, 60:311–313, 1992.
112. R. D. Skeel and J. Izaguirre. The five femtosecond time step barrier. In P. Deuflhard, J. Hermans, B. Leimkuhler, A. Mark, S. Reich, and R. D. Skeel, editors, *Algorithms for Macromolecular Modelling*, volume 4 of *Lecture Notes in Computational Science and Engineering*, pages 303–318. Springer-Verlag, 1998.
113. R. D. Skeel, G. Zhang, and T. Schlick. A family of symplectic integrators: stability, accuracy, and molecular dynamics applications. *SIAM J. Sci. Comput.*, 18(1):203–222, Jan. 1997.
114. D. M. Stoffer. *Some Geometric and Numerical Methods for Perturbed Integrable Systems.* PhD thesis, Swiss Federal Institute of Technology, Zürich, 1988.
115. C. Störmer. Sur les trajectoires des corpuscles életrisés. *Arch. Sci.*, 24:5–18, 113–158, 221–247, 1907.

116. W. B. Streett, D. Tildesley, and G. Saville. Multiple time step methods in molecular dynamics. *Mol. Phys.*, 35:639–648, 1978.
117. S. J. Stuart, R. H. Zhou, and B. J. Berne. Molecular dynamics with multiple time scales—the selection of efficient reference system propagators. *J. Chem. Phys.*, 105(4):1426—1436, July 22, 1996.
118. M. Suzuki. General theory of fractal path integrals with applications to many-body theories and statistical physics. *J. Math. Phys.*, 32(2), Feb. 1991.
119. M. Suzuki. Improved Trotter-like formula. *Physics Letters A*, 180(3), June 1993.
120. D. J. Tobias, G. J. Martyna, and M. L. Klein. Molecular dynamics simulations of a protein in the canonical ensemble. *J. Phys. Chem.*, 97(47):12959–12966, 1993.
121. S. Toxvaerd. Hamiltonians for discrete dynamics. *Phys. Rev. E*, 1995.
122. M. Tuckerman, B. J. Berne, and G. J. Martyna. Reversible multiple time scale molecular dynamics. *J. Chem. Phys*, 97(3):1990–2001, 1992.
123. W. F. van Gunsteren and H. J. C. Berendsen. Algorithms for macromolecular dynamics and constraint dynamics. *Molecular Phys*, 34:1311–1327, 1977.
124. W. F. van Gunsteren and H. J. C. Berendsen. A leap-frog algorithm for stochastic dynamics. *Molecular Simulation*, 1987.
125. W. F. van Gunsteren and H. J. C. Berendsen. *GROMOS Manual*. BIOMOS b. v., Lab. of Phys. Chem., Univ. of Groningen, 1987.
126. L. Verlet. Computer 'experiments' on condensed fluids I. thermodynamical properties of Lennard–Jones molecules. *Phys. Rev.*, 159:98–103, 1967.
127. R. F. Warming and B. J. Hyett. The modified equation approach to the stability and accuracy analysis of finite difference methods. *J. Comput. Phys*, 14:159–179, 1974.
128. M. Watanabe and M. Karplus. Simulation of macromolecules by multiple-time-step methods. *J. Phys. Chem.*, 99(15):5680–5697, Apr. 13, 1995.
129. J. Wisdom. The origin of the Kirkwood gaps: A mapping for asteroidal motion near the 3/1 commensurability. *Astr. J.*, 87:577–593, 1982.
130. J. Wisdom, M. Holman, and J. Touma. Symplectic correctors. In J. E. Marsden, G. W. Patrick, and W. F. Shadwick, editors, *Integration Algorithms and Classical Mechanics*, pp. 217–244, volume 10 of *Fields Institute Communications*. Fields Institute, American Mathematical Society, July 1996.
131. H. Yoshida. Construction of higher order symplectic integrators. *Phys. Lett. A*, 150:262–268, 1990.
132. M. Q. Zhang and R. D. Skeel. Symplectic integrators and the conservation of angular momentum. *J. Comput. Chem.*, 16:365–369, Mar. 1995.

Numerical Methods for Bifurcation Problems

Alastair Spence and Ivan G. Graham

Department of Mathematical Sciences, University of Bath
Claverton Down, Bath BA2 7AY, U.K.

1. Introduction

This set of lecture notes provides an introduction to the numerical solution
of bifurcation problems. The lectures are pitched at UK MSc level and the
theory is given for finite dimensional operators – so we shall require only
matrix theory, finite dimensional calculus, etc. Only the basic principles for
three of the most common bifurcations will be discussed, but the hope is
that after reading these notes a student should be able to tackle the original
journal papers. Almost all the results extend to infinite dimensional operators
defined in an appropriate setting, e.g. Banach or Hilbert Spaces.

There are many books on bifurcation theory – for example Chow & Hale
[2] gives an all-round treatment, Vanderbauwhede [48] gives an early account
of bifurcation in the presence of symmetries, and the important books by
Golubitsky & Schaeffer [9], and Golubitsky, Stewart & Schaeffer [10] look at
multiparameter bifurcation problems using singularity theory. An early conference proceedings is Rabinowitz [37], which contains one of the first papers
on the numerical *analysis* (as compared with numerical methods) for bifurcation problems written by H. B. Keller [25]. As might be expected, early books
about the numerical analysis of bifurcations were conference proceedings, see
Mittelmann & Weber [32], Küpper et al. [28], Küpper et al. [29], Roose et
al.[40] and Seydel et al. [46].

H.B. Keller's book [26] is a published version of lectures on Numerical
Methods in Bifurcation Problems delivered at the Indian Institute of Science, Bangalore. Rheinboldt's book [39] is a collection of his papers and also
gives information and listing of the code PITCON for numerical continuation
of parameter dependent nonlinear problems. The code AUTO, developed by
Doedel [7], but with recent extensions by several others, is now the lead-
ing piece of software for nonlinear systems, and can handle steady and time
dependent problems, and discretized boundary value problems. Seydel [45]
contains discussion of numerical methods and many interesting examples. A
comprehensive treatment, including a full discussion of numerical methods
using singularity theory, is to appear in the forthcoming book by Govaerts
[13]. Also, at a previous Summer School in Lancaster, Beyn [1] gave a survey
on numerical methods for dynamical systems, including methods for homo-
clinic and heteroclinic orbits. In fact, AUTO now has an option to compute
and follow paths of these orbits.

The plan of these notes is as follows. Section 2 contains three generic examples, namely a fold bifurcation, a Hopf bifurcation, and bifurcation from the trivial solution. Section 3 contains a brief account of Newton's method and the Implicit Function Theorem. Section 4 discusses the ideas behind Keller's pseudo-arclength numerical continuation algorithm [25]. Sections 5, 6 and 8 provide an introduction to the numerical analysis of the three types of bifurcation phenomena introduced in Section 2. Section 7 discusses bifurcation theory in nonlinear ODEs using results in Sections 5 and 6.

There are many phenomena not considered here, for example, bifurcation in the presence of symmetry (see for example [51] and the books [9], [10], [48]) and high order singularities in multiparameter problems (see [9], [10], [13], [23]).

This article is based on lectures given by the authors at the University of Bath, 1987-1998.

2. Examples

Bifurcation is the study of nonlinear problems with parameters, with the main interest being the determination of changes in solution behaviour as a parameter varies. In particular, interest centres on how to detect, calculate and classify points where there is a change in the type of solution of the nonlinear problem. This section contains some examples of some typical bifurcation phenomena.

In these notes we shall consider systems of the form

$$\boldsymbol{F}(\boldsymbol{x}, \lambda) = \boldsymbol{0} \tag{2.1}$$

where $\boldsymbol{F} : \mathbb{R}^{n+1} \to \mathbb{R}^n$, $\boldsymbol{x} \in \mathbb{R}^n$ is the *state* variable, and $\lambda \in \mathbb{R}$ is a *parameter*. We shall study the behaviour of $\boldsymbol{x}$ as λ varies, in fact, loosely speaking, λ may be thought of as the independent variable and $\boldsymbol{x}$ as the dependent variable.

Problems like (2.1) arise when studying autonomous systems of ordinary differential equations

$$\frac{d\boldsymbol{x}}{dt} = \boldsymbol{F}(\boldsymbol{x}, \lambda), \quad \boldsymbol{x}(0) \text{ given}, \ \boldsymbol{x}(t) \in \mathbb{R}^n. \tag{2.2}$$

Steady states (equilibria) of (2.2) are given by $\frac{d\boldsymbol{x}}{dt} = \boldsymbol{0}$, and hence satisfy (2.1). An important topic in the study of systems like (2.2) is the analysis of how the solutions change as λ varies and the determination of changes in stability. Often a first step is to find the steady states by solving (2.1) for a range of λ values and then determine any changes in stability of these steady states. This theme is described in the first example.

Example 2.1. Consider $f : \mathbb{R}^2 \to \mathbb{R}$ defined by

$$f(x, \lambda) = \lambda - x^2 = 0.$$

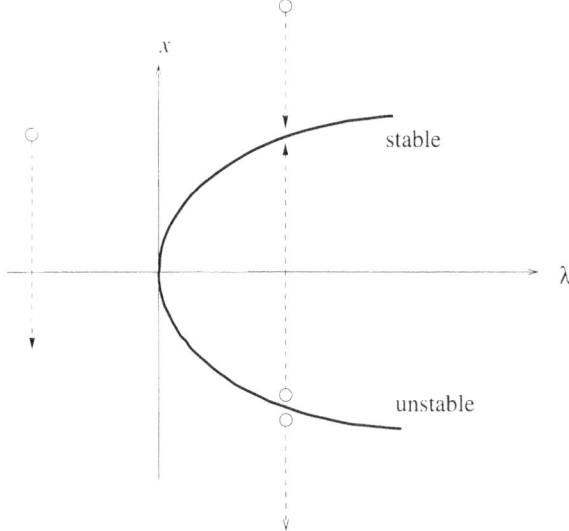

Fig. 2.1. Equilibria

For $\lambda < 0$ there are no real solutions; for $\lambda = 0$, $x = 0$ (twice); and for $\lambda > 0$ there are two solutions, $x = \pm\sqrt{\lambda}$. The steady solutions are shown by the solid line in Fig. 2.1. Consider now the solutions of the ODE

$$x_t = \lambda - x^2, \ x(t) \in \mathbb{R}, \ x(0) \text{ given.}$$

If $\lambda - x(0)^2 < 0$, then initially $x_t < 0$ and x decreases in time. If $\lambda - x(0)^2 > 0$ then $x_t > 0$ and x increases in time. The trajectories (dashed lines) for 4 different initial values (denoted ∘) are shown in Fig. 2.1. Thus for $\lambda > 0, x = \sqrt{\lambda}$ is a stable equilibrium, and for $\lambda < 0, x = -\sqrt{\lambda}$ is an unstable equilibrium. □

It is clear that even in this simple example, knowledge of the zeros of $f(x, \lambda) = 0$ helps us understand the behaviour of solutions of $x_t = f(x, \lambda)$. It is instructive before reading on to carry out a similar analysis for $x_t = \lambda x - x^3$.

The type of solution behaviour exhibited in Example 2.1 occurs in many physical examples. The point $(x, \lambda) = (0,0)$ is called a fold point (turning point or, in the dynamical systems literature, a saddle node) and we return to this kind of phenomenon in Section 5.

A very clear account of the stability of nonlinear systems is given in Chapter 9 of [20], where it is proved (p.187) that if x is a stable steady state for a given λ, then the Jacobian matrix $F_x(x, \lambda)$ (i.e. the matrix with the (i, j)th component $\frac{\partial F_i}{\partial x_j}(x, \lambda)$) has no eigenvalues with positive real part. Hence stability of a steady state of (2.2) is lost when one or more eigenvalues of $F_x(x, \lambda)$ moves into the right half-plane as λ varies. It is an instructive exercise to see in Example 2.1 how the eigenvalue of the (1×1) Jacobian matrix changes along the path of steady state solutions.

In Example 2.1 it was trivial to find the path of steady states analytically. In general the solutions to a nonlinear problem $\boldsymbol{F}(\boldsymbol{x}, \lambda) = \boldsymbol{0}$ will not be known analytically. In the following sections we shall describe how to compute such solution paths and recognise the parameter values at which the number of solutions changes.

The following example is two dimensional, and it is more convenient to use (x, y) rather than $\boldsymbol{x}$.

Example 2.2. Consider the pair of ODEs

$$
\begin{pmatrix} x_t \\ y_t \end{pmatrix} = \begin{pmatrix} \lambda & 1 \\ -1 & \lambda \end{pmatrix} \begin{pmatrix} x \\ y \end{pmatrix} - (x^2 + y^2) \begin{pmatrix} x \\ y \end{pmatrix}.
$$

Clearly $(x(t), y(t)) = \boldsymbol{0}$ is a solution for any $\lambda \in \mathbb{R}$. For any $\lambda > 0$ each $(x(t), y(t))^T$ satisfying $x(t) = \sqrt{\lambda} \sin t, y(t) = \sqrt{\lambda} \cos t$, is a nontrivial periodic solution. The Jacobian of the right hand side is

$$
\begin{pmatrix} \lambda & 1 \\ -1 & \lambda \end{pmatrix} - \begin{pmatrix} 3x^2 + y^2 & 2xy \\ 2xy & x^2 + 3y^2 \end{pmatrix}.
$$

At $(x, y) = (0, 0)$ the eigenvalues of this matrix are $\lambda \pm i$ and so the trivial solution is stable for $\lambda < 0$ and unstable for $\lambda > 0$. On the other hand a short calculation shows that the eigenvalues of this matrix on the above nontrivial solution are $-\lambda \pm \sqrt{\lambda^2 - 1}$, which always have negative real part when $\lambda > 0$ and so the periodic solution is stable. (See also §1 in [26].)

In summary: As λ passes through zero there is a birth of periodic orbits in (2.2). The eigenvalues of $\begin{pmatrix} \lambda & 1 \\ -1 & \lambda \end{pmatrix}$ are $\lambda \pm i$, and these are purely imaginary at $\lambda = 0$. This is a simple example of a *Hopf* bifurcation. We discuss this topic in §8.   □

Example 2.3. Consider the differential equation

$$
\frac{d^2 y}{dt^2} + \lambda \sin y = 0 \tag{2.3}
$$

where $\lambda > 0$ is given and $y(t)$ is to be found on $t \in [0, l]$ subject to the boundary conditions

$$
\frac{dy}{dt}(0) = 0 = \frac{dy}{dt}(l). \tag{2.4}
$$

This models the behaviour of an elastic rod occupying $0 \le t \le l$ which is fixed at each end and subject to a force λ in the direction of the rod (see [2], Chap. 1 for a figure and more detailed discussion).

Here $y(t)$ represents the angle the tangent to the rod at a distance t along the rod makes to the horizontal. Physically, as λ increases the rod can buckle. Obviously the trivial solution $y \equiv 0$ solves (2.3), (2.4) for all λ.

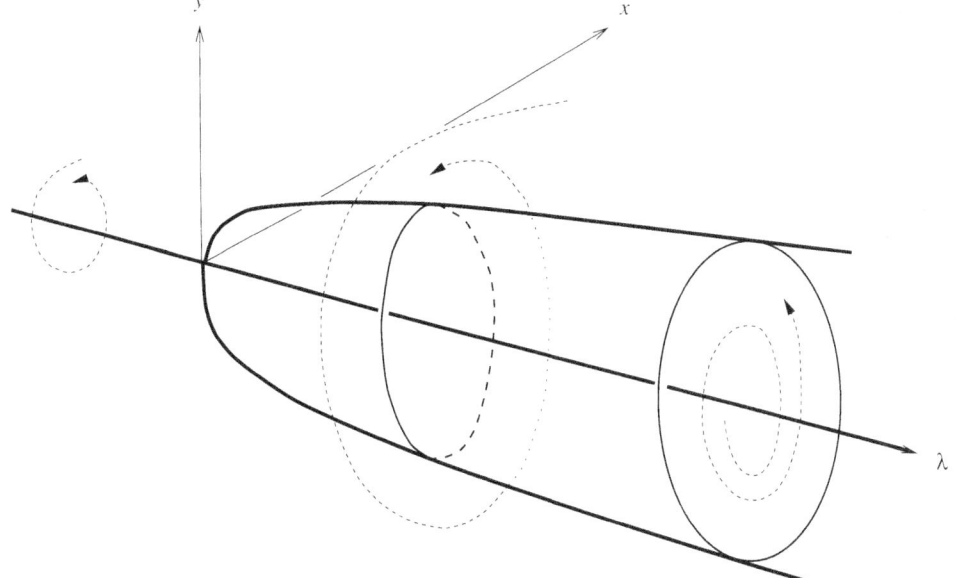

Fig. 2.2.

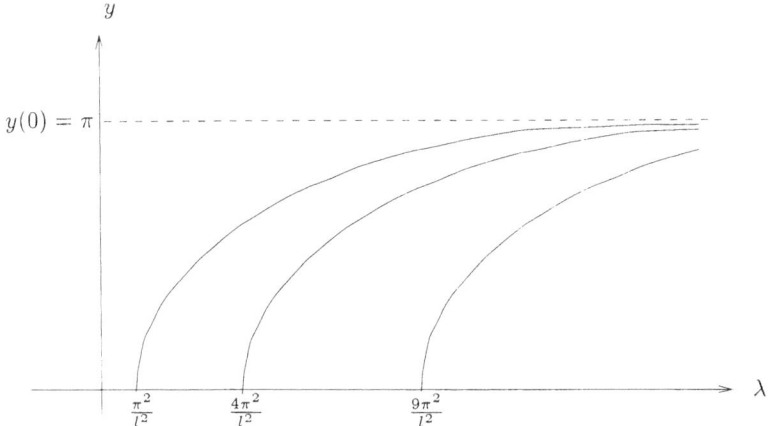

Fig. 2.3.

182   A. Spence and I.G. Graham

The interesting solution is the buckled state $y \not\equiv 0$. The differential equation is tractable to theoretical analysis (the first step is to multiply (2.3) by $\frac{dy}{dt}$ and integrate) and it is shown in [2] that nontrivial solutions emanate from $(y, \lambda) = (0, m^2 \pi^2 / l^2)$, $m = 1, 2, 3, \ldots$ as shown in Fig. 2.3. (The value of $y(0)$ is plotted on the vertical axis). The nontrivial branches correspond to buckled states. This is classical "bifurcation from the trivial solution" and it was the analysis of this and similar buckling problems that prompted the initial studies in bifurcation theory.

In practise we may compute buckled states by approximating (2.3), (2.4). One way to do this is to set $h = l/n$ ($n \in \mathbb{Z}$) and introduce the mesh

$$t_i = ih , \quad i = 0, \cdots, n$$

of equally spaced points on $[0, l]$. Then set $y_i = y(t_i)$ and approximate

$$\frac{d^2 y}{dt^2}(t_i) \quad \text{by} \quad \frac{1}{h^2} \{y_{i-1} - 2y_i + y_{i+1}\} .$$

Substituting in (2.3) and forcing equality gives the approximation

$$\frac{1}{h^2} \{Y_{i-1} - 2Y_i + Y_{i+1}\} + \lambda \sin Y_i = 0, \quad i = 1, \ldots, (n-1), \qquad (2.5)$$

where $Y_i \simeq y(t_i)$. We can approximate (2.4) by

$$\frac{Y_1 - Y_0}{h} = 0 = \frac{Y_n - Y_{n-1}}{h} . \qquad (2.6)$$

Now using (2.5) for $i = 1, .., (n-1)$ together with (2.6) yields the $(n-1)$ dimensional nonlinear system

$$
\begin{aligned}
\boldsymbol{F}(\boldsymbol{Y}, \lambda) \quad &=: \quad \boldsymbol{AY} - \lambda \sin(\boldsymbol{Y}) \\
&= \frac{1}{h^2}
\begin{bmatrix}
-1 & 1 & & & \\
1 & -2 & 1 & & \\
& & & & \\
& & 1 & -2 & 1 \\
& & & 1 & -1
\end{bmatrix}
\begin{bmatrix}
Y_1 \\
\\
\\
Y_{n-1}
\end{bmatrix}
+ \lambda
\begin{bmatrix}
\sin Y_1 \\
\\
\\
\sin Y_{n-1}
\end{bmatrix}
= \boldsymbol{0}. \; (2.7)
\end{aligned}
$$

Clearly $\boldsymbol{Y} = \boldsymbol{0}$ is always a solution. It is of interest to find out (i) For what λ do nontrivial $\boldsymbol{Y}$ exist? (ii) Do they approximate the λ given by the ordinary differential equation theory outlined above? (iii) What is the corresponding $\boldsymbol{Y}$? (iv) How would we compute $\boldsymbol{Y}$ as λ varies?   □

3. Newton's Method and the Implicit Function Theorem

The main computational tool to solve systems like (2.1) is *Newton's method*, which we discuss in §3.1. The main theoretical tool, which also has important numerical implications is the *Implicit Function Theorem* which we discuss in §3.2. Two applications of the Implicit Function Theorem are discussed in §3.3.

Recall that if $G : D \subset \mathbb{R}^n \to \mathbb{R}^n$, and $\| \cdot \|$ denotes a norm on $\mathbb{R}^n$ then G is called *Lipschitz continuous* (with respect to $\| \cdot \|$) if there exists $\gamma \in \mathbb{R}$, such that for all $x, y \in D$

$$\|G(x) - G(y)\| \le \gamma \|x - y\|, \tag{3.1}$$

and we write $G \in \mathrm{Lip}_\gamma(D)$. Throughout these lectures $\| \cdot \|$ will denote the Euclidean norm $\|x\| = \{x^T x\}^{1/2}$ on $\mathbb{R}^n$ and also the matrix norm induced by the Euclidean norm. With respect to this norm, $B(x, r)$ will denote the open ball in $\mathbb{R}^n$ with centre x and radius r, while $\bar{B}(x, r)$ denotes its closure.

3.1. Newton's Method for Systems

A very nice treatment of Newton's method for systems of nonlinear equations is given in [6]. To find a root, x_0 say, of $F(x) = 0$, Newton's method for a given starting guess x^0 is

$$x^{k+1} = x^k + d^k, \quad \text{where } F_x(x^k)d^k = -F(x^k), \qquad k \ge 0. \tag{3.2}$$

Theorem 3.1. *Assume*

(a) $F : \mathbb{R}^n \to \mathbb{R}^n$ *is continuously differentiable in an open convex set $D \subset \mathbb{R}^n$, and $F_x \in \mathrm{Lip}_\gamma(B(x_0, r))$, for some $r > 0$,*
(b) $F_x(x_0)$ *is nonsingular.*

Then provided x^0 satisfies $x^0 \in B(x_0, \epsilon)$, for small enough $\epsilon > 0$, Newton's method is well defined and $x^k \to x_0$ quadratically. (See [6] for a fuller account.)

In fact ϵ can be given explicitly as $\min\{r, (2\beta\gamma)^{-1}\}$, where $\beta = \|(F_x(x_0))^{-1}\|$, and this approaches 0 as $r \to 0$ or $\beta \to \infty$ or $\gamma \to \infty$.

It is interesting to formulate the matrix eigenvalue problem as a system of $(n+1)$ equations in $(n+1)$ unknowns, as is done in the following example.

Example 3.1. Let A be a real symmetric matrix, with simple eigenvalue μ_0 and corresponding eigenvector ϕ_0 satisfying $\phi_0^T \phi_0 = 1$. Consider the problem of computing (ϕ_0, μ_0) by Newton's method. Define $F : \mathbb{R}^{n-1} \to \mathbb{R}^{n-1}$ by

$$F(y) = \begin{pmatrix} A\phi - \mu\phi \\ \phi^T \phi - 1 \end{pmatrix} = 0, \qquad \text{where } y = \begin{pmatrix} \phi \\ \mu \end{pmatrix} \in \mathbb{R}^{n+1}. \tag{3.3}$$

If we apply Newton's method to compute the eigenpair $(\boldsymbol{\phi}^T, \mu)^T$ then the first step in verifying the convergence would be to check that the Jacobian matrix $\boldsymbol{F_y}$ is nonsingular. A simple calculation shows

$$\boldsymbol{F_y} = \begin{pmatrix} A - \mu I & -\boldsymbol{\phi} \\ 2\boldsymbol{\phi}^T & 0 \end{pmatrix}.$$

The proof that this is nonsingular at $\boldsymbol{y}_0^T = (\boldsymbol{\phi}_0^T, \mu_0)$ may be obtained directly or by application of case *(ii)* of the ABCD Lemma, which we now state.

Lemma 3.1 ("ABCD Lemma" (Keller [25])). *Given an $n \times n$ real matrix A, $\boldsymbol{c}$, $\boldsymbol{b} \in \mathbb{R}^n$, $d \in \mathbb{R}$, consider the $(n+1) \times (n+1)$ bordered matrix*

$$M = \begin{pmatrix} A & \boldsymbol{b} \\ \boldsymbol{c}^T & d \end{pmatrix}.$$

(i) If A is nonsingular then M is nonsingular if and only if $d - \boldsymbol{c}^T A^{-1} \boldsymbol{b} \neq 0$.
(ii) If $\mathrm{rank}(A) = n - 1$, M is nonsingular if and only if

$$\boldsymbol{\psi}^T \boldsymbol{b} \neq 0 \text{ for all } \boldsymbol{\psi} \in \ker(A^T) \setminus \{\boldsymbol{0}\},$$

and

$$\boldsymbol{c}^T \boldsymbol{\phi} \neq 0 \text{ for all } \boldsymbol{\phi} \in \ker(A) \setminus \{\boldsymbol{0}\}.$$

(iii) If $\mathrm{rank}(A) \leq n - 2$, then M is singular.

Proof. The proofs of parts *(i)* and *(iii)* are left as an exercise. In part *(ii)* to prove the "if" part consider $M \begin{pmatrix} \boldsymbol{x} \\ y \end{pmatrix} = \begin{pmatrix} \boldsymbol{0} \\ 0 \end{pmatrix}$. The first equation is $A\boldsymbol{x} + \boldsymbol{b}y = \boldsymbol{0}$, and multiplication on the left by $\boldsymbol{\psi}^T$ gives $y = 0$. Thus $A\boldsymbol{x} = \boldsymbol{0}$ and so $\boldsymbol{x} = \alpha\boldsymbol{\phi}$, for some non zero α. The second equation provides $\boldsymbol{c}^T(\alpha\boldsymbol{\phi}) = 0$ and hence $\alpha = 0$. Thus $\boldsymbol{x} = \boldsymbol{0}$. To prove the "only if" part assume $\boldsymbol{c}^T \boldsymbol{\phi} = 0$. Thus $(\boldsymbol{\phi}^T, 0)^T \in \ker(M)$ and so M is singular. Similarly if $\boldsymbol{\psi}^T \boldsymbol{b} = 0$ then $(\boldsymbol{\psi}^T, 0)^T \in \ker(M^T)$. □

Clearly different normalisations for the eigenvectors are possible. Replacing $\boldsymbol{\phi}^T \boldsymbol{\phi} = 1$ in (3.3) with $\boldsymbol{e}_r^T \boldsymbol{\phi} = 1$, where $\boldsymbol{e}_r$ is the unit vector with $(\boldsymbol{e}_r)_i = \delta_{ir}$, one can show that Newton's method applied to the eigenvalue problem can be interpreted as a version of inverse iteration (see [47] for more details).

3.2. The Implicit Function Theorem

The Implicit Function Theorem is obtained as an application of the Contraction Mapping Theorem to a nonlinear system with a parameter. So, let us first recall the Contraction Mapping Theorem.

Theorem 3.2 (Contraction Mapping Theorem). *Suppose*

(i) $\boldsymbol{G} \in \mathrm{Lip}_\alpha(\bar{B}(\boldsymbol{x}_0, r))$ *for some* $r > 0$, *with* $0 \le \alpha < 1$;
(ii) $\|\boldsymbol{x}_0 - \boldsymbol{G}(\boldsymbol{x}_0)\| \le (1 - \alpha)r$.

Then

(a) For all $\boldsymbol{x}^0 \in \bar{B}(\boldsymbol{x}_0, r)$, *the sequence* $\boldsymbol{x}^k$ *defined by* $\boldsymbol{x}^{k+1} = \boldsymbol{G}(\boldsymbol{x}^k)$ *converges to a limit* $\boldsymbol{x}^\star \in \bar{B}(\boldsymbol{x}_0, r)$;
(b) $\boldsymbol{x}^\star$ *is the unique fixed point of* $\boldsymbol{G}$ *in* $\bar{B}(\boldsymbol{x}_0, r)$.

The proof is in most books on nonlinear equations. Other versions of this theorem replace the assumption *(ii)* with the requirement that $\boldsymbol{G}(\bar{B}(\boldsymbol{x}_0, r)) \subseteq \bar{B}(\boldsymbol{x}_0, r)$. For numerical analysis purposes the present version is better since checking *(ii)* requires only checking that $\boldsymbol{G}(\boldsymbol{x}_0)$ should not be too far from $\boldsymbol{x}_0$. The proof of this version of the Contraction Mapping Theorem is in [38].

The contraction mapping theorem has many uses. One example of its use is in the analysis of the modified Newton method, given by the fixed point iteration $\boldsymbol{x}^{k+1} = \boldsymbol{G}(\boldsymbol{x}^k)$, where

$$\boldsymbol{G}(\boldsymbol{x}) = \boldsymbol{x} - \boldsymbol{F}_{\boldsymbol{x}}(\boldsymbol{x}^0)^{-1} \boldsymbol{F}(\boldsymbol{x}). \tag{3.4}$$

Using the Contraction Mapping Theorem it can be shown that if $\boldsymbol{x}^0$ is sufficiently close to a solution $\boldsymbol{x}_0$ of $\boldsymbol{F}(\boldsymbol{x}) = \boldsymbol{0}$ then $\boldsymbol{x}^k \to \boldsymbol{x}_0$ linearly as $k \to \infty$.

Another example of its use is in the proof of the Implicit Function Theorem. Consider the parameter dependent problem

$$\boldsymbol{F}(\boldsymbol{x}, \lambda) = \boldsymbol{0},$$

for $(\boldsymbol{x}, \lambda) \in D$, where $\boldsymbol{F} : D \subset \mathbb{R}^{n+1} \to \mathbb{R}^n$. Let S be the solution set

$$S = \{(\boldsymbol{x}, \lambda) \in D : \quad \boldsymbol{F}(\boldsymbol{x}, \lambda) = \boldsymbol{0}\}.$$

It is natural to ask the following question. If $(\boldsymbol{x}_0, \lambda_0) \in S$ and λ is near λ_0, is there a corresponding unique $\boldsymbol{x}(\lambda)$ such that $(\boldsymbol{x}(\lambda), \lambda) \in S$ and $\boldsymbol{x}(\lambda_0) = \boldsymbol{x}_0$? If so, we say $\boldsymbol{x}$ is *parametrised* by λ near $(\boldsymbol{x}_0, \lambda_0)$, written "$\boldsymbol{x} = \boldsymbol{x}(\lambda)$ near $(\boldsymbol{x}_0, \lambda_0)$". The Implicit Function Theorem provides the answer, but first consider a simple example.

Example 3.2. Consider

$$f(x, \lambda) = x^2 + \lambda^2 - 1.$$

Clearly if $|x_0| < 1$ and $(x_0, \lambda_0) \in S$ then $x = x(\lambda)$ near (x_0, λ_0).

Since $\boldsymbol{F}$ now depends on $\boldsymbol{x}$ and λ, we use the notation $\boldsymbol{F}_{\boldsymbol{x}}$ to mean the $n \times n$ matrix with (i, j)th element $\dfrac{\partial F_i}{\partial x_j}$, and by $\boldsymbol{F}_\lambda$ we mean the $n \times 1$ vector with elements $\dfrac{\partial F_i}{\partial \lambda}$. If $(\boldsymbol{x}_0, \lambda_0) \in D$ we write

$$\begin{aligned} \boldsymbol{F}^0 &= \boldsymbol{F}(\boldsymbol{x}_0, \lambda_0), \quad \boldsymbol{F}_{\boldsymbol{x}}^0 = \boldsymbol{F}_{\boldsymbol{x}}(\boldsymbol{x}_0, \lambda_0), \\ \boldsymbol{F}_\lambda^0 &= \boldsymbol{F}_\lambda(\boldsymbol{x}_0, \lambda_0), \quad \text{etc}. \end{aligned}$$

The Implicit Function Theorem

In the proof of this theorem we assume that for all $(\boldsymbol{x}, \lambda), (\boldsymbol{x}, \mu), (\boldsymbol{y}, \lambda) \in D$,

$$(A1) \quad \|\boldsymbol{F}(\boldsymbol{x}, \lambda) - \boldsymbol{F}(\boldsymbol{x}, \mu)\| \leq \sigma_2 |\lambda - \mu|$$
$$(A2) \quad \|\boldsymbol{F}_{\boldsymbol{x}}(\boldsymbol{x}, \lambda) - \boldsymbol{F}_{\boldsymbol{x}}(\boldsymbol{y}, \lambda)\|_2 \leq \gamma_1 \|\boldsymbol{x} - \boldsymbol{y}\|$$
$$(A3) \quad \|\boldsymbol{F}_{\boldsymbol{x}}(\boldsymbol{x}, \lambda) - \boldsymbol{F}_{\boldsymbol{x}}(\boldsymbol{x}, \mu)\|_2 \leq \gamma_2 |\lambda - \mu|.$$

Clearly (A1–A3) hold if $\boldsymbol{F}$ has two continuous derivatives with respect to $(\boldsymbol{x}, \lambda) \in D$. In many applications in fact $\boldsymbol{F}$ will be infinitely continuously differentiable on D, which we write as $\boldsymbol{F} \in C^\infty(D)$.

Theorem 3.3. *(Implicit Function Theorem) Suppose (A1–A3) hold and suppose there exists $(\boldsymbol{x}_0, \lambda_0) \in D$ such that*

(A4) $\boldsymbol{F}(\boldsymbol{x}_0, \lambda_0) = \boldsymbol{0}$,
(A5) $\boldsymbol{F}_{\boldsymbol{x}}(\boldsymbol{x}_0, \lambda_0)$ is nonsingular.

Then there exist neighbourhoods $B(\lambda_0, \varepsilon_\lambda)$, $B(\boldsymbol{x}_0, \varepsilon_x)$ of λ_0, $\boldsymbol{x}_0$ such that for all $\lambda \in B(\lambda_0, \varepsilon_\lambda)$ there exists $\boldsymbol{x}(\lambda) \in B(\boldsymbol{x}_0, \varepsilon_x)$ with

(a) $\boldsymbol{F}(\boldsymbol{x}(\lambda), \lambda) = \boldsymbol{0}$,
(b) $\boldsymbol{x}(\lambda)$ is the unique solution of $\boldsymbol{F}(\boldsymbol{x}, \lambda) = \boldsymbol{0}$ in $B(\boldsymbol{x}_0, \varepsilon_x)$,
(c) $\boldsymbol{x}(\lambda_0) = \boldsymbol{x}_0$,
(d) $\boldsymbol{F}_{\boldsymbol{x}}(\boldsymbol{x}(\lambda), \lambda)$ is nonsingular for all $\lambda \in B(\lambda_0, \varepsilon_\lambda)$,
(e) $\boldsymbol{x}(\lambda)$ is continuous with respect to $\lambda \in B(\lambda_0, \varepsilon_\lambda)$.

Remark If $\boldsymbol{F} \in C^\infty(D)$ then (e) can be replaced by $\boldsymbol{x} \in C^\infty(B(\lambda_0, \varepsilon_\lambda))$.

Proof. The proof of *(a),(b)* and *(c)* uses the Contraction Mapping Theorem applied to $\boldsymbol{K}(\boldsymbol{x}, \lambda) := \boldsymbol{x} - \left(\boldsymbol{F}_{\boldsymbol{x}}^0\right)^{-1} \boldsymbol{F}(\boldsymbol{x}, \lambda)$, which is very like the form of the mapping $\boldsymbol{G}$ used in the theory of the modified Newton method (3.4). Part *(d)* follows using 3.1.4 in [6], and *(e)* by standard manipulation. $\square$

With respect to the parameter dependent problem we make the following definition.

Definition 3.1. $(\boldsymbol{x}_0, \lambda_0) \in S$ *is called a regular point of S if $\boldsymbol{F}_{\boldsymbol{x}}^0$ is nonsingular. The Implicit Function Theorem can then be applied to show $\boldsymbol{x} = \boldsymbol{x}(\lambda)$ near $(\boldsymbol{x}_0, \lambda_0)$. If a point $(\boldsymbol{x}_0, \lambda_0) \in S$ is not regular it is called a singular point.*

Example 3.3. Consider

$$\boldsymbol{F}(\boldsymbol{x}, \lambda) = \begin{bmatrix} x_1^2 + x_2^2 - \lambda \\ x_2^2 - 2x_1 + 1 \end{bmatrix}.$$

It is clear that the solution set S is the intersection of the circle centred on the origin with radius $\sqrt{\lambda}$, and a parabola. (It is helpful to draw a sketch.)

Clearly $\boldsymbol{F} \in \mathcal{C}^{\infty}(\mathbb{R}^{n+1})$ and $\boldsymbol{F_x} = \begin{bmatrix} 2x_1 & 2x_2 \\ -2 & 2x_2 \end{bmatrix}$. Consider $(\boldsymbol{x}_0, \lambda_0) :=$
$(0.73, 0.68, 1)$. Then $(\boldsymbol{x}_0, \lambda_0) \in S$ and $\det(\boldsymbol{F_x^0}) = 4(x_0)_1(x_0)_2 + 4(x_0)_2 \neq 0$.
So $(\boldsymbol{x}_0, \lambda_0)$ is a regular point and the Implicit Function Theorem shows that
$\boldsymbol{x} = \boldsymbol{x}(\lambda)$ near $(\boldsymbol{x}_0, \lambda_0)$. Consider instead $(\boldsymbol{x}_0, \lambda_0) = (1/2, 0, 1/4) \in S$. Then
$\boldsymbol{F_x^0}$ is found to be singular. So $(\boldsymbol{x}_0, \lambda_0)$ is a singular point and we cannot
conclude that $\boldsymbol{x} = \boldsymbol{x}(\lambda)$ near $(\boldsymbol{x}_0, \lambda_0)$. (Plotting the path of solutions $\boldsymbol{x}(\lambda)$
against λ shows why not.) The difficulty here is simply that the solution set
turns around at $\lambda = \lambda_0 = \frac{1}{4}$.

This is a special type of singular point called a fold or turning point.

Definition 3.2. *If $(\boldsymbol{x}_0, \lambda_0) \in S$ is a singular point and if $\text{Rank}(\boldsymbol{F_x^0}) = n - 1$
then $(\boldsymbol{x}_0, \lambda)$ is called a fold point (or turning point) if $\boldsymbol{F}_\lambda^0 \notin \text{Image}(\boldsymbol{F_x^0})$.
In this case the $n \times (n+1)$ augmented Jacobian $[\boldsymbol{F_x^0}|\boldsymbol{F_\lambda^0}]$ must have rank n
and hence has a subset of n linearly independent columns. By selecting the
variables corresponding to these columns as the dependent variables we can
still apply the Implicit Function Theorem.*

Example 3.4. Consider again Example 3.3. $(\boldsymbol{x}_0, \lambda_0) = (1/2, 0, 1/4)$, and hence

$$[\boldsymbol{F_x^0}|\boldsymbol{F_\lambda^0}] = \begin{bmatrix} 1 & 0 & -1 \\ -2 & 0 & 0 \end{bmatrix},$$

which has full rank. The first and third columns are linearly independent so
if we write

$$\boldsymbol{G}(\boldsymbol{y}, x_2) = \begin{bmatrix} x_1^2 - \lambda + x_2^2 \\ -2x_1 + 1 + x_2^2 \end{bmatrix}.$$

The solution set for $\boldsymbol{G} = \boldsymbol{0}$ is identical to the solution of $\boldsymbol{F} = \boldsymbol{0}$ but x_2 is now
considered to be a parameter and $\boldsymbol{y} = (x_1, \lambda)$. Then $\boldsymbol{G_y} = \begin{bmatrix} 2x_1 & -1 \\ -2 & 0 \end{bmatrix}$,
which is nonsingular at $(\boldsymbol{y}_0, (x_2)_0) = (\frac{1}{2}, \frac{1}{4}, 0)$ so the Implicit Function Theorem shows that $\boldsymbol{y} = \boldsymbol{y}(x_2)$ near $(\boldsymbol{y}_0, (x_2)_0)$. $\qquad \square$

This example shows that change of parametrisation can remove the problems of a fold point. If a singular point is not a fold point, further analysis is
required (see §6).

3.3. Two Examples

We now give two examples.

Example 3.5. (See Example 2.3) Consider the $n - 1$ dimensional nonlinear
system with $\boldsymbol{F} : \mathbb{R}^n \to \mathbb{R}^{n-1}$ given in (2.7). Clearly $(\boldsymbol{Y}_0, \lambda_0) := (\boldsymbol{0}, \lambda_0) \in S$,
for all $\lambda_0 \in \mathbb{R}$, and

$$[\boldsymbol{F_Y}|\boldsymbol{F_\lambda}] \;=\; [A - \lambda \mathrm{diag}(\cos \boldsymbol{Y})|\sin \boldsymbol{Y}]$$
$$=\; [A - \lambda_0 I|\boldsymbol{0}] \text{ at } (\boldsymbol{Y}_0, \lambda_0).$$

Now A has the $(n-1)$ eigenvalues $\mu_k = \frac{n^2}{l^2}\left(2 - 2\cos\frac{k\pi}{(n-1)}\right)$, (with corresponding eigenfunctions $\boldsymbol{x}^k$, with $x_j^k = \cos(k\pi(2j-1)/2(n-1))$) for $k = 0, ..., (n-2)$, which are distinct and hence simple. So if $\lambda_0 \neq -\mu_k$ for any k then $(\boldsymbol{Y}_0, \lambda_0)$ is a regular point, and so near λ_0 we have $\boldsymbol{Y} = \boldsymbol{Y}(\lambda)$. But if $\lambda_0 = \mu_k$, then $\mathrm{Rank}(A - \lambda_0 I) = n - 2$, so $(\boldsymbol{Y}_0, \lambda_0)$ is a singular point. In addition $\boldsymbol{F}_\lambda^0 = \boldsymbol{0} \in \mathrm{Image}(A - \lambda_0 I) = \mathrm{Image}(\boldsymbol{F}_{\boldsymbol{Y}}^0)$, so $(\boldsymbol{Y}_0, \lambda_0)$ is not a fold point either.

Example 3.6. (Perturbation theory for algebraically simple eigenvalues.)
Let A be a real $n \times n$ matrix with a simple eigenvalue μ_0 (i.e. algebraic multiplicity is 1) and corresponding eigenvector ϕ_0. If A is perturbed to $A + \epsilon B$, one question is to find the dominant term in the perturbation of μ_0.

Start the perturbation theory by considering the nonlinear system (cf. Example 3.1 but without the assumption that A is symmetric)

$$\boldsymbol{F}(\boldsymbol{y}, \epsilon) := \left(\begin{array}{c} (A + \epsilon B)\phi - \mu\phi \\ \phi^T\phi - 1 \end{array} \right) = \left(\begin{array}{c} \boldsymbol{0} \\ 0 \end{array} \right), \quad \boldsymbol{y} = \left(\begin{array}{c} \phi \\ \mu \end{array} \right). \tag{3.5}$$

Clearly with $\boldsymbol{y}_0 = (\phi_0^T, \mu_0)^T$, $\boldsymbol{F}(\boldsymbol{y}_0, 0) = \boldsymbol{0}$, and $\boldsymbol{F_y}(\boldsymbol{y}_0, 0)$ is nonsingular. (This is proved using part *(ii)* of Lemma 3.1, though note that the condition of algebraic simplicity is needed since A is no longer assumed symmetric.) Thus using the Implicit Function Theorem, for small $|\epsilon|$ there exists a unique $\boldsymbol{y}(\epsilon)$ such that $\boldsymbol{F}(\boldsymbol{y}(\epsilon), \epsilon) = 0$ and $\boldsymbol{F_y}(\boldsymbol{y}(\epsilon), \epsilon)$ is nonsingular. The latter result ensures that $\mu(\epsilon)$ is simple. Since $\mu(\epsilon) \in C^\infty(\mathbb{R})$ we can write $\mu(\epsilon) = \mu_0 + \epsilon\mu'(0) + \mathcal{O}(\epsilon^2)$. To find the dominant term in the error we need to find $\mu'(0)$. To do this we differentiate $(A + \epsilon B)\phi(\epsilon) = \mu(\epsilon)\phi(\epsilon)$ with respect to ϵ, set $\epsilon = 0$, and multiply on the left by $\psi_0 \in \ker((A - \mu_0 I)^T) \setminus \{\boldsymbol{0}\}$. This leads to

$$\mu'(0) = \psi_0^T B\phi_0 / \psi_0^T \phi_0.$$

(Note that $\psi_0^T\phi_0 \neq 0$. If $\psi_0^T\phi_0 = 0$, then $\phi_0 \in \mathrm{Image}(A - \mu_0 I)$ and $\dim(\ker(A - \mu_0 I)^2) > 1$, contradicting the assumption of algebraic simplicity.) If A is symmetric then $\mu'(0) = \phi_0^T B\phi_0 / \phi_0^T \phi_0$.

4. Computation of Solution Paths

In this section we consider the general problem

$$\boldsymbol{F}(\boldsymbol{x}, \lambda) = \boldsymbol{0}, \tag{4.1}$$

where $\boldsymbol{F} : \mathbb{R}^{n+1} \to \mathbb{R}$, $\boldsymbol{F} \in C^\infty(\mathbb{R}^{n+1})$. Set

$$S = \{(\boldsymbol{x}, \lambda) \in \mathbb{R}^{n+1} : \boldsymbol{F}(\boldsymbol{x}, \lambda) = \boldsymbol{0}\}. \tag{4.2}$$

Often in applications one is interested in computing the whole set S or a continuous portion of it. For example in fluid dynamics $\boldsymbol{x}$ may represent the velocity and pressure of a flow, whereas λ is some physical parameter such as the Reynolds number. In practice S is computed by finding a discrete set of points on S and then using some graphics package to interpolate. So the basic numerical question to consider is: Given a point $(\boldsymbol{x}_0, \lambda_0) \in S$ how would we compute a nearby point on S? Throughout we use the notation $\boldsymbol{F}^0 = \boldsymbol{F}(\boldsymbol{x}_0, \lambda_0)$, $\boldsymbol{F}_{\boldsymbol{x}}^0 = \boldsymbol{F}_{\boldsymbol{x}}(\boldsymbol{x}_0, \lambda_0)$, etc.

If $\boldsymbol{F}_{\boldsymbol{x}}^0$ is nonsingular then the Implicit Function Theorem implies that for λ near λ_0 the solutions of $\boldsymbol{F}(\boldsymbol{x}, \lambda) = \boldsymbol{0}$ satisfy $\boldsymbol{x} = \boldsymbol{x}(\lambda)$ with $\boldsymbol{F}_{\boldsymbol{x}}(\boldsymbol{x}(\lambda), \lambda)$ nonsingular. Hence Theorem 3.1 implies that Newton's method for finding the solution $\boldsymbol{x}(\lambda)$ of $\boldsymbol{F}(\boldsymbol{x}(\lambda), \lambda) = \boldsymbol{0}$ with starting value $\boldsymbol{x}_0$ will converge in some ball centred on $\boldsymbol{x}(\lambda)$ for small enough $\lambda - \lambda_0$.

A simple strategy for computing a point of S near $(\boldsymbol{x}_0, \lambda_0)$ is to choose a steplength $\Delta\lambda$, set $\lambda_1 = \lambda_0 + \Delta\lambda$ and solve

$$\boldsymbol{F}(\boldsymbol{x}, \lambda_1) = \boldsymbol{0}$$

by Newton's method with starting guess $\boldsymbol{x}^0 = \boldsymbol{x}_0$. We then know this will work if $\Delta\lambda$ is sufficiently small. However this method will fail (or at best require repeated reduction of step $\Delta\lambda$) as a turning point is approached. For this reason the pseudo-arclength method described in the next section was introduced.

4.1. Keller's Pseudo-Arclength Continuation [25]

Ideally we would like a method that has no difficulties near, or passing round, a fold point. This isn't unreasonable since at a fold point there is nothing geometrically "wrong" with the curve, though λ is the wrong parameter to use to describe the curve. In this section we shall assume that there is an arc of S such that at all points in the arc

$$\text{Rank } [\boldsymbol{F}_{\boldsymbol{x}} | \boldsymbol{F}_\lambda] = n, \tag{4.3}$$

and so any point in the arc is either a regular or fold point of S. The Implicit Function Theorem thus implies that the arc is a smooth curve in $\mathbb{R}^{n-1}$, and so there is a unique tangent direction at each point of the arc.

Let t denote any parameter used to describe the arc. Then along the arc $(\boldsymbol{x}, \lambda) = (\boldsymbol{x}(t), \lambda(t))$. Suppose $(\boldsymbol{x}_0, \lambda_0) = (\boldsymbol{x}(t_0), \lambda(t_0))$ and denote the tangent at $(\boldsymbol{x}_0, \lambda_0)$ by $\boldsymbol{\tau}_0 = (\dot{\boldsymbol{x}}_0, \dot{\lambda}_0)$ where

$$\dot{\boldsymbol{x}} = \frac{d\boldsymbol{x}}{dt}, \qquad \dot{\lambda} = \frac{d\lambda}{dt}.$$
$$\dot{\boldsymbol{x}}_0 = \dot{\boldsymbol{x}}(t_0), \qquad \dot{\lambda}_0 = \dot{\lambda}(t_0).$$

The tangent $\boldsymbol{\tau}_0$ is well defined even if $(\boldsymbol{x}_0, \lambda_0)$ is a fold point and can be computed in practice using the following result.

Lemma 4.1. *Assume (4.3). Then the tangent at* $(\boldsymbol{x}_0, \lambda_0) \in S$ *satisfies*

$$\boldsymbol{\tau}_0 \in \ker [\boldsymbol{F}_{\boldsymbol{x}}^0 | \boldsymbol{F}_{\lambda}^0]. \tag{4.4}$$

Proof. Since $\boldsymbol{F}(\boldsymbol{x}(t), \lambda(t)) = \boldsymbol{0}$, differentiating with respect to t gives

$$\boldsymbol{F}_{\boldsymbol{x}}(\boldsymbol{x}(t), \lambda(t))\dot{\boldsymbol{x}}(t) + \boldsymbol{F}_{\lambda}(\boldsymbol{x}(t), \lambda(t))\dot{\lambda}(t) = \boldsymbol{0}.$$

Put $t = t_0$ and we have $\begin{bmatrix} \dot{\boldsymbol{x}}_0 \\ \dot{\lambda}_0 \end{bmatrix} \in \ker [\boldsymbol{F}_{\boldsymbol{x}}^0 | \boldsymbol{F}_{\lambda}^0]$ and so the result follows.

Suppose now that $\boldsymbol{\tau}_0 = [\boldsymbol{s}_0, \sigma_0]$ denotes the unit tangent i.e. $\boldsymbol{\tau}_0^T \boldsymbol{\tau}_0 = 1$. We can use this vector to devise an extended system which can be solved by Newton's method without fail for a point $(\boldsymbol{x}_1, \lambda_1)$ on S near $(\boldsymbol{x}_0, \lambda_0)$. The appropriate extended system is

$$\boldsymbol{H}(\boldsymbol{y}, t) = \boldsymbol{0} \tag{4.5}$$

where $\boldsymbol{y} = (\boldsymbol{x}, \lambda) \in \mathbb{R}^{n+1}$ and $\boldsymbol{H} : \mathbb{R}^{n+2} \to \mathbb{R}^{n+1}$. is given by

$$\boldsymbol{H}(\boldsymbol{y}, t) = \begin{bmatrix} \boldsymbol{F}(\boldsymbol{x}, \lambda) \\ \boldsymbol{s}_0^T(\boldsymbol{x} - \boldsymbol{x}_0) + \sigma_0(\lambda - \lambda_0) - (t - t_0) \end{bmatrix}. \tag{4.6}$$

The last equation in system (4.5) is the equation of the plane perpendicular to $\boldsymbol{\tau}_0$ a distance $\Delta t = (t - t_0)$ from t_0 (see Fig. 4.1). So in (4.5) we in fact implement a specific parametrisation local to $(\boldsymbol{x}_0, \lambda_0)$, namely parametrisation by the length of the projection of $(\boldsymbol{x}, \lambda)$ onto the tangent direction at $(\boldsymbol{x}_0, \lambda_0)$.

With $\boldsymbol{y}_0 = (\boldsymbol{x}_0, \lambda_0)$, we have $H(\boldsymbol{y}_0, t_0) = \boldsymbol{0}$ and $\boldsymbol{H}_{\boldsymbol{y}}(\boldsymbol{y}_0, t_0) = \begin{bmatrix} \boldsymbol{F}_{\boldsymbol{x}}^0 & \boldsymbol{F}_{\lambda}^0 \\ \boldsymbol{s}_0^T & \sigma_0 \end{bmatrix}$. Since $(\boldsymbol{s}_0^T, \sigma_0)^T$ is orthogonal to each of the rows of $[\boldsymbol{F}_{\boldsymbol{x}}^0, \boldsymbol{F}_{\lambda}^0]$, the matrix $\boldsymbol{H}_{\boldsymbol{y}}(\boldsymbol{y}_0, t_0)$ is nonsingular and so by the Implicit Function Theorem solutions of (4.5) satisfy $\boldsymbol{y} = (\boldsymbol{x}, \lambda) = (\boldsymbol{x}(t), \lambda(t))$ for t near t_0. For $t_1 = t_0 + \Delta t$ and Δt sufficiently small we know that $\boldsymbol{F}(\boldsymbol{y}, t_1) = \boldsymbol{0}$ has a unique solution $\boldsymbol{y} = \boldsymbol{y}(t_1) = (\boldsymbol{x}_1, \lambda_1)$ and $\boldsymbol{H}_{\boldsymbol{y}}(\boldsymbol{y}(t_1), t_1)$ is nonsingular. Thus Newton's method will converge for small enough Δt. If we take as starting guess $\boldsymbol{y}^0 = \boldsymbol{y}_0 = (\boldsymbol{x}_0, \lambda_0)$, it is a straightforward exercise to show, (i) $\boldsymbol{y}^1$, the first Newton iterate, is given by $\boldsymbol{y}^1 = (\boldsymbol{x}_0, \lambda_0) + \Delta t(\boldsymbol{s}_0, \sigma_0)$, that is, the first iterate "steps out" along the tangent, as one might expect, and (ii) $\boldsymbol{\tau}_0^T(\boldsymbol{y}^k - \boldsymbol{y}_0) = \Delta t \quad \forall \, k \geq 1$, which means that all the Newton iterates lie in the plane shown in Fig. 4.1.

Since length along the tangent at $(\boldsymbol{x}_0, \lambda_0)$ is used as parameter this technique is called *pseudo-arclength continuation* [25].

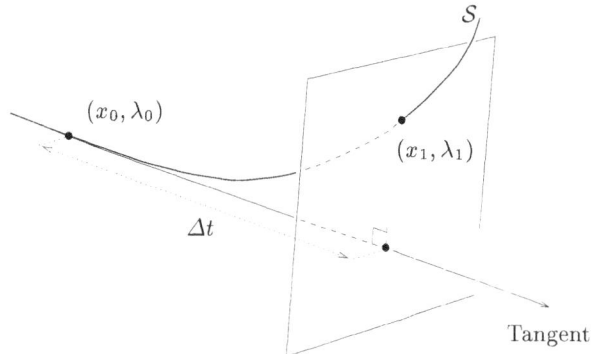

Fig. 4.1.

Another Interpretation

Since we know that solutions of (4.5) satisfy $y = y(t)$, t near t_0, where t is pseudo-arclength, we have $H(y(t), t) = 0$ and we can differentiate with respect to t to get $H_y(y(t), t)\dot{y}(t) + H_t(y(t), t) = 0$. Since H_y is nonsingular for t near t_0,

$$\dot{y}(t) = -H_y(y(t), t)^{-1} H_t(y(t), t), \tag{4.7}$$

which is an ordinary differential equation for $y(t)$ with initial condition

$$y(t_0) = y_0. \tag{4.8}$$

We can use Euler's method to solve (4.7), (4.8) in which case the first step is

$$y_{\text{Euler}}^1 = y_0 - \Delta t [H_y^0]^{-1} H_t^0 \tag{4.9}$$

Since $H_t^0 = \begin{bmatrix} 0 \\ -1 \end{bmatrix}$, (4.9) is equivalent to

$$
\begin{aligned}
y_{\text{Euler}}^1 &= y_0 + \begin{bmatrix} F_x^0 & F_\lambda^0 \\ s_0^T & \sigma_0 \end{bmatrix}^{-1} \begin{bmatrix} 0 \\ \Delta t \end{bmatrix} \\
&= y^0 + \Delta t \begin{bmatrix} s_0 \\ \sigma_0 \end{bmatrix}.
\end{aligned}
$$

So the first step of Newton's method for (4.5) is equivalent to one step of Euler's method applied to (4.7), (4.8). We can think of this as using Euler's method to provide a "predicted guess" for $y(t_1) = (x_1, \lambda_1)$. Then continuing with Newton's method can be thought of as "correcting" this initial guess.

The choice of an appropriate step control strategy for Δt seems to be harder than in the ODE case, perhaps because the real problem is $F(x(t), \lambda(t)) = 0$ and not the differential equation derived from it. This topic is discussed in §4.6 of [45] or §7.4 of [39] but experience indicates that simple techniques often work just as well as sophisticated approaches.

Practical Implementation of Pseudo-Arclength Continuation

The following is a suggested algorithm for implementing the pseudo-arclength continuation method introduced above.

Step 1 Suppose F_x^0 is nonsingular, solve

$$F_x^0 z_0 = -F_\lambda^0 \qquad (4.10)$$

for z_0. Then set $\begin{bmatrix} s_0 \\ \sigma_0 \end{bmatrix} = \dfrac{1}{(z_0^T z_0 + 1)^{1/2}} \begin{bmatrix} z_0 \\ 1 \end{bmatrix}.$

Step 2 (Euler predictor) Choose a step length Δt and set

$$\begin{bmatrix} x^1 \\ \lambda^1 \end{bmatrix} = \begin{bmatrix} x_0 \\ \lambda_0 \end{bmatrix} + \Delta t \begin{bmatrix} s_0 \\ \sigma_0 \end{bmatrix}. \qquad (4.11)$$

Step 3 (Newton's method) For $k \geq 1$ iterate

$$\begin{bmatrix} x^{k+1} \\ \lambda^{k+1} \end{bmatrix} = \begin{bmatrix} x^k \\ \lambda^k \end{bmatrix} + \begin{bmatrix} d^k \\ \delta^k \end{bmatrix}$$

with

$$\begin{bmatrix} F_x^k & F_\lambda^k \\ s_0^T & \sigma_0 \end{bmatrix} \begin{bmatrix} d^k \\ \delta^k \end{bmatrix} = - \begin{bmatrix} F^k \\ s_0^T(x^k - x_0) + \sigma_0(\lambda^k - \lambda_0) - \Delta t \end{bmatrix} \qquad (4.12)$$

Note that if F_x^0 is singular then the continuation method will still work provided the condition Rank $[F_x^0|F_\lambda^0] = n$ holds, but Step 1 will fail to find the tangent vector. In practise this problem usually does not arise since F_x^0 only becomes singular at isolated points on the solution set S and effectively the probability of landing precisely on such a point is 0. However, care is needed when F_x^0 is nearly singular as is discussed in the next subsection. As a further precaution many continuation methods monitor the determinant of F_x^0.

If the tangent at a singular point is required then the null vector z_0 of F_x^0 can be computed (say, by the inverse power method) and then the tangent vector can be taken as

$$\begin{bmatrix} s_0 \\ \sigma_0 \end{bmatrix} = \frac{1}{(z_0^T z_0)^{1/2}} \begin{pmatrix} z_0 \\ 0 \end{pmatrix}.$$

4.2. Block Elimination

As seen in (4.12) it is repeatedly necessary to solve "bordered systems" with coefficient matrix

$$M^k = \begin{bmatrix} F_x^k & F_\lambda^k \\ s_0^T & \sigma_0 \end{bmatrix}.$$

In many applications, where $F(x, \lambda)$ arises from a solution of a differential equation, F_x^k may have some special structure (e.g. tridiagonal, banded, sparse) which makes systems with matrix F_x^k easy to solve, but this structure is not present in M^k. Then the "block elimination method" (see [25]) is useful for quickly solving such systems.

"Block elimination" is merely Gaussian Elimination performed blockwise. If A is an $n \times n$ matrix, $b, c \in \mathbb{R}^n$ and $d \in \mathbb{R}$ then the block matrix

$$M := \begin{pmatrix} A & b \\ c^t & d \end{pmatrix} = \begin{pmatrix} I & 0 \\ l_n^T & 1 \end{pmatrix} \begin{pmatrix} A & b \\ 0^T & u_{n+1} \end{pmatrix}, \tag{4.13}$$

where $l_n = c^T A^{-1}$, $u_{n+1} = d - c^T A^{-1} b$. If l_{n+1} and u_{n+1} are computed, then the system

$$\begin{pmatrix} A & b \\ c^T & d \end{pmatrix} \begin{pmatrix} x \\ y \end{pmatrix} = \begin{pmatrix} f \\ g \end{pmatrix} \tag{4.14}$$

is readily solved using block forward and back substitution. One algorithm to accomplish this is

(i) Solve $Az = b$, and $Aw = f$, and then set
(ii) $y = (g - c^T w)/(d - c^T z)$, $x = w - yz$.

If A and M are both well conditioned then this algorithm for (4.14) works well, but if A is poorly conditioned, as occurs in pseudo-arclength continuation near a fold point, then it may fail to produce reliable results (in linear algebra terms, the algorithm is not "backward stable"). A complete analysis of why the algorithm fails in the latter case was first given by Moore [34] using a deflation argument. The account by Govaerts [11] avoids deflation but provides a stable algorithm based on combining the decomposition (4.13) with the alternative decomposition

$$\begin{pmatrix} A & b \\ c^T & d \end{pmatrix} = \begin{pmatrix} A & 0 \\ c^T & \lambda_{n+1} \end{pmatrix} \begin{pmatrix} I_n & u_n \\ 0^T & 1 \end{pmatrix}. \tag{4.15}$$

Roughly speaking the improved algorithm of Govaerts uses an iterative refinement approach. The idea is as follows. In actual calculation the step (ii) above produces a good approximation, $\hat{y}$ say, for y, but the approximation for x is often worthless. So the x approximation is discarded. To compute the residual after the first solve the approximate solution $(x_0, y_0) = (0, \hat{y})$ is used. Then the approximation is corrected using the block LU decomposition (4.15) in a second solve. The analysis of why this works is fairly technical [11]. The main work in the resulting stable algorithm involves two solves with A and one solve with A^T. Moore [34] provides a stable algorithm also using only 3 solves (see quoted papers for analysis and algorithmic details).

5. The Computation of Fold (Turning) Points

Let (x_0, λ_0) be a point on S satisfying

$$\boldsymbol{F}_{\boldsymbol{x}}^0 \text{ is singular,} \quad \text{rank}[\boldsymbol{F}_{\boldsymbol{x}}^0 | \boldsymbol{F}_\lambda^0] = n. \tag{5.1}$$

Such a point is a fold point (see Example 2.1 and Definition 3.2). In a general one parameter problem $\boldsymbol{F}(\boldsymbol{x}, \lambda) = \boldsymbol{0}$, fold points are the generic singular points and there are many examples of their occurrence in applications. It is important to understand this type of nonlinear phenomenon in its own right, but also because the theoretical analysis and numerical methods for more complicated singularities are often extensions of fold point techniques.

Before attempting a discussion of the n-dimensional case in §5.1 it is a useful exercise to analyse first the scalar case: this is the subject of the following example.

Example 5.1. Assume $f : \mathbb{R}^2 \to \mathbb{R}$ with $f \in C^\infty(\mathbb{R}^2)$ and set $S = \{(x, \lambda) \in \mathbb{R}^2 : f(x, \lambda) = 0\}$. Assume rank$[f_x, f_\lambda] = 1$ on S (i.e. either $f_x \neq 0$ or $f_\lambda \neq 0$ at all points $(x, \lambda) \in S$). (i) Analyse the behaviour of S near a fold point (x_0, λ_0) i.e. where $f_x^0 = 0$. (ii) Derive a 2×2 system for the accurate calculation of a fold point and determine when the fold point is a regular solution of this system.

Sketch of solution: Recall the usual notation $f_x^0 = f_x(x_0, \lambda_0)$, etc. By the rank condition, at a fold point $f_\lambda^0 \neq 0$ and so the Implicit Function Theorem implies $\lambda = \lambda(x)$ near x_0. Repeated differentiation of $f(x, \lambda(x)) = 0$ provides $\lambda'(x_0) = 0$, $\lambda''(x_0) = -f_{xx}^0/f_\lambda^0$, and so the first three terms of the Taylor expansion of $\lambda(x)$ can be found. (Sketch $\lambda(x)$ when $f_{xx}^0 \neq 0$.) Thus if $f_{xx}^0 \neq 0$ we see that $\lambda'(x)$ changes sign at $x = x_0$. To accurately compute (x_0, λ_0), consider the 2×2 system $\boldsymbol{G}(\boldsymbol{y}) = [f, f_x]^T = \boldsymbol{0}$ to be solved for $\boldsymbol{y} = (x, \lambda)$. It is easy to show that (x_0, λ_0) is a regular point of $\boldsymbol{G}(\boldsymbol{y}) = \boldsymbol{0}$ if and only if $f_{xx}^0 \neq 0$.

We shall see in the following subsection that the theory for fold points of the n-dimensional system $\boldsymbol{F}(\boldsymbol{x}, \lambda) = \boldsymbol{0}$ is very similar to the one dimensional example above.

5.1. Analysis of Fold Points

Consider (4.1) with S as in (4.2). Let $(\boldsymbol{x}(t), \lambda(t))$ denote a smooth arc of S with rank$[\boldsymbol{F}_{\boldsymbol{x}} | \boldsymbol{F}_\lambda] = n$ and let $(\boldsymbol{x}_0, \lambda_0) = (\boldsymbol{x}(t_0), \lambda(t_0))$ satisfy the fold point condition (5.1). Let

$$\phi_0 \in \ker(\boldsymbol{F}_{\boldsymbol{x}}^0) \setminus \{\boldsymbol{0}\}, \quad \psi_0 \in \ker(\boldsymbol{F}_{\boldsymbol{x}}^{0\,T}) \setminus \{\boldsymbol{0}\}. \tag{5.2}$$

Note that $\psi_0^T \boldsymbol{F}_\lambda^0 \neq 0$ (since otherwise $\psi_0 \in \ker[\boldsymbol{F}_{\boldsymbol{x}}^0, \boldsymbol{F}_\lambda^0]^T = \{\boldsymbol{0}\}$). Differentiation of $\boldsymbol{F}(\boldsymbol{x}(t), \lambda(t)) = \boldsymbol{0}$ with respect to t, evaluation at $t = t_0$, and left multiplication by ψ_0^T yields

$$\dot{\lambda}(t_0) = 0, \quad \dot{\boldsymbol{x}}(t_0) = \alpha\phi_0, \quad \ddot{\lambda}(t_0) = -\alpha^2 \psi_0^T(\boldsymbol{F}_{\boldsymbol{xx}}^0 \phi_0)\phi_0/\psi_0^T \boldsymbol{F}_\lambda^0, \tag{5.3}$$

for some $\alpha \neq 0$, where $\boldsymbol{F}_{\boldsymbol{xx}}^0 \phi_0$ denotes the Jacobian matrix of the n-vector $\boldsymbol{F}_{\boldsymbol{x}} \phi_0$.

Definition 5.1. *We call $(\boldsymbol{x}_0, \lambda_0)$ a* quadratic *fold point if* $\ddot{\lambda}(t_0) \neq 0$.

To compute the fold point it is natural to set up the system

$$\boldsymbol{T}(\boldsymbol{y}) := \begin{pmatrix} \boldsymbol{F}(\boldsymbol{x}, \lambda) \\ \boldsymbol{F_x}(\boldsymbol{x}, \lambda)\boldsymbol{\phi} \\ \boldsymbol{\phi}^T \boldsymbol{\phi} - 1 \end{pmatrix} = \boldsymbol{0}, \quad \text{to be solved for } \boldsymbol{y} = \begin{pmatrix} \boldsymbol{x} \\ \boldsymbol{\phi} \\ \lambda \end{pmatrix} \in \mathbb{R}^{2n+1},$$

(5.4)

where $\boldsymbol{T} : \mathbb{R}^{2n+1} \to \mathbb{R}^{2n+1}$. Here the second and third equations say that $\boldsymbol{F_x}$ has a zero eigenvalue with corresponding normalised eigenvector $\boldsymbol{\phi}$, (cf. Example 3.1). We shall see in §5.2 that there are alternative choices of system to compute a fold point, but (5.4) is very convenient for analysis. In fact, the following theorem shows that (5.4) characterises a quadratic fold point.

Theorem 5.1. *Let $(\boldsymbol{x}_0, \lambda_0)$ satisfy (5.1).*

(a) The point $(\boldsymbol{x}_0, \boldsymbol{\phi}_0, \lambda_0) \in \mathbb{R}^{2n+1}$ is a regular solution of $\boldsymbol{T}(\boldsymbol{y}) = \boldsymbol{0}$ if and only if $\ddot{\lambda}(t_0) \neq 0$, i.e. $(\boldsymbol{x}_0, \lambda_0)$ is a quadratic fold point.

(b) In addition, assume $\mu_0 = 0$ is an algebraically simple eigenvalue of $\boldsymbol{F_x^0}$. Let $\mu(t)$ denote the eigenvalue of $\boldsymbol{F_x}(\boldsymbol{x}(t), \lambda(t))$ near $t = t_0$ with $\mu(t_0) = 0$. Then $\dot{\mu}(t_0) \neq 0$ if and only if $(\boldsymbol{x}_0, \lambda_0)$ is a quadratic fold point.

The second result says that at a quadratic fold point a simple eigenvalue passes smoothly through zero. Thus there is a smooth change in sign of $\det(\boldsymbol{F_x})$ through a quadratic fold point and this fact is often used in continuation codes.

Proof. The proof of Theorem 5.1 is straightforward. One way is to consider $\boldsymbol{T_y}(\boldsymbol{y}_0)$, given by

$$\boldsymbol{T_y}(\boldsymbol{y}_0) = \begin{bmatrix} \boldsymbol{F_x^0} & 0 & \boldsymbol{F_\lambda^0} \\ \boldsymbol{F_{xx}^0}\boldsymbol{\phi}_0 & \boldsymbol{F_x^0} & \boldsymbol{F_{x\lambda}^0}\boldsymbol{\phi}_0 \\ 0 & 2\boldsymbol{\phi}_0^T & 0 \end{bmatrix}$$

and use part *(ii)* of Keller's ABCD Lemma (Lemma 3.1). In the notation of the lemma, $A = \begin{pmatrix} \boldsymbol{F_x^0} & 0 \\ \boldsymbol{F_{xx}^0}\boldsymbol{\phi}_0 & \boldsymbol{F_x^0} \end{pmatrix}$ with corresponding choices for $\boldsymbol{b}$ and $\boldsymbol{c}$. It can then be shown that $\boldsymbol{T_y}(\boldsymbol{y}_0)$ is nonsingular if and only if $\boldsymbol{\psi}_0^T (\boldsymbol{F_x^0}\boldsymbol{\phi}_0)\boldsymbol{\phi}_0 \neq 0$, and part *(a)* follows from (5.3). The proof of part *(b)* follows by formulating the eigenvalue problem as

$$\boldsymbol{G}(\boldsymbol{z}, t) = \begin{pmatrix} \boldsymbol{F_x}(\boldsymbol{x}(t), \lambda(t))\boldsymbol{\phi} - \mu\boldsymbol{\phi} \\ \boldsymbol{\phi}^T\boldsymbol{\phi} - 1 \end{pmatrix} = \boldsymbol{0}, \quad \boldsymbol{z} = \begin{pmatrix} \boldsymbol{\phi} \\ \mu \end{pmatrix} \in \mathbb{R}^{n+1}. \quad (5.5)$$

Now $\boldsymbol{G}(\boldsymbol{z}_0, t_0) = \boldsymbol{0}$ and $\boldsymbol{G_z}(\boldsymbol{z}_0, t_0)$ is nonsingular (cf. Example 3.1), and the Implicit Function Theorem shows that $\mu = \mu(t)$ is a simple eigenvalue of $\boldsymbol{F_x}(\boldsymbol{x}(t), \lambda(t))$ near $t = t_0$. Differentiation of $\boldsymbol{F_x}(\boldsymbol{x}(t), \lambda(t))\boldsymbol{\phi}(t) = \mu(t)\boldsymbol{\phi}(t)$, evaluation at $t = t_0$, and left multiplication by $\boldsymbol{\psi}_0^T$ provides $\dot{\mu}(t_0) \neq 0$ iff $\ddot{\lambda}(t_0) \neq 0$. (Note that $\boldsymbol{\psi}_0^T\boldsymbol{\phi}_0 \neq 0$ since zero is an algebraically simple eigenvalue of $\boldsymbol{F_x^0}$.) □

5.2. Numerical Calculation of Fold Points

The system (5.4) can easily be used to compute fold points. It was so used by Seydel [44], [43] and by Moore and Spence [35]. It is important to realise that when solving (5.4) by Newton's method one need not solve the $(2n + 1) \times (2n + 1)$ linear systems directly. In [35] an efficient solution procedure is described using only solves with an $n \times n$ nonsingular matrix, which is formed from F_λ and $n - 1$ linearly independent columns of F_x. The details are not given here, but the main work involves 4 linear solves with the *same* matrix i.e. only one LU factorisation of an $n \times n$ matrix is needed per Newton step to solve (5.4).

The fact that the solution of the $(2n + 1) \times (2n + 1)$ linear Jacobian systems is accomplished using solves of $n \times n$ systems has been used many times since, and another example of this technique is in the calculation of Hopf bifurcation points (see §8 and [16]).

Griewank and Reddien [17, 18] (and with improvements Govaerts [12]) suggested an alternative way of calculating fold points (and other higher order singularities). This involves setting up a "minimal" defining system

$$T(y) = \begin{bmatrix} F(x, \lambda) \\ g(x, \lambda) \end{bmatrix} = 0, \qquad y \in \mathbb{R}^{n+1}, \tag{5.6}$$

where $g(x, \lambda) : \mathbb{R}^n \times \mathbb{R} \to \mathbb{R}$ is implicitly defined through the equations

$$M(x, \lambda) \begin{bmatrix} v(x, \lambda) \\ g(x, \lambda) \end{bmatrix} = \begin{bmatrix} 0 \\ 1 \end{bmatrix}, \tag{5.7}$$

and

$$(w^T(x, \lambda), g(x, \lambda)) M(x, \lambda) = (0^T, 1), \tag{5.8}$$

where

$$M(x, \lambda) = \begin{bmatrix} F_x(x, \lambda) & b \\ c^T & d \end{bmatrix}, \tag{5.9}$$

for some $b, c \in \mathbb{R}^n$, $d \in \mathbb{R}$. (The fact that $g(x, \lambda)$ is defined uniquely by *both* (5.7) and (5.8) may be seen since both equations imply that $g(x, \lambda) = [M^{-1}(x, \lambda)]_{n+1,n+1}$.) Note that $M(x, \lambda)$ is a bordering of F_x, as arises in the numerical continuation method (§4). Assuming b, c, and d are chosen so that $M(x, \lambda)$ is nonsingular (see the ABCD Lemma 3.1) then $g(x, \lambda)$ and $v(x, \lambda)$ in (5.7) are uniquely defined. (Note, if S is parametrised by t near (x_0, λ_0), i.e. $(x(t), \lambda(t))$ near $t = t_0$, then $v = v(x(t), \lambda(t))$ and $g = g(x(t), \lambda(t))$, and these functions may be differentiated with respect to t.) Also, if we apply Cramer's Rule in (5.7) (an idea due to Govaerts) we have (with $M(x, \lambda)$ nonsingular)

$$g(x, \lambda) = \det(F_x(x, \lambda))/\det(M(x, \lambda)) \tag{5.10}$$

and so

$$g(x, \lambda) = 0 \iff F_x(x, \lambda) \text{ is singular.}$$

It is easily shown that quadratic fold points are regular solutions of (5.6). To apply Newton's method to (5.6) derivatives of $g(x, \lambda)$ are required and these can be found by differentiation of (5.7). When the details of an efficient implementation of Newton's method applied to (5.6) are worked out then the main cost is two linear solves with M and one with M^T. A nice summary of these ideas is given in Beyn [1]. A complete account is in the forthcoming book by Govaerts [13].

6. Bifurcation from the Trivial Solution

As usual we consider the problem $F(x, \lambda) = 0$ and we assume that $F \in C^\infty(\mathbb{R}^{n+1})$. We shall also assume that

$$F(0, \lambda) = 0, \quad \text{for all } \lambda \in \mathbb{R}, \tag{6.1}$$

with $(0, \lambda)$ being the path of trivial solutions. (As an example, consider $f \in \mathbb{R}^2$, $f(x, \lambda) = x\lambda - x^3$.) In this section we use the G^0 to denote the value of any function G at $(0, \lambda_0)$.

A formal definition of bifurcation from the trivial solution is as follows.

Definition 6.1. *A point* $(0, \lambda) \in \mathbb{R}^{n+1}$ *is said to be a bifurcation point for* $F(x, \lambda) = 0$ *if and only if there exists a sequence* (x_k, λ_k), *with* $x_k \to 0$, $\lambda_k \to \lambda_0$, *as* $k \to \infty$, *such that* $F(x_k, \lambda_k) = 0$ *and* $x_k \neq 0$ *for all* k.

Application of the Implicit Function Theorem shows immediately that if $(0, \lambda_0)$ is a bifurcation point then F_x^0 must be singular. We shall assume that F_x^0 has an algebraically simple zero eigenvalue: that is, there exist $\phi_0, \psi_0 \in \mathbb{R}^n \setminus \{0\}$ such that

$$\begin{aligned}
\ker(F_x^0) &= \operatorname{span}\{\phi_0\}, \\
\ker(F_x^{0\,T}) &= \operatorname{span}\{\psi_0\}, \\
\text{and} \quad \psi_0^T \phi_0 &\neq 0.
\end{aligned} \tag{6.2}$$

In §6.1 we shall consider the scalar case since it is simpler and the scalar result is used in the n-dimensional case. We shall see in §7 that the scalar case is interesting in its own right. In §6.2 we then discuss the n-dimensional case.

6.1. Scalar Case

Consider the problem $f(x, \lambda) = 0$, with $f(0, \lambda) = 0$ for all λ. Then we have the following theorem.

Theorem 6.1. *Suppose* $f : D \subseteq \mathbb{R} \times \mathbb{R} \to \mathbb{R}$ *where* D *is an open subset of* $\mathbb{R}^2$. *Suppose* $f \in C^\infty(D)$. *Let* $S = \{(x, \lambda) \in D : f(x, \lambda) = 0\}$, *and assume* $(0, \lambda) \in S$, *for all* $\lambda \in \mathbb{R}$. *Also assume for some* $\lambda_0 \in \mathbb{R}$

(i) $f_x^0 = 0$,

(ii) $f_{x\lambda}^0 \neq 0$,

(where $f_x^0 = f_x(0, \lambda_0)$, etc). Then nontrivial solutions bifurcate from the trivial solution $x = 0$ at $(0, \lambda_0)$. Moreover, near λ_0,

$$\lambda(x) = \lambda_0 + xv(x) \tag{6.3}$$

where $v(x)$ is a smooth function of x with $v(0) = -f_{xx}^0 / 2f_{x\lambda}^0$.

Proof. Note that $f_\lambda^0 = f_{\lambda\lambda}^0 = f_{\lambda\lambda\lambda}^0 = \cdots = 0$ since $(0, \lambda) \in S$ for all λ. Thus $\mathrm{rank}[f_x^0, f_\lambda^0] = 0$ and the Implicit Function Theorem cannot be applied directly on $f(x, \lambda) = 0$. However, we can introduce a new problem on which the Implicit Function Theorem can be applied. Define $h(x, v) : \mathbb{R}^2 \to \mathbb{R}$ by, for $x \neq 0$,

$$h(x, v) = \frac{1}{x^2} f(x, \lambda), \quad v = (\lambda - \lambda_0)/x. \tag{6.4}$$

Expanding $f(x, \lambda_0 + xv)$ in powers of x using Taylor's theorem, and using $f^0 = f_x^0 = f_\lambda^0 = f_{\lambda\lambda}^0 = 0$, gives, for $x \neq 0$,

$$h(x, v) = \frac{1}{2!}[f_{xx}^0 + 2f_{x\lambda}^0 v] + x[\text{smooth function of } v].$$

Define $h(x, v)$ at $x = 0$ by

$$h(0, v) = \frac{1}{2!}[f_{xx}^0 + 2f_{x\lambda}^0 v]. \tag{6.5}$$

Then h is a smooth function of (x, v). Moreover by setting

$$v_0 = -f_{xx}^0 / 2f_{x\lambda}^0, \tag{6.6}$$

then, at $(x, v) = (0, v_0)$ we have, from (6.5), $h^0 = h(0, v_0) = 0$. Moreover, by assumption (ii), $h_v^0 \neq 0$, and the Implicit Function Theorem gives that for $|x|$ sufficiently small there exists $v(x)$ such that $h(x, v(x)) = 0$. Using this function v, recall (6.4) and define

$$\lambda(x) = \lambda_0 + xv(x), \tag{6.7}$$

from which $f(x, \lambda(x)) = 0$ and the existence of nontrivial bifurcating solutions is proved.

Differentiating (6.7) and evaluation at $x = 0$ gives

$$\lambda'(0) = v(0) = v_0 = -f_{xx}^0 / 2f_{x\lambda}^0, \tag{6.8}$$

which tells us the slope of the tangent to the nontrivial solution branch at the bifurcation point (see Fig. 6.1).   □

To compute $(x, \lambda(x))$ near $(0, \lambda_0)$ with $x \neq 0$, consider solving the system

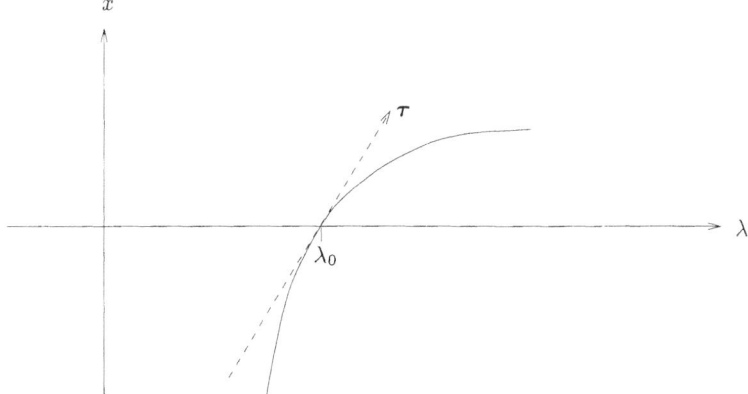

Fig. 6.1. Here τ is the tangent to the bifurcating branches at $(0, \lambda_0)$. The slope of the tangent is $(-2f_{x\lambda}^0 / f_{xx}^0)$.

$$\boldsymbol{G}(\boldsymbol{y}, t) := \left(\begin{array}{c} f(x, \lambda) \\ x - t \end{array} \right) = \boldsymbol{0}, \qquad \boldsymbol{y} = \left(\begin{array}{c} x \\ \lambda \end{array} \right) \in \mathbb{R}^2 . \qquad (6.9)$$

If $(\boldsymbol{y}, t)$ solves (6.9) with sufficiently small $t \neq 0$, then $x = t$ and $(x, \lambda) = (t, \lambda(t)) =: \boldsymbol{y}(t)$. Moreover,

$$\det \left(\boldsymbol{G_y}(\boldsymbol{y}(t), t) \right) = -f_\lambda(t, \lambda(t)) = (-f_{x\lambda}^0)t + \mathcal{O}(t^2)$$

and so the Newton theory (Theorem 5.2.1 in [6]) shows that convergence of Newton's method can only be guaranteed for starting guesses in a ball of radius $\mathcal{O}(t)$. If we take as starting guess the following point on the tangent τ depicted in Fig. 6.2:

$$\left(\begin{array}{c} x^0 \\ \lambda^0 \end{array} \right) = \left(\begin{array}{c} 0 \\ \lambda_0 \end{array} \right) + t \left(\begin{array}{c} 1 \\ \lambda'(0) \end{array} \right)$$

with $\lambda'(0)$ given by (6.8), then

$$\left(\begin{array}{c} t \\ \lambda(t) \end{array} \right) - \left(\begin{array}{c} x^0 \\ \lambda^0 \end{array} \right) = \left(\begin{array}{c} 0 \\ \lambda(t) - \lambda(0) - t\lambda'(0) \end{array} \right) = \mathcal{O}(t^2)$$

and one can show that Newton's method will converge for sufficiently small t. Notice that in this case t is *not* the pseudo-arclength parameter.

6.2. n-Dimensional Case

For the general case we have the classical theorem by Crandall and Rabinowitz on bifurcation from a simple eigenvalue [4] (which holds for general operators on Banach spaces under minimal smoothness requirements).

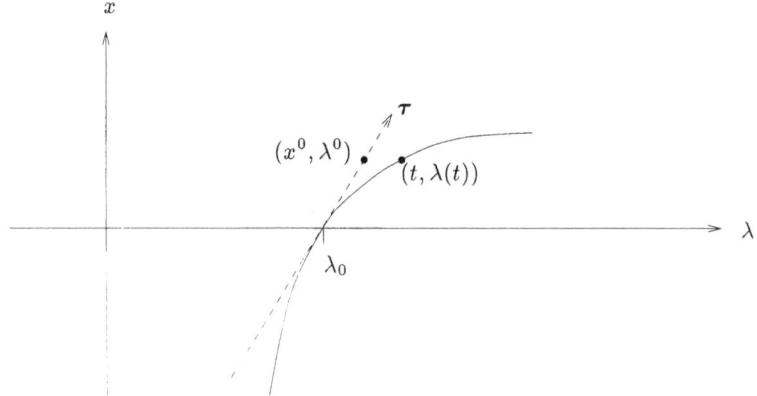

Fig. 6.2. Numerical continuation away from the bifurcation point. Here $x^0 = t$ and in this case t is merely the value of the (non-zero) x component.

Theorem 6.2. *Suppose $\boldsymbol{F} \in C^{\infty}(\mathbb{R}^{n+1})$ and (6.1), (6.2) hold. If*

$$\boldsymbol{\psi}_0^T \boldsymbol{F}_{\boldsymbol{x},\lambda}^0 \boldsymbol{\phi}_0 \neq 0, \tag{6.10}$$

then $(\boldsymbol{0}, \lambda_0)$ is a bifurcation point, and there exist smooth functions $(\boldsymbol{x}(t), \lambda(t))$ parametrised by t near $t = 0$ such that $\boldsymbol{F}(\boldsymbol{x}(t), \lambda(t)) = \boldsymbol{0}$ with $\lambda(0) = \lambda_0$, $\boldsymbol{x}(0) = \boldsymbol{0}$, and $\boldsymbol{x}'(0) = \boldsymbol{\phi}_0$.

Proof. We present a proof which we believe is within the scope of a typical UK research student just starting a PhD. More elegant proofs are given in most text books on bifurcation theory. The method of proof is an example of the "Lyapunov-Schmidt reduction" [2].

We give the proof only under the assumption that $\boldsymbol{F}_{\boldsymbol{x}}^0$ has distinct real eigenvalues, $\mu_0, \ldots, \mu_{n-1}$, with linearly independent eigenvectors $\{\boldsymbol{\phi}_0, \boldsymbol{\phi}_1, \ldots, \boldsymbol{\phi}_{n-1}\}$, and $\{\boldsymbol{\psi}_0, \boldsymbol{\psi}_1, \ldots, \boldsymbol{\psi}_{n-1}\}$ the corresponding linearly independent eigenvectors of $(\boldsymbol{F}_{\boldsymbol{x}}^0)^T$. Recall that $\mu_0 = 0$ and all the other eigenvalues of $\boldsymbol{F}_{\boldsymbol{x}}^0$ are nonzero. Also, because the eigenvalues are distinct and simple, $\boldsymbol{\psi}_i^T \boldsymbol{\phi}_j = 0$, $(i \neq j)$ and $\boldsymbol{\psi}_i^T \boldsymbol{\phi}_i \neq 0$.

Now since the $\boldsymbol{\psi}_i$ span $\mathbb{R}^n$, the equation $\boldsymbol{F}(\boldsymbol{x}, \lambda) = \boldsymbol{0}$ can be written as

$$\boldsymbol{\psi}_0^T \boldsymbol{F}(\boldsymbol{x}, \lambda) = 0 \tag{6.11}$$

$$\text{and} \quad \boldsymbol{\psi}_i^T \boldsymbol{F}(\boldsymbol{x}, \lambda) = 0, \quad i = 1, \ldots, (n-1). \tag{6.12}$$

(The proof of this is an elementary exercise.) Also we write $\boldsymbol{x} \in \mathbb{R}^n$ in the form

$$\boldsymbol{x} = \boldsymbol{x}(\boldsymbol{y}, t) = t\boldsymbol{\phi}_0 + V\boldsymbol{y}, \tag{6.13}$$

where V is the $n \times (n-1)$ matrix with ith column $\boldsymbol{\phi}_i$, for $i = 1, \ldots, (n-1)$. Thus we decompose $\mathbb{R}^n$ into $\mathbb{R}^n = \text{span}\{\boldsymbol{\phi}_0\} \oplus \mathcal{R}$, where $\mathcal{R} = \text{Image}(\boldsymbol{F}_{\boldsymbol{x}}^0) =$

span$\{\phi_1, \ldots, \phi_{n-1}\}$. Note that $\psi_0^T v = 0$ for all $v \in \mathcal{R}$. Now consider the $n - 1$ equations given by (6.12) in the form

$$\tilde{F}(y, t, \lambda) = 0, \quad \tilde{F} : \mathbb{R}^{n-1} \times \mathbb{R} \times \mathbb{R} \to \mathbb{R}^{n-1} \tag{6.14}$$

where $(\tilde{F}(y, t, \lambda))_i = \psi_i^T F(t\phi_0 + Vy, \lambda)$. We shall use the Implicit Function Theorem to parametrise the solution y of (6.14) as a function of (t, λ). To do this observe that

$$\tilde{F}(0, 0, \lambda) = 0 \in \mathbb{R}^{n-1},$$

and the matrix $\tilde{F}_y(0, 0, \lambda)$ is given by

$$\left[\tilde{F}_y(0, 0, \lambda)\right]_{i,j} = \psi_i^T F_x(0, \lambda)\phi_j, \quad i, j = 1, \ldots, (n-1).$$

Thus $[\tilde{F}_y(0, 0, \lambda_0)]_{ij} = \psi_i^T F_x^0 \phi_j = \mu_i \psi_i^T \phi_i$, and since $\mu_i \neq 0$, $i = 1, \ldots, (n-1)$, $\tilde{F}_y(0, 0, \lambda_0)$ is nonsingular on $\mathbb{R}^{n-1}$, and so the Implicit Function Theorem (extended to the case of a two-dimensional parameter [2]) shows that solutions y of (6.14) may be parametrised by (t, λ) i.e. $y = y(t, \lambda)$ near $(0, \lambda_0)$. Thus $\tilde{F}(y(t, \lambda), t, \lambda) = 0$ for (t, λ) near $(0, \lambda_0)$ and (by uniqueness) $y(0, \lambda) = 0$. In addition, $y_t(0, \lambda_0) = 0$ since $\tilde{F}_t(0, 0, \lambda_0) = 0$, and $y_\lambda(0, \lambda) = 0$ since $\tilde{F}_\lambda(0, 0, \lambda) = 0$.

Having solved (6.14) (i.e. the equations (6.12)), we return to eqn. (6.11) and write it in the form

$$f(t, \lambda) := \psi_0^T F(t\phi_0 + Vy(t, \lambda), \lambda) = 0. \tag{6.15}$$

This is in the form of a scalar nonlinear problem, and all that remains to prove the existence of nontrivial solutions is to check the conditions of Theorem 6.1. Not surprisingly (6.15) (or (6.11)) is called the *bifurcation equation* in the Lyapunov-Schmidt reduction of $F(x, \lambda) = 0$. It is essentially a projection of the original n dimensional system into the one dimensional space spanned by the null eigenvector. Observe that

$$\begin{aligned} f(0, \lambda) &= \psi_0^T F(Vy(0, \lambda), \lambda) \\ &= \psi_0^T F(0, \lambda) \\ &= 0 \end{aligned}$$

and

$$f_t(0, \lambda_0) = \psi_0^T F_x^0 \phi_0 = 0.$$

since $\phi_0 \in \ker(F_x(0, \lambda_0))$. Finally

$$\begin{aligned} f_{t\lambda}^0 &= \psi_0^T (F_{xx}^0(\phi_0 + Vy_t^0)(Vy_\lambda^0) + F_{x\lambda}^0(\phi_0 + Vy_t^0) + F_x^0 Vy_{t\lambda}^0) \\ &= \psi_0^T F_{x\lambda}^0 \phi_0 \neq 0 \end{aligned}$$

by (6.10). Here we have used $y_t^0 = 0$, $y_\lambda^0 = 0$, and $Vy_{t\lambda}^0 \in \mathcal{R}$ to simplify the expression for $f_{t\lambda}^0$. So the conclusions of Theorem 6.1 apply. Hence nontrivial solutions of $f(t, \lambda) = 0$ bifurcate at $\lambda = \lambda_0$. The corresponding $x(t) = x(y(t, \lambda), t) = t\phi_0 + Vy(t, \lambda)$, with $x(0) = 0$ and $x'(0) = \phi_0$, provides the nontrivial solution of $F(x, \lambda) = 0$. $\qquad \square$

Detection of bifurcation points is relatively easy for this case. We seek points where $F_x(0, \lambda)$ is singular. If $\mu(\lambda)$ denotes the eigenvalue of $F_x(0, \lambda)$ along the trivial solution path with $\mu(\lambda_0) = 0$ then it is a simple exercise to show that $\mu'(\lambda_0) \neq 0$ if and only if (6.10) holds. Hence (6.10) is another example of a nondegeneracy condition which can be interpreted as an eigenvalue going through zero with nonzero speed. Note that $\det(F_x(0, \lambda))$ changes sign at the bifurcation point as λ passes through λ_0.

Example 6.1. In Example 2.3, eqn. (2.7) is

$$F(Y, \lambda) = AY + \lambda \sin(Y).$$

(It is an instructive exercise to go through the proof of Theorem 6.2 for this example.) In this case $F(0, \lambda) = 0$, for all $\lambda \in \mathbb{R}$, $F_Y(0, \lambda) = (A + \lambda I)$, and $F_{Y\lambda}(0, \lambda) = I$. Now A has $(n-1)$ algebraically simple eigenvalues

$$\mu_k = -h^{-2}(2 - 2\cos k\pi/(n-1)), \quad k = 0, 1, \ldots, n-2.$$

The zero eigenvalue is ruled out of the buckling example on physical grounds. so Theorem 6.2 shows that bifurcation from the trivial solution occurs at $\lambda = -\mu_k$, $k = 0, \ldots, (n-2)$ since the nondegeneracy condition (6.5) reduces to the condition that the eigenvalue be algebraically simple. □

If the n-dimensional problem arises from a discretization of an ODE or PDE then the eigenvalues and eigenfunctions of the linearisation of the continuous problem might be known analytically. To find λ_0 and ϕ_0 in the discretized problem a simple inverse iteration approach applied to $F_x(0, \lambda)$ with the exact value for λ at the bifurcation point used would almost certainly work very quickly.

To move off the trivial branch a technique similar to that in §6.1 may be used. At $(0, \lambda_0)$ $[F_x^0 | F_\lambda^0]$ has a two dimensional kernel spanned by $(\phi_0^T, 0)^T$ and $(0^T, 1)^T$. It is straightforward to show that the tangent to the bifurcating nontrivial branch has the form

$$\tau_0^T = ((-2\psi_0^T F_{x\lambda}^0 \phi_0)\phi_0^T, \psi_0^T (F_{xx}^0 \phi_0)\phi_0)$$

(cf. the scalar case in the previous section). The fact that the ordinary pseudo-arclength method works in these circumstances is proved in [22]. However, the direct analogue of the approach in §6.1 is merely to set equal to t one component of x. The best component to choose is the rth, where $(\phi_0)_r$ is the component of maximum modulus of ϕ_0 (see [39]).

Remarks

1. It is important to note that bifurcation from the trivial solution is a rather special case but nonetheless a very important case in applications. The trivial solution forms an invariant subspace under the action of F in $\mathbb{R}^{n+1}$ and bifurcating nontrivial solutions break the subspace. Werner [49]

gives a general theory of subspace-breaking bifurcation. A different case but with comparable results arises when $F(x, \lambda)$ satisfies a symmetry condition (see [48], [51]).

2. In the absence of any special features (for example, symmetry) a bifurcation where two (nontrivial) solution curves intersect will not typically arise in a one parameter problem $F(x, \lambda) = 0$. In this case one needs *two* parameters to detect and compute bifurcation points (see [33], [18],[23]).

7. Bifurcation in Nonlinear ODEs

The bifurcation theory in these notes is given for finite dimensional problems. However several of the theoretical results on bifurcation can be applied to infinite dimensional problems involving nonlinear ODEs by use of the shooting method, which also provides a computational tool. In fact Poincaré's analysis of bifurcation of periodic orbits in ODEs using the Poincaré section is the first example of the use of a shooting method to prove analytical results. The use of shooting to study steady bifurcations in nonlinear boundary value problems (BVPs) seems to have been considered first by J.B.Keller in 1960. The treatment here is based on Keller's article [27]. An account of the numerical analysis of shooting methods for nonlinear BVPs is given in [24].

First recall a standard theorem of existence, uniqueness and continuity with respect to initial data for systems of ordinary differential equations (ODEs) of the form

$$u' = f(t, u), \quad t > a \tag{7.1}$$

with initial condition

$$u(a) = \alpha \tag{7.2}$$

where $u(t) \in \mathbb{R}^n$ is to be found for $t > a$, $f : \mathbb{R} \times \mathbb{R}^n \to \mathbb{R}^n$ is given, $\alpha \in \mathbb{R}^n$ is given, and $a \in \mathbb{R}$ is given.

Theorem 7.1. *Suppose f is continuous on $[a, b] \times \mathbb{R}^n$, and suppose*

$$\|f(t, u) - f(t, v)\| \leq L\|u - v\| \tag{7.3}$$

for some $L > 0$ and for $t \in [a, b]$ and all $u, v \in \mathbb{R}^n$. Then for any $\alpha \in \mathbb{R}^n$ the IVP (7.1), (7.2) has a unique solution $u = u(t, \alpha)$ defined for $t \in [a, b]$. Moreover u is Lipschitz continuous in α, and in fact

$$\|u(t, \alpha) - u(t, \beta)\| \leq e^{L(t-a)}\|\alpha - \beta\|$$

for all $\alpha, \beta \in \mathbb{R}^n$.

Remarks

1. In many problems of interest (7.3) will not be true over all $u, v \in \mathbb{R}^n$, but rather over all $u, v \in B(\alpha, r)$ for some $\alpha \in \mathbb{R}^n$, $r > 0$ fixed. In this case

a similar theorem holds, but the solution may exist only for $t \in [a, b_0]$ with $b_0 = \min\{b, r/L\}$.

2. The numerical solution of (7.1), (7.2) over finite ranges of $t > a$ is now well understood. Many codes exist in which a user specifies $\boldsymbol{f}, a, b$ and $\boldsymbol{\alpha}$ and a required tolerance, and the program returns the value of $\boldsymbol{u}(b)$ at b or at any intermediate points between a and b. We will assume that (7.1), (7.2) have a unique solution which can be found numerically for $t \in [a, b]$, with $b > a$.

7.1. The Shooting Method for ODEs

Consider the second order ODE

$$- y'' - g(t, y, y') = 0, \quad t \in [a, b] \tag{7.4}$$

subject to the boundary conditions

$$
\begin{aligned}
a_0 y(a) - a_1 y'(a) &= \alpha & (7.5) \\
b_0 y(b) + b_1 y'(b) &= \beta & (7.6)
\end{aligned}
$$

with $|a_0| + |a_1| \neq 0$, $|b_0| + |b_1| \neq 0$. Here it is assumed that $g(t, y_1, y_2)$ is continuous on $D := \{(t, y_1, y_2) : t \in [a, b], y_1^2 + y_2^2 < \infty\}$ and satisfies a uniform Lipschitz condition in y_1 and y_2.

To solve this BVP consider the associated initial value problem (IVP)

$$- u'' - g(t, u, u') = 0 \quad (\text{as in } (7.4)) \tag{7.7}$$

subject to the initial conditions

$$a_0 u(a) - a_1 u'(a) = \alpha \quad (\text{as in } (7.5)) \tag{7.8}$$

and

$$c_0 u(a) - c_1 u'(a) = s, \tag{7.9}$$

where s is a parameter which will be determined below. We choose c_0, c_1 s.t.

$$d := a_1 c_0 - a_0 c_1 \neq 0. \tag{7.10}$$

Then (7.8), (7.9) are independent initial conditions and the matrix $\begin{bmatrix} a_0 & -a_1 \\ c_0 & -c_1 \end{bmatrix}$ is invertible. From Theorem 7.1 we know that (7.7)–(7.9) has a unique solution, which we denote by $u(t; s)$, $t > 0$, $s \in \mathbb{R}$. To solve (7.4)–(7.6) we need to find s such that

$$f(s) = 0 \tag{7.11}$$

where $f(s)$ is defined by the right-hand boundary condition (7.6)

$$f(s) = \{b_0 u(b; s) + b_1 \frac{\partial u}{\partial t}(b; s) - \beta\}. \tag{7.12}$$

Thus the solution of (7.4)–(7.6) is reduced to solving the nonlinear problem (7.11) where f is implicitly defined in terms of solutions of (7.7)–(7.9).

The numerical analysis of shooting methods for solving (7.4)–(7.6) where solutions of (7.7)–(7.9) are evaluated numerically is given in [24].

The equivalence of (7.4)–(7.6) to (7.11) is given in the following lemma (see [24],[27]).

Lemma 7.1. *(i) If s_0 solves (7.11) then $y(t) = u(t; s_0)$ solves (7.4)–(7.6). If $y(t)$ solves (7.4)–(7.6) then $s_0 = c_0 y(a) - c_1 y'(a)$ solves (7.11).*

(ii) s_0 is the unique solution of (7.11) if and only if $y(t)$ is the unique solution of (7.4)–(7.6).

Proof. (i) Suppose $f(s_0) = 0$ then $y(t) = u(t; s_0)$ solves (7.4)–(7.6). Conversely if $y(t)$ solves (7.4)–(7.6) then set $s_0 = c_0 y(a) - c_1 y'(a)$. By uniqueness of solutions to initial value problems $u(t; s_0) = y(t)$ and $f(s_0) = 0$.

(ii) See Theorem 2 [27] where a more general problem is considered.

To analyse and solve (7.7)–(7.9) we reduce to a first order system. To do this we set $u_1 = u$, $u_2 = u' = u_1'$. Then (7.7) becomes the 2×2 system

$$\begin{bmatrix} u_1' \\ u_2' \end{bmatrix} = \begin{bmatrix} u_2 \\ -g(t, u_1, u_2) \end{bmatrix} =: \boldsymbol{f}(t, \boldsymbol{u}) \tag{7.13}$$

and (7.8), (7.9) become (using (7.10)),

$$\begin{bmatrix} u_1(a) \\ u_1(a) \end{bmatrix} = \frac{1}{d} \begin{bmatrix} -c_1 & a_1 \\ -c_0 & a_0 \end{bmatrix} \begin{bmatrix} \alpha \\ s \end{bmatrix}, \tag{7.14}$$

and so (7.13), (7.14) has a unique solution using Theorem 7.1.

To implement Newton's method for (7.11) we need to be able to evaluate not only $f(s)$ but also $f_s(s) = \left\{ b_0 w(b; s) + b_1 \dfrac{\partial w}{\partial t}(b; s) \right\}$ where $w = \dfrac{\partial u}{\partial s}$. We obtain an equation for w by differentiating (7.7)–(7.9) with respect to s to get the IVP

$$\left. \begin{aligned} -w'' - g_u(t, u, u')w - g_{u'}(t, u, u')w' &= 0, \\ a_0 w(a) - a_1 w'(a) &= 0, \\ c_0 w(a) - c_1 w'(a) &= 1. \end{aligned} \right\} \tag{7.15}$$

(Note $'$ always means differentiation with respect to t.) This system together with (7.7)–(7.9) can be reduced to a first order system of dimension 4 which we can solve using standard ODE software. We illustrate this by means of an example.

Example 7.1. Consider the boundary value problem

$$\left. \begin{aligned} -y'' + e^y &= 0 \\ y(0) = 0, \; y(1) &= 0, \end{aligned} \right\} \tag{7.16}$$

and the corresponding IVP

$$\left.\begin{aligned} -u'' + e^u = 0, \\ u(0) = 0, \ u'(0) = s. \end{aligned}\right\} \tag{7.17}$$

Denote the solution by $u(t; s)$. The nonlinear problem (7.11) is

$$f(s) := u(1; s) = 0. \tag{7.18}$$

To find f_s, set $w = \dfrac{\partial u}{\partial s}$, then differentiate (7.17) with respect to s

$$\left.\begin{aligned} -w'' + e^u w &= 0, \\ w(0) &= 0, \\ w'(0) &= 1. \end{aligned}\right\} \tag{7.19}$$

From the solution of this we obtain $f_s(s) = w(1; s)$.

We solve (7.17), (7.19) simultaneously by the substitutions

$$u_1 = u, \qquad u_2 = u' = u_1',$$
$$u_3 = w, \qquad u_4 = w' = u_3',$$

to obtain a first order system of four equations:

$$\begin{bmatrix} u_1 \\ u_2 \\ u_3 \\ u_4 \end{bmatrix}' = \begin{bmatrix} u_2 \\ \exp(u_1) \\ u_4 \\ \exp(u_1)u_3 \end{bmatrix}$$

with initial condition

$$\begin{bmatrix} u_1 \\ u_2 \\ u_3 \\ u_4 \end{bmatrix}_{t=0} = \begin{bmatrix} 0 \\ s \\ 0 \\ 1 \end{bmatrix}.$$

We solve this system up to $t = 1$, from which we obtain

$$f(s) = u_1(1), \quad f_s(s) = u_3(1).$$

If s is a guess to the solution of (7.11) then the values of $f(s)$ and $f_s(s)$ can be used to generate a new guess for s using Newton's method, and this process can be iterated.

7.2. Analysis of Parameter Dependent ODEs

As mentioned at the start of this section, the shooting method can be used as a technique for theoretical analysis as well as solving problems numerically.

Example 7.2. (Recall Example 2.3). Prove that nontrivial solutions exist for the following BVP

$$\left.\begin{array}{l} -y'' - \lambda \sin y = 0, \\ y'(0) = 0, \quad y'(l) = 0. \end{array}\right\} \tag{7.20}$$

To prove this we use again the shooting approach and consider the IVP

$$\left.\begin{array}{l} -u'' - \lambda \sin u = 0, \\ u'(0) = 0, \quad u(0) = s. \end{array}\right\} \tag{7.21}$$

Note that the solution u depends on t, s and also λ, ie. $u = u(t; s, \lambda)$. In view of the right hand boundary conditions in (7.20), we consider

$$f(s, \lambda) = \frac{\partial u}{\partial t}(l; s, \lambda) = 0. \tag{7.22}$$

By uniqueness for the IVP (7.21) we have

$$f(0, \lambda) = 0 \qquad \text{for all } \lambda \in \mathbb{R}.$$

By Lemma 7.1, (7.22) is equivalent to (7.20), so nontrivial solutions y of (7.20) bifurcate at $\lambda = \lambda_0$ if and only if nontrivial solutions s of (7.22) bifurcate at $\lambda = \lambda_0$. To prove the latter assertion we use Theorem 6.1, for which we need f_s and $f_{s\lambda}$. So set $w = \dfrac{\partial u}{\partial s}$ and differentiate (7.21) with respect to s to obtain

$$\left.\begin{array}{l} -w'' - \lambda(\cos u)w = 0, \\ w'(0) = 0, \quad w(0) = 1, \end{array}\right\} \tag{7.23}$$

with solution $w = w(t; s, \lambda)$. Now when $s = 0$ $u = 0$ (by uniqueness for (7.21)) and (7.23) implies that $w(t) = w(t; 0, \lambda)$ satisfies the linear 2nd order ODE

$$- w'' - \lambda w = 0. \tag{7.24}$$

Thus $w(t) = A \sin \sqrt{\lambda} t + B \cos \sqrt{\lambda} t$, and to satisfy the boundary conditions, we have $A = 0$, $B = 1$. Then, by (7.22)

$$f_s(0, \lambda) = \frac{\partial w}{\partial t}(l; 0, \lambda) = -\sqrt{\lambda} \sin(\sqrt{\lambda} l). \tag{7.25}$$

This vanishes for $\lambda = \lambda_0 = \frac{m^2 \pi^2}{l^2}$, $m = 0, 1, 2....$ (In the application of the buckling of a rod the case $\lambda = 0$ is ruled out on physical grounds.) To check if bifurcation occurs at $\lambda = \lambda_0$, we have to compute $f_{s\lambda}(0, \lambda_0)$. To do this set

$$v = \frac{\partial w}{\partial \lambda},$$

and differentiate (7.23) with respect to λ to get

$$\left. \begin{array}{c} -v'' - (\cos u)w + \lambda(\sin u)\frac{\partial u}{\partial \lambda}w - \lambda(\cos u)v = 0, \\ v'(0) = 0, \quad v(0) = 0. \end{array} \right\} \qquad (7.26)$$

At $s = 0$, $u = 0$, and so $v(t) = v(t; 0, \lambda)$ satisfies

$$\left. \begin{array}{c} -v'' - \lambda v = w, \\ v'(0) = 0, \quad v(0) = 0. \end{array} \right\} \qquad (7.27)$$

Bifurcation occurs at λ_0 if

$$f_{s\lambda}(0, \lambda_0) = \frac{\partial v}{\partial t}(l, 0, \lambda_0) \neq 0. \qquad (7.28)$$

Suppose (7.28) does not hold. Then $v(t) = v(t; 0, \lambda_0)$ satisfies (7.27) with $\lambda = \lambda_0$, together with

$$v'(l) = 0. \qquad (7.29)$$

Then (7.27) implies

$$\begin{aligned} \int_0^l w^2 &= -\int_0^l v''w - \lambda_0 \int_0^l vw \\ &= +\int_0^l v'w' - \lambda_0 \int_0^l vw \quad \text{by (7.29)} \\ &= -\int_0^l vw'' - \lambda_0 \int_0^l vw \quad \text{since } w'(0) = 0 = w'(l) \\ &= \int_0^l v(-w'' - \lambda_0 w) = 0 \quad \text{by (7.24),} \end{aligned}$$

which is impossible since $w(t) = w(t; 0, \lambda_0) = \sqrt{\lambda_0} \cos \sqrt{\lambda_0} t$ and $\lambda_0 \neq 0$. So bifurcation from the trivial solution occurs at $\lambda = \lambda_0 = \frac{m^2 \pi^2}{l^2}$, $m = 1, 2, \ldots$

7.3. Calculation of Fold Points in ODEs Using Shooting

We consider this technique via an example. See also Seydel [45].

Example 7.3. Consider the following nonlinear ODE:

$$-y'' - \lambda \exp(y) = 0, \quad y(0) = 0 = y(1).$$

Using the development in Example 7.2 we set up an associated IVP and $f(s) := u(1; s, \lambda)$. To calculate the values of f and its derivatives in the shooting method, we have the three initial value problems:

$-u''$	$=$	λe^u	$-w''$	$=$	$\lambda e^u w$	$-v''$	$=$	$\lambda e^u v + e^u$
$u(0)$	$=$	0	$w(0)$	$=$	0	$v(0)$	$=$	0
$u'(0)$	$=$	s	$w'(0)$	$=$	1	$v'(0)$	$=$	0

Then

$$f(s, \lambda) = u(1; s, \lambda); \quad f_s(s, \lambda) = w(1; s, \lambda); \quad f_\lambda(s, \lambda) = v(1; s, \lambda).$$

We may follow the solution curve of $f(s, \lambda) = 0$ numerically by continuation with respect to s or λ, with a check on size of $|f_s|$, $|f_\lambda|$. If one of these becomes small then we use the other as a parameter. If they are not both zero then the curve has only turning points. Observe also that when $s = 0$, $\lambda = 0$ we have

$$u = 0, \quad \text{so} \quad f(0, 0) = 0;$$
$$w = t, \quad \text{so} \quad f_s(0, 0) = 1;$$
$$v = -\frac{1}{2}t^2, \quad \text{so} \quad f_\lambda(0, 0) = -\frac{1}{2}.$$

We may use $(0, 0)$ as the starting point for continuation. To solve for u, v, w simultaneously, set $u_1 = u$, $u_2 = u'$, $u_3 = w$, $u_4 = w'$, $u_5 = v$, $u_6 = v'$. Then the three problems become:

$$\begin{bmatrix} u_1 \\ u_2 \\ u_3 \\ u_4 \\ u_5 \\ u_6 \\ u_7 \end{bmatrix}' = \begin{bmatrix} u_2 \\ -u_7 \exp(u_1) \\ u_4 \\ -u_7 \exp(u_1)u_3 \\ u_6 \\ -u_7 \exp(u_1)u_5 - \exp(u_1) \\ 0 \end{bmatrix}, \quad \text{with} \quad \begin{bmatrix} u_1 \\ u_2 \\ u_3 \\ u_4 \\ u_5 \\ u_6 \\ u_7 \end{bmatrix} = \begin{bmatrix} 0 \\ s \\ 0 \\ 1 \\ 0 \\ 0 \\ \lambda \end{bmatrix} \quad \text{at } t = 0.$$

For any (s, λ) we solve this system by any numerical routine, and then

$$f(s, \lambda) = u_1(1), \quad f_s(s, \lambda) = u_3(1), \quad f_\lambda(s, \lambda) = u_5(1). \tag{7.30}$$

These may be used in a continuation method as described in §4.

8. Hopf Bifurcation

As was seen in Example 2.2, one way a steady state of $\dot{x} = F(x, \lambda)$ can lose stability as λ varies is when a complex pair of eigenvalues of $F_x(x, \lambda)$ crosses the imaginary axis. This situation is described by the classical Hopf bifurcation theorem [19].

Theorem 8.1 (Hopf Bifurcation). *Let* $F \in C^2(\mathbb{R}^{n+1})$ *and assume*

(i) $F(x_0, \lambda_0) = 0$,
(ii) $F_x(x_0, \lambda_0)$ *has a simple, purely imaginary eigenvalue* $\mu(\lambda_0) = +i\beta_0$, $\beta_0 \neq 0$, *with eigenvector* $\phi_0 + i\psi_0$, *and no other eigenvalues on the imaginary axis apart from at* $-i\beta_0$,

(iii) $\left\{ \frac{d}{d\lambda} \mathrm{Re}(\mu) \right\} \Big|_{\lambda=\lambda_0} \neq 0.$

Then there exists an $a_0 > 0$ and a parameter a such that $\dot{x} = F(x, \lambda)$ has a smooth branch of $T(a)$-periodic solutions $(x(t, a; \lambda(a)), \lambda(a))$ for $0 \leq t \leq T(a)$, for all $|a| < a_0$ with the following properties

$$
\begin{aligned}
x(t, a; \lambda(a)) &= x^s(\lambda(a)) + a(\cos(\beta_0 t)\phi_0 - \sin(\beta_0 t)\psi_0) + \mathcal{O}(a^2), \\
\lambda(a) &= \lambda_0 + \mathcal{O}(a^2), \\
T(a) &= \frac{2\pi}{\beta_0} + \mathcal{O}(a^2),
\end{aligned}
$$

where $x^s(\lambda(a))$ denotes the steady solution at $\lambda = \lambda(a)$.

The theorem states that at (x_0, λ_0) there is a birth of periodic solutions that may be parametrised by the amplitude a. If conditions *(i),(ii)* and *(iii)* hold then (x_0, λ_0) is called a *Hopf bifurcation point*.

Note that since F_x^0 is nonsingular, the Implicit Function Theorem ensures that S (the solution set of the steady problem) may be parametrised by λ near $\lambda = \lambda_0$. Using the approach in the proof of part (b) of Theorem 5.1 extended to $\mathbb{C}^{n+1}$ shows that a (real or complex) simple eigenvalue of $F_x(x(\lambda), \lambda)$ is a smooth function of λ near λ_0. Hence we can write $\mathrm{Re}(\mu) = \mathrm{Re}(\mu(\lambda))$.

Condition *(iii)* in Theorem 8.1 is another example of a nondegeneracy condition where an eigenvalue smoothly crosses the imaginary axis.

Theorem 8.1 is due to Hopf in a famous paper in 1942, though in 1929 Andronov was the first to formulate a theorem and Poincaré's work in 1892 contained examples of this type of bifurcation. So this phenomenon is now often called Poincaré/Andronov/Hopf bifurcation. A nice treatment of the theory of Hopf bifurcation with references to the work of Andronov and Poincaré is given in Wiggins [52].

8.1. Calculation of a Hopf Bifurcation Point

If a good estimate of the Hopf bifurcation point is known then it may be computed exactly by setting up and solving an appropriate extended system (cf. the fold point system in (5.4).) Consider the nonlinear system

$$
H(y) = 0 \tag{8.1}
$$

where

$$
H(y) := \begin{pmatrix} F(x, \lambda) \\ F_x(x, \lambda)\phi - \beta\psi \\ c^T\phi - 1 \\ F_x(x, \lambda)\psi + \beta\phi \\ c^T\psi \end{pmatrix}, \quad y := \begin{pmatrix} x \\ \phi \\ \lambda \\ \psi \\ \beta \end{pmatrix} \in \mathbb{R}^{3n+2} \tag{8.2}
$$

with $\boldsymbol{H} : \mathbb{R}^{3n+2} \to \mathbb{R}^{3n+2}$. This is the obvious system to write down as can be seen from conditions *(i)* and *(ii)* in Theorem 8.1. There are two conditions on the eigenvector $\boldsymbol{\phi} + i\boldsymbol{\psi}$ since a complex vector requires two real normalisations.

The following theorem is readily proved (see [16]).

Theorem 8.2. *Let $(\boldsymbol{x}_0, \lambda_0)$ be a Hopf bifurcation point (i.e. (i), (ii) and (iii) of Theorem 8.1 hold) and assume $\boldsymbol{c}$ has non-zero projection on* span $\{\boldsymbol{\phi}_0\}$. *Then $\boldsymbol{y}_0 := (\boldsymbol{x}_0^T, \boldsymbol{\phi}_0^T, \lambda_0, \boldsymbol{\psi}_0^T, \beta_0)^T \in \mathbb{R}^{3n+2}$ is a regular solution of (8.1).*

Note that fold points also satisfy (8.1) since if $(\boldsymbol{x}_0, \lambda_0)$ is a fold point and $\boldsymbol{\phi}_0 \in \ker(\boldsymbol{F_x}(\boldsymbol{x}_0, \lambda_0))$ then $\boldsymbol{y}_0 = (\boldsymbol{x}_0, \boldsymbol{\phi}_0, \lambda_0, \boldsymbol{0}, 0)$ satisfies $\boldsymbol{H}(\boldsymbol{y}_0) = \boldsymbol{0}$. In fact $\boldsymbol{y}_0$ is a regular solution if the conditions of Theorem 5.1 hold.

System (8.1) was first introduced by Jepson [21] and independently by Griewank and Reddien [16] who showed that the linearisation of (8.1) could be reduced to solving systems with a bordered form of $\boldsymbol{F_x^2}(\boldsymbol{x}, \lambda) + \beta^2 I$. This is natural since an alternative system for a Hopf bifurcation can be derived by using the fact that the second and fourth equations of (8.1) can be written as $(\boldsymbol{F_x}(\boldsymbol{x}, \lambda) + \beta^2 I)\boldsymbol{v} = \boldsymbol{0}$ with $\boldsymbol{v} = \boldsymbol{\phi}$ or $\boldsymbol{\psi}$.

To eliminate the possibility of computing a fold point rather than a Hopf bifurcation point, Werner and Janovsky [50] used the system

$$\boldsymbol{R}(\boldsymbol{y}) = \boldsymbol{0} \tag{8.3}$$

where

$$\boldsymbol{R}(\boldsymbol{y}) = \begin{pmatrix} \boldsymbol{F}(\boldsymbol{x}, \lambda) \\ (\boldsymbol{F_x^2}(\boldsymbol{x}, \lambda) + \nu I)\boldsymbol{\phi} \\ \boldsymbol{c}^T \boldsymbol{\phi} \\ \boldsymbol{c}^T \boldsymbol{F_x}(\boldsymbol{x}, \lambda)\boldsymbol{\phi} - 1 \end{pmatrix}, \quad \boldsymbol{y} = \begin{pmatrix} \boldsymbol{x} \\ \boldsymbol{\phi} \\ \lambda \\ \nu \end{pmatrix} \in \mathbb{R}^{2n+2} \tag{8.4}$$

in which $\boldsymbol{R} : \mathbb{R}^{2n+2} \to \mathbb{R}^{2n+2}$, and where $\boldsymbol{c}$ is a constant vector. The last equation in (8.3) ensures that the solution cannot be a fold point. The system $\boldsymbol{R}(\boldsymbol{y}) = \boldsymbol{0}$ is closely related to a system derived by Roose and Hlavacek [41], but (8.3) has several advantages when computing paths of Hopf bifurcations if a second parameter is varying (see [50]).

The fact that (8.1) or (8.3) is regular at a Hopf bifurcation is important since Newton's method (or some variant) will probably be used to solve the system. Just as is the case in §5 for the computation of a fold point, there are efficient ways of solving the Jacobian systems in Newton's method. In [16] an efficient procedure is described for the solution of the $(3n + 2) \times (3n + 2)$ Jacobian systems arising from (8.1) by solving systems with a bordering of $(\boldsymbol{F_x^2}(\boldsymbol{x}, \lambda) + \beta^2 I)$. We do not give the details here. A nice summary is given in [1].

8.2. The Detection of Hopf Bifurcations in Large Systems

The extended systems in §8.1 can only be used when we know we are near a Hopf point. The following section describes how this might be determined in practice.

When computing a path of steady solutions of $\dot{x} = F(x, \lambda)$ using a numerical continuation method it is easy to pass over a Hopf bifurcation point without "noticing" it, since when a complex pair of eigenvalues crosses the imaginary axis there is no easy detection test based on the linear algebra of the continuation method. In particular the sign of the determinant of F_x does not change. If n is small then the simplest test is merely to compute all the eigenvalues of F_x during the continuation. For large n, say when F arises from a discretized PDE, such an approach will usually be out of the question. The efficient detection of Hopf bifurcations in large systems is an important and, as yet, unsolved problem. The review article [8] discusses in detail both classical techniques from complex analysis and linear algebra-based methods. It is natural to try to use classical ideas from complex analysis for this problem since then one seeks an *integer*, namely the number of eigenvalues in the unstable half-plane, and counting algorithms are applicable. This is explored for large systems in [14] but there is still work to be done in this area.

The *rightmost* eigenvalues of $F_x(x, \lambda)$ determine the (linearised) stability of the steady solutions of $\dot{x} = F(x, \lambda)$ and one strategy for the detection of Hopf bifurcation points is to monitor a few of the rightmost eigenvalues as the path of steady state solutions is computed. (Note that the rightmost eigenvalue is not a continuous function of λ, see [36].) Standard iterative methods, e.g. Arnoldi's method and simultaneous iteration, compute extremal or dominant eigenvalues, and there is no guarantee that the rightmost eigenvalue will be computed by direct application of these methods to F_x. The approach in [8] and [3] is to first transform the eigenvalue problem using the Generalised Cayley Transform

$$C(A) = (A - \alpha_1 I)^{-1}(A - \alpha_2 I), \quad \alpha_1, \alpha_2 \in \mathbb{R},$$

which has the key property that if $\mu \neq \alpha_1$ is an eigenvalue of A then $\theta := (\mu - \alpha_1)^{-1}(\mu - \alpha_2)$ is an eigenvalue of $C(A)$. Also, $\mathrm{Re}(\mu) \leq (\geq)(\alpha_1 + \alpha_2)/2$ if and only if $|\theta| \leq (\geq)1$. Thus eigenvalues to the right of the line $\mathrm{Re}(\mu) = (\alpha_1 + \alpha_2)/2$ are mapped outside the unit circle and eigenvalues to the left of the line mapped inside the unit circle. In [8] and [3] algorithms based on computing dominant eigenvalues of $C(F_x)$ using Arnoldi or simultaneous iteration are presented, with consequent calculation of rightmost eigenvalues of F_x. These algorithms were tested on a variety of problems, including systems arising from mixed finite element discretizations of the Navier-Stokes equations. Quite large problems can in fact be tackled. Indeed, in [15] the problem of the stability of flow over a backward facing step is discussed in detail and the rightmost eigenvalues of a system with over 3×10^5 degrees of freedom are found using the Generalised Cayley transform allied with simultaneous iteration.

However it was later noted (see [31]) that since

$$C(A) = I + (\alpha_1 - \alpha_2)(A - \alpha_1 I)^{-1},$$

Arnoldi's method applied to $C(A)$ builds the same Krylov subspace as Arnoldi's method applied to the shift-invert transformation $(A - \alpha_1 I)^{-1}$. Thus if Arnoldi's method is the eigenvalue solver there is no advantage in using the Cayley transform, which needs two parameters, over the standard shift-invert transformation (see [31]).

One can think of the approach in [3] as the computation of the subspace containing the eigenvectors corresponding to the rightmost eigenvalues of F_x. A similar theme, derived using a completely different approach, is described by [42] and refined by [5]. In these papers the subspace corresponding to a set of (say rightmost) eigenvalues is computed using a hybrid iterative process based on a splitting technique. Roughly speaking a small subspace is computed using a Newton-type method and the solution in the larger complementary space is found using a Picard (contraction mapping) approach. One advantage is that the Jacobian matrix F_x need never be evaluated.

When detecting Hopf bifurcations in the Navier-Stokes equations using mixed finite elements, a generalised eigenvalue problem of the form $A\phi = \mu B\phi$ arises where B is singular. A common method is to apply Arnoldi's method to the shifted-inverted matrix $(A - \alpha B)^{-1}B$, which is singular since B is singular. In [30] it is noted that great care is needed here when using Arnoldi's method because of the generation of spurious eigenvalues due to perturbation of the zero eigenvalue. The details are quite technical and are omitted.

Finally we note that chapter 5 of [45] contains an overview of Hopf detection techniques.

Bibliography

1. W. J. Beyn. Numerical methods for dynamical systems. In W. Light, editor, *Advances in Numerical Analysis*, pages 175–227. Clarendon Press, Oxford, 1991.
2. S-N. Chow and J. K. Hale. *Methods of Bifurcation Theory*. Springer-Verlag, New York, 1982.
3. K. A. Cliffe, T. J. Garratt, and A. Spence. Eigenvalues of the discretized Navier-Stokes equation with application to the detection of Hopf bifurcations. *Advances in Computational Maths.*, 1:337–356, 1993.
4. M. G. Crandall and P. H. Rabinowitz. Bifurcation from a simple eigenvalue. *J. Functional Analysis*, 8:321–340, 1971.
5. Bryan D. Davidson. Large-scale continuation and numerical bifurcation for partial differential equations. *SIAM J. Numer. Anal.*, 34:2008–2027, October 1997.

6. J.E Dennis Jr and Robert B. Schnabel. *Numerical Methods for Unconstrained Optimization and Nonlinear Equations.* Prentice Hall, New Jersey, 1983.
7. E. J. Doedel and J. P. Kernevez. AUTO: Software for continuation and bifurcation problems in ordinary differential equations. Technical report, Caltech, Pasadena, 1986.
8. T. J. Garratt, G. Moore, and A. Spence. Two methods for the numerical detection of Hopf bifurcations. In T. Küpper R. Seydel, F. W. Schneider and H. Troger, editors, *Bifurcation and Chaos; Analysis, Algorithms, Applications,* volume **97**. Birkhäuser, Basel, 1991.
9. M. Golubitsky and D. G. Schaeffer. *Singularities and Groups in Bifurcation Theory, Vol I,* volume **51**. Springer-Verlag, New York, 1985.
10. M. Golubitsky, I. Stewart, and D. G. Schaeffer. *Singularities and Groups in Bifurcation Theory, Vol II,* volume **69**. Springer-Verlag, New York, 1988.
11. W. Govaerts. Stable solvers and block elimination for bordered systems. *SIAM J Matrix Anal. and Applications,* **12**:469–483, 1991.
12. W. Govaerts. Bordered matrices and singularities of large nonlinear systems. *Int. J. of Bifurcation and Chaos,* **5**:243–250, 1995.
13. W. Govaerts. *Numerical Methods for Bifurcations of Dynamic Equilibria.* SIAM, To be published, 1999.
14. W. Govaerts and A. Spence. Detection of Hopf points by counting sectors in the complex plane. *Numer. Math.,* **75**:43–58, 1996.
15. P. M. Gresho, D. K. Gartling, J. R. Torczynski, K. A. Cliffe, K. H. Winters, T. J. Garratt, A. Spence, and J. W. Goodrich. Is the steady viscous incompressible 2D flow over a backward facing step ar Re=800 stable? *Int. J. Numer. Meth. Fluids,* **17**:501–541, 1993.
16. A. Griewank and G. Reddien. The calculation of Hopf points by a direct method. *IMA J. Numer. Anal.,* **3**:295–303, 1983.
17. A. Griewank and G. Reddien. Characterization and computation of generalized turning points. *SIAM J. Numer. Anal.,* **21**:176–185, 1984.
18. A. Griewank and G. Reddien. Computation of cusp singularities for operator equations and their discretizations. *J. Comp. Appl Maths,* **16**:133–153, 1989.
19. B. D. Habard, N. D. Kazarinoff, and Y-H Wan. Theory and applications of Hopf bifurcation. *London Math. Soc. Lecture Note Series,* **41**, 1981.
20. M. W. Hirsch and S. Smale. *Differential Equations, Dynamical Systems, and Linear Algebra.* Academic Press, New York, 1974.
21. A. D. Jepson. *Numerical Hopf Bifurcation.* PhD thesis, Caltech, Pasadena, 1981.
22. A. D. Jepson and D. W. Decker. Convergence near bifurcation. *SIAM J. Numer. Anal.,* **23**:959–975, 1986.
23. A. D. Jepson and A. Spence. The numerical solution of nonlinear equations having several parameters, I: scalar equations. *SIAM J. Numer. Anal.,* **22**:736–759, 1985.
24. H. B. Keller. *Numerical methods for two-point boundary value problems.* Ginn-Blaisdell, Waltham, 1968.
25. H. B. Keller. *Numerical solution of bifurcation and nonlinear eigenvalue problems,* pages 359–384. Academic Press, New York, 1977.
26. H. B. Keller. *Numerical methods in bifurcation problems.* Springer-Verlag, 1987.
27. J. B. Keller. Bifurcation Theory for Ordinary Differential Equations. In J. B. Keller and S. Antmann, editors, *Bifurcation Theory and Nonlinear Eigenvalue Problems.* Benjamin, New York, 1969.
28. T. Küpper, H. D. Mittelmann, and H. Weber. *Numerical Methods for Bifurcation Problems,* volume **70**. Birkhäuser, Basel, 1984.

29. T. Küpper, R. Seydel, and H. Troger. Bifurcation: Analysis, Algorithms, Applications. *Proceedings of a conference in Dortmund*, **79**, 1987.

30. K. Meerbergen and A. Spence. Implicitly Restarted Arnoldi with purification for the shift-invert transformation. *Mathematics of Computation*, **67**:667–689, 1997.

31. K. Meerbergen, A. Spence, and D. Roose. Shift-Invert and Cayley transforms for detection of rightmost eigenvalues of nonsymmetric matrices. *BIT*, **34**:409–423, 1994.

32. H. D. Mittelmann and H. Weber. *Bifurcation problems and their Numerical Solution*, volume **54**. Birkhaüser, Basel, 1980.

33. G. Moore. Numerical Treatment of nontrivial bifurcation points. *Numer. Funct. Anal. and Optimiz.*, **2**:441–472, 1980.

34. G. Moore. Some remarks on the deflated block elimination method. In T. Küpper, R. Seydel, and H. Troger, editors, *Bifurcation: Analysis, Algorithms and Applications*, International Series in Numerical Mathematics, pages 222–234. Birkhäuser, 1987.

35. G. Moore and A. Spence. The calculation of Turning Points of nonlinear equations. *SIAM J. Numer. Anal.*, **17**:567–576, 1980.

36. R. Neubert. Predictor-Corrector techniques for detecting Hopf bifurcation points. *International J. Bifurcation and Chaos*, **3**:1311–1318, 1993.

37. P. H. Rabinowitz. *Applications of Bifurcation Theory*. Academic Press, New York, 1977.

38. L. B. Rall. *Computational solution of nonlinear operator equations*. Wiley, New York, 1969.

39. W. C. Rheinboldt. *Numerical Analysis of parametrized nonlinear equations*. Wiley-Interscience, New York, 1986.

40. D. Roose, B. De Dier, and A. Spence. *Continuation and Bifurcations: Numerical Techniques and Applications*, volume **313**. Kluwer, Dordrecht, 1990.

41. D. Roose and V. Hlavacek. A direct method for the computation of Hopf bifurcation points. *SIAM J. Appl. Math.*, **45**:879–894, 1985.

42. G. Schroff and H. B. Keller. Stabilization of unstable procedures: The recursive projection method. *SIAM J. Numer. Analysis*, **30**:1099–1120, 1993.

43. R. Seydel. Numerical Computation of branch points in nonlinear equations. *Numer. Math.*, **32**:339–352, 1979.

44. R. Seydel. Numerical Computation of branch points in ordinary differential equations. *Numer. Math.*, **32**:51–68, 1979.

45. R. Seydel. *Practical Bifurcation and Stability Analysis: From equilibrium to Chaos*. Springer-Verlag, New York, 1994.

46. R. Seydel, T. Küpper, F. W. Schneider, and H. Troger. *Bifurcation and Chaos: Analysis, Algorithms, Applications*, volume **97** of *International Series in Numerical Mathematics*. Birkhäuser, Basel, 1991.

47. G. Symm and J. H. Wilkinson. Realistic Error Bounds for a Simple Eigenvalue and its Associated Eigenvector. *Numer. Math.*, **35**:113–126, 1980.

48. A. Vanderbauwhede. Local bifurcation and symmetry. In *Research notes in mathematics*, volume **75**. Pitman, London, 1982.

49. B. Werner. Regular systems for bifurcation points with underlying symmetries. In T. Küpper, H. D. Mittelmann, and H. Weber, editors, *Numerical Methods for Bifurcation Problems*, volume **70**, pages 562–574. Birkhäuser, Basel, 1984.

50. B. Werner and V. Janovsky. Computation of Hopf branches bifurcating from Takens-Bogdanov points for problems with symmetry. In R. Seydel, T. Küpper, F. W. Schneider, and H. Troger, editors, *Bifurcation and Chaos: Analysis, Algorithms, Applications*, pages 377–388. Birkhauser, 1991.

51. B. Werner and A. Spence. The computation of symmetry breaking bifurcation points. *SIAM J. Numer. Anal.*, **21**:388–399, 1984.
52. S. Wiggins. *Introduction to Applied Nonlinear Dynamical Systems and Chaos.* Springer-Verlag, New York, 1990.

Spectra and Pseudospectra

The Behaviour of Non-Normal Matrices and Operators

Lloyd N. Trefethen

Oxford University Computing Laboratory, Wolfson Building, Parks Road, Oxford OX1 3QD, UK

Preface

The five sections of these notes will one day be the first five chapters of a book, to appear some time after 2001.

1. Eigenvalues

Eigenvalues are among the most successful tools of applied mathematics. Here are some of the fields where they are important, with representative citations.

acoustics [47]	chemistry
control theory [40]	earthquake engineering [8]
ecology [48]	economics
fluid mechanics [20]	functional analysis
helioseismology [29]	magnetohydrodynamics [7]
Markov chains [52]	matrix iterations [31]
numerical soln. of differential eqns. [61]	physics of music [23]
quantum mechanics	spectroscopy
partial differential eqs. [11]	structural analysis [9]
vibration analysis	

Figures 1.1 and 1.2 present images of eigenvalues in two quite different scientific fields.

In the simplest context of matrices, the definitions are as follows. Let A be an $N \times N$ matrix with real or complex coefficients; we write $A \in \mathbb{C}^{N \times N}$. Let v be a nonzero real or complex column vector of length N, and let λ be a real or complex scalar; we write $v \in \mathbb{C}^N$ and $\lambda \in \mathbb{C}$. Then v is an **eigenvector** of A, and $\lambda \in \mathbb{C}$ is its corresponding **eigenvalue**, if

$$Av = \lambda v. \tag{1.1}$$

(Even if A is real, its eigenvalues are in general complex unless A is symmetric.) The set of all the eigenvalues of A is the **spectrum** of A, a nonempty

Fig. 1.1. Spectroscopic image of light from the sun. The black 'Fraunhofer lines' correspond to various differences of eigenvalues of the Schrödinger operator for atoms such as H, Fe, Ca, Na, and Mg that are present in the cooler outer layers of the sun. Light at these frequencies resonates with frequencies of the transitions between energy states in these atoms and is absorbed. Spectroscopic measurements such as these are a crucial tool in chemical analysis, not only of astronomical bodies, and by making possible the measurement of redshifts, they led to the discovery of the expanding universe. This figure is reproduced by permission of Prof. J.M. Malherbe of the Observatoire de Paris.

subset of the complex plane $\mathbb{C}$ which we denote by $\Lambda(A)$. The spectrum can also be defined as the set of points $z \in \mathbb{C}$ where the **resolvent** matrix,

$$(z - A)^{-1}, \tag{1.2}$$

does not exist. Here and throughout these notes, $z - A$ is an abbreviation for $zI - A$, where I is the identity.

Unlike singular values [70], eigenvalues make sense only for a matrix that is square. This reflects the fact that in applications, they are generally used where a matrix is to be compounded iteratively, for example, as a power A^k or an exponential $e^{tA} = I + tA + (tA)^2/2 + \cdots$.

For most matrices A, there exists a **complete set of eigenvectors**, a set of N linearly independent eigenvectors $v_1, \ldots, v_N$ with $Av_j = \lambda_j v_j$. If A has N distinct eigenvalues, then it is guaranteed to have a complete set of eigenvectors, and they are unique up to normalization by scalar factors. For any matrix A with a complete set of eigenvectors $\{v_j\}$, let V be the $N \times N$ matrix whose jth column is v_j, a **matrix of eigenvectors**. Then we can write all N eigenvalue conditions at once by the matrix equation

$$AV = V\Lambda, \tag{1.3}$$

where Λ is the diagonal $N \times N$ matrix whose jth diagonal entry is λ_j. Pictorially,

$$
A \begin{bmatrix} | & | & & | \\ v_1 & v_2 & \cdots & v_N \\ | & | & & | \end{bmatrix} = \begin{bmatrix} | & | & & | \\ v_1 & v_2 & \cdots & v_N \\ | & | & & | \end{bmatrix} \begin{pmatrix} \lambda_1 & & & \\ & \lambda_2 & & \\ & & \ddots & \\ & & & \lambda_N \end{pmatrix}.
$$

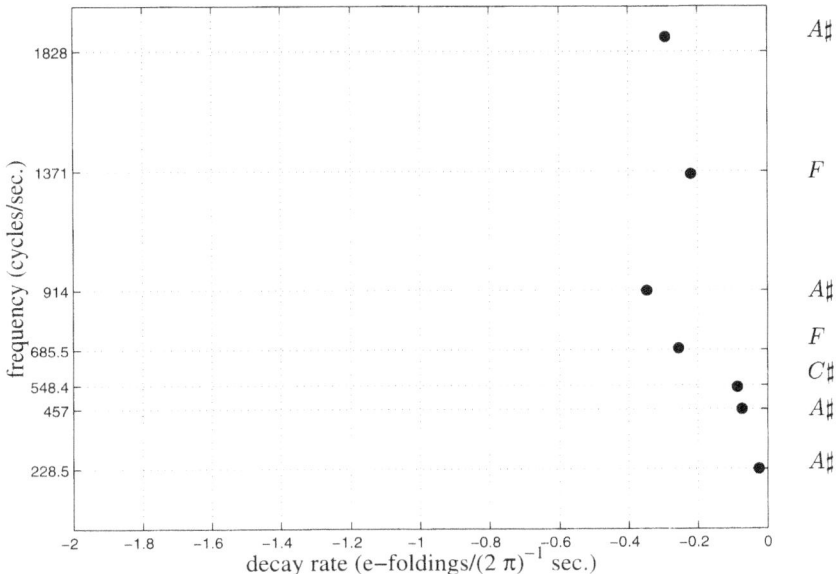

Fig. 1.2. Measured eigenvalues in the complex plane of a minor third $A_4\sharp$ carillon bell (figure from [38] based on data from [63]). The grid lines show the positions of the frequencies corresponding to a minor third chord at 456.8 Hz, together with two octaves above the fundamental and one below. Shortly after the bell is struck, the ear hears all seven of the frequencies portrayed; a little later, the higher four have decayed and mostly the lowest three are heard; still later the lowest mode, the "hum", dominates. The simple rational relationships among these frequencies would not hold for arbitrarily shaped bells but are the result of generations of evolution of the shapes of bells to achieve a pleasing effect.

Since the eigenvectors v_j are linearly independent, V is nonsingular, and thus we can multiply (1.3) on the right by V^{-1} to obtain the factorization

$$A = V \Lambda V^{-1}, \tag{1.4}$$

known as an **eigenvalue decomposition** or a **diagonalization** of A. In view of this formula, a matrix with a complete set of eigenvectors is said to be **diagonalizable.** An equivalent term is **nondefective.**

The eigenvalue decomposition expresses a change of basis to "eigenvector coordinates," i.e., coefficients in an expansion in eigenvectors. If $A = V \Lambda V^{-1}$, for example, then we have

$$V^{-1}(A^k x) = V^{-1}(V \Lambda V^{-1})^k x = \Lambda^k (V^{-1} x). \tag{1.5}$$

Now the product $V^{-1}(A^k x)$ is equal to the vector c of coefficients in an expansion $A^k x = Vc = \sum c_j v_j$ of $A^k x$ as a linear combination of the eigenvectors $\{v_j\}$, and similarly, $V^{-1}x$ is the vector of coefficients in an expansion of x. Thus, (1.5) asserts that to compute $A^k x$, we can expand x in the basis of eigenvectors, apply the diagonal matrix Λ^k, and interpret the result as a vector of coefficients for another expansion in the basis of eigenvectors. In other words, the change of basis has rendered the problem diagonal and hence trivial. For $e^{tA}x$, similarly, we have

$$V^{-1}(e^{tA}x) = V^{-1}(Ve^{t\Lambda}V^{-1})x = e^{t\Lambda}(V^{-1}x), \tag{1.6}$$

so diagonalization makes this problem trivial too, and likewise for other functions $f(A)$.

So far we have taken A to be a matrix, but eigenvalues are equally important when A is a more general linear operator such as an infinite matrix, a differential operator, or an integral operator. Indeed, eigenvalue problems for matrices often come about through discretization of linear operators. The **spectrum** $\Lambda(A)$ of a closed operator A defined in a Banach (or Hilbert) space is defined as the set of numbers $z \in \mathbb{C}$ for which the resolvent (1.2) does not exist as a bounded operator defined on the whole space [41]. It can be any closed set in the complex plane, including the empty set. Eigenvalues and eigenvectors (also called **eigenfunctions** or **eigenmodes**) are still defined by (1.1), but among the new features that arise in the operator case is the phenomenon that not every $z \in \Lambda(A)$ is necessarily an eigenvalue. These notes avoid fine points of spectral theory wherever possible, for the main issues to be investigated are orthogonal to the differences between matrices and operators. In particular, the distinction between spectra and pseudospectra has little to do with the distinction between point and continuous spectra. In certain contexts, of course, it will be necessary for us to be more precise.

Our topic is the limitations of eigenvalues, and alternatives to eigenvalues. In the remainder of this introductory section, let us accordingly consider the question: What are eigenvalues useful for? Why are eigenvalues and eigenfunctions—more generally, spectra and spectral theory—among the standard tools of applied mathematics?

We begin with a one-paragraph history [4],[18],[64]. It is not too great an oversimplification to say that a major part of eigenvalue analysis originated early in the nineteenth century with Fourier's solution of the heat equation by series expansions. Fourier's ideas were extended by Poisson, and other highlights of the nineteenth century include Sturm and Liouville's treatment of more general second-order differential equations in the 1830s, Sylvester and Cayley's diagonalization of symmetric matrices in the 1850s (the origins of this idea go back to Cauchy, Jacobi, Lagrange, Euler, Fermat, and Descartes), Weber and Schwarz's treatment of a vibrating membrane in 1869 and 1885 (whose origins in vibrating strings go back to D. Bernoulli, Euler, d'Alembert,..., Pythagoras), Lord Rayleigh's treatise *The Theory of Sound* in 1877 [55], and further developments by Poincaré around 1890. By 1900, eigenvalues and eigenfunction expansions were well known, especially in the context of differential equations. The new century brought the mathematical theory of linear operators due to Fredholm, Hilbert, Schmidt, von Neumann and others; the terms "eigenvalue" and "spectral theory" appear to have been coined by Hilbert. The influential book by Courant and Hilbert in 1924 surveyed a large amount of material concerning eigenvalues of differential equations and vibration problems [10]. Just two years later came the explosive ideas of quantum mechanics, which in a short time, at the hands of Heisenberg, Jordan, Schrödinger, Dirac, and others, moved matrices and operators to center stage of the scientific world. Quantum "matrix mechanics" revealed that energy states of atoms and molecules could be viewed as eigenfunctions of Schrödinger operators, thereby explaining Fig. 1.1, the periodic table of the elements, and countless other scientific observations besides, and from that time on, every mathematical scientist has known the basics of matrices, operators, eigenvalues, and eigenfunctions.

What exactly do eigenvalues offer that makes them useful for so many problems? I believe that there are three principal answers to this question, more than one of which may be important in a particular application.

1. DIAGONALIZATION AND SEPARATION OF VARIABLES: USE OF THE EIGENFUNCTIONS AS A BASIS. One thing eigenvalues may accomplish is the decoupling as in (1.4)–(1.6) of a problem involving vectors or functions into a collection of problems involving scalars, which may make subsequent computations easier. For example, in Fourier's problem of heat conduction in a solid bar with zero temperature at both ends, the eigenmodes are sine waves that decay independently as a function of t. If an arbitrary initial temperature distribution is expanded as a sum of these sine waves, then the solution at a later time can be calculated by summing the components of the expansion.

2. RESONANCE: HEIGHTENED RESPONSE TO SELECTED INPUTS. Diagonalization is an algorithmic idea; the other uses of eigenvalues are more physical. One is the analysis of the phenomenon of resonance, perhaps most familiar in the context of vibrating strings, drums, and mechanical structures. Any visitor to science museums has seen demonstrations that certain systems respond preferentially to vibrations at special frequencies. These frequencies are the eigenvalues of the linear or linearized operator that governs the system in question, and the form of the response is associated with the corresponding eigenfunctions. Examples of resonance are familiar—one thinks of soldiers breaking step as they cross bridges, or of the less fortunate Tacoma Narrows Bridge in the 1940s, whose collapse was driven by a wind-induced flow oscillation too close to a structural eigenfrequency; of buildings and their response to the vibrations of earthquakes—an application where eigenvalues are written into legal codes; of that old cartoon standby, the soprano whose high E shatters windows. In other examples resonance is desired rather than feared: examples include AM radio, where the signal from a far-off station is selected

from a sea of background noise by a finely tuned resonant circuit, and the cochlea of the human ear, whose basilar membrane resonates preferentially in different locations according to the frequency of the sound input and thus in a sense tunes in all stations at once. These last two examples, incidentally, illustrate the wide range of complexity in applications of eigenvalue ideas, for the radio problem is straightforward and almost perfectly linear, whereas the ear is a complicated nonlinear system, not yet fully understood, for which eigenmodes are only a first step.

3. ASYMPTOTICS AND STABILITY: DOMINANT RESPONSE TO GENERAL INPUTS. A related application of eigenvalues is to questions of the form, what will happen as time elapses (or in the extreme, $t \to \infty$) to a system that has experienced some more or less random disturbance? Fourier's heat problem again affords an example: whatever the shape of the initial temperature distribution, the higher sine waves decay faster than the lowest one, and therefore almost any initial distribution will eventually come to look like the half-wavelength sine with zeros just at the two ends of the interval. Similarly, what makes a church bell as in Fig. 1.2 chime musically? As the clapper strikes, all frequencies are excited, but differential decay rates soon filter out all but a few dominant ones, and the result is a pleasing sound. Kettledrums operate on the same principle, as do Markov chains in probability theory. Sometimes the crucial issue is a question of stability: are there modes that grow rather than decay with t? For example, in fluid mechanics a standard technique to determine whether small perturbations to a laminar flow will be amplified into large ones—which may then trigger the onset of turbulence—is to calculate whether the eigenvalues of the system all lie in the left half of the complex plane. (This technique is not always successful.) Similar questions arise in control theory and in numerical analysis, where time is discrete and stability depends on eigenvalues being less than one in complex modulus. Problems of convergence of matrix iterations in numerical analysis are also related, the convergence rate being determined by how close certain eigenvalues are to zero.

Principles (1), (2) and (3) account for most applications of eigenvalues. (Sometimes the latter two are hard to distinguish, as for example in the operation of bowed or blown musical instruments. The significance of eigenvalues in quantum mechanics also has special features, not well captured by (1)–(3) and indeed not really understood at all.) In view of the ubiquity of vibrations, oscillations, and linear or approximately linear processes in the physical world, they amply justify the great attention that has been given to eigenvalues over the years.

And I think there is a fourth reason, too, for the success of eigenvalues.

4. THEY GIVE A MATRIX A PERSONALITY. We humans like images; our brains are specially adapted to interpret them. Eigenvalues enable us to take

the abstraction of a matrix or linear operator, for whose analysis we possess no hardwired talent, and portray it as a picture.

These notes are about a class of problems for which eigenvalue methods may fail: problems involving matrices or operators for which the matrix V^{-1} of (1.4)–(1.6), if it exists, is very large:

$$\|V^{-1}\| \gg 1. \tag{1.7}$$

(This often turns out to mean exponentially large with respect to some parameter.) This formulation of the matter assumes that the matrix V itself is in some sense reasonably scaled, with $\|V\|$ roughly of order 1. If no assumptions are made about the scaling of $\|V\|$, then (1.7) should be replaced by a statement about the **condition number** of V in the norm $\|\cdot\|$,

$$\|V\|\,\|V^{-1}\| \gg 1, \tag{1.8}$$

and to be still more precise we should require that (1.8) hold not just for some eigenvector matrix V, whose eigenvector columns might be badly scaled relative to one another, but for any eigenvector matrix V. For operators as opposed to matrices, a suitable generalization of (1.8) is still a good starting point, though such simple formulations do not always work.

The conditions (1.7) and (1.8) depend upon the choice of norm $\|\cdot\|$. Though sometimes it is essential to consider other possibilities, most of our examples will be based on the use of the 2-norm $\|\cdot\| = \|\cdot\|_2$, defined by $\|v\|_2 = (\sum |v_j|^2)^{1/2}$ for a vector v and then by $\|A\| = \max_x \|Ax\|/\|x\|$ for a matrix A. This choice of norm corresponds mathematically to formulation in a Hilbert space and physically to consideration of energy defined by a sum of squares, and in this important special case, (1.8) amounts to the condition that the eigenvectors of A, if they exist, are far from orthogonal. At the other extreme is a **normal** matrix, one that has a complete set of orthogonal eigenvectors; symmetric or hermitian matrices fall in this category. In this case, if each v_j is normalized by $\|v_j\| = 1$, then V is a **unitary** matrix (in the real case we say **orthogonal**), with $V^{-1} = V^*$ (V^* denotes the conjugate transpose) and $\|V\| = \|V^{-1}\| = 1$. Thus for $\|\cdot\| = \|\cdot\|_2$, (1.8) is a statement that A is in some sense far from normal. In this norm it is the non-normal matrices for which eigenvalue analysis may fail, and in these notes, starting with the subtitle on the title page, we often speak of problems that are "non-normal" or "far from normal" when a more careful statement would refer to the more general condition (1.8).

The majority of the familiar applications of eigenvalue analysis involve matrices or operators that are normal or close to normal, having eigenfunctions orthogonal or nearly so. Among the examples mentioned so far, all of the physical ones are in this category except certain problems of fluid mechanics. The familiar mechanical oscillations are governed by normal operators, for example, and so are the oscillations of quantum mechanics, at least in their standard formulation. As a consequence, our intuition about eigenvalues has been formed by the normal case. Two centuries of successes have generated confidence that the eigenvalue idea is both powerful in practice and fundamental in concept. It has not always been noted that as most of these successes involve problems governed by normal or near-normal operators, our grounds for confidence in the non-normal case are less solid.

With this in mind, I will now briefly indicate what can go wrong with (1). (2), and (3) in certain applications.

Consider first (2). If a linear operator is normal, then the degree of resonant amplification that may occur in response to an input at frequency ω is equal to the inverse of the distance in the complex plane between ω and the nearest eigenvalue. (This formula can be found in first-year physics textbooks, usually without the word eigenvalue.) For a non-normal operator. however, the resonant amplification may be orders of magnitude greater. *The resonances of a non-normal system are not determined just by the eigenvalues.* This phenomenon is at the heart of the topic known as "receptivity" in fluid mechanics.

Next, consider (3). It is true that for a purely linear, constant-coefficient, homogeneous problem, eigenvalues govern the asymptotic behavior as $t \to \infty$. If the problem is normal, this is a robust statement; the eigenvalues also have relevance to short-time or transient behavior, and moreover, their influence tends to persist if the problem is altered in small ways. If the problem is far from normal, however, conclusions based on eigenvalues are in general not robust. First, there may be a long transient that looks quite different from the asymptote and has no connection to the eigenvalues. Second, even the asymptote may change beyond recognition if the problem is modified slightly. *Eigenvalues do not always govern the transient behavior of a non-normal system, nor the asymptotic behavior in the presence of nonlinear terms, variable coefficients, lower-order terms, inhomogeneous forcing data. or other perturbations.* Few applied problems are free of all these complications. For those that are, it is rare that one is interested so purely in the limit $t \to \infty$ as one may at first imagine. These issues are at the heart of convergence and stability investigations in numerical analysis.

This brings us to (1). Unlike (2) and (3), the algorithmic idea of diagonalization is not in general invalidated if $\|V\| \|V^{-1}\|$ is large (although in extreme cases there may be difficulties caused by rounding errors on a computer). On the other hand, there is a different difficulty that sometimes makes diagonalization less useful than one might expect, even for normal problems. In practice, for differential or other operators one works with truncated expansions; an infinite series is approximated by finite sum. The difficulty that arises sometimes is that the choice of the basis of eigenfunctions for such an expansion may necessitate taking an unacceptably large number of terms in the expansion to achieve the required accuracy. *Eigenfunction expansions may be exceedingly inefficient.* This fact was publicized by Orszag around 1970 in the context of spectral methods for the numerical solution of differential equations [24],[53]. Spectral methods, by contrast, are based on expansions in functions that have nothing to do with the eigenfunctions of the problem at hand, but which may converge geometrically where an expansion in eigenfunctions converges only linearly. Thirty Chebyshev polynomials may resolve a problem as well as ten thousand eigenfunctions.

What about (4), a matrix or operator's personality? In the highly non-normal case, vivid though the image may be, the location of the eigenvalues may be as fragile an indicator of underlying character as the hair color of a Hollywood actress. We shall see that pseudospectra provide equally compelling images that may capture the spirit underneath more robustly.

In summary, eigenvalues and eigenfunctions have a distinguished history of application throughout the mathematical sciences; we could not get along without them. Their clearest successes, however, are associated with problems that involve well-behaved systems of eigenvectors, which in most contexts means matrices or operators that are normal or nearly so. This class of problems encompasses the majority of applications, but not all of them. For non-normal problems, the record is less clear, and even the conceptual significance of eigenvalues is open to question.

2. Pseudospectra

In certain applications, eigenvalue analysis proves to be misleading. Here are some examples, with a single citation in each case. (Listing a field like "ecology," of course, does not mean that eigenvalue analysis is always misleading in that field, just that it sometimes is.)

hydrodynamic stability [71]	lasers [62]
control theory [35]	rounding error analysis [6]
ecology [51]	non-hermitian quantum mechanics [32]
meteorology [22]	matrix iterations [49]
numer. soln. of diffl. eqs. [19]	magnetohydrodynamics [2]
Markov chains [39]	operator theory [3]

These notes are about phenomena that arise in these applications where eigenvalues are troublesome, and about ways to treat them mathematically. Specifically, they are about the mathematical tool known as pseudospectra, and the present section is devoted to describing this tool. [1]

Let $\|\cdot\|$ denote a norm on $\mathbb{C}^N$, the space of complex N-vectors, and also the associated induced norm on $\mathbb{C}^{N \times N}$, the space of complex $N \times N$ matrices. (Vector and matrix norms are described, for example, in [36] and [70].) After Theorem 2.1 we shall usually assume that $\|\cdot\|$ is the 2-norm $\|\cdot\|_2$ defined by

---

[1] In this section we consider pseudospectra of matrices. For linear operators, many of the definitions are analogous, but new technical details arise.

the Euclidean inner product. Here and throughout these notes, A denotes a matrix in $\mathbb{C}^{N \times N}$ or a linear operator on an infinite-dimensional space.

One way to motivate the idea of pseudospectra is as follows. It is a familiar observation of applied mathematics that the question "Is A singular?" is not robust, for an arbitrarily small perturbation can change the answer from yes to no. A better question for applied purposes is, "Is $\|A^{-1}\|$ large?" Now the condition that defines eigenvalues is a condition of singularity of a matrix. To ask "Is z an eigenvalue of A?" is the same as to ask,

$$\text{Is } z - A \text{ singular?}$$

(As always in these notes, $z - A$ is shorthand for $zI - A$.) Therefore, the property of being an eigenvalue of a matrix is also not robust. A better question may be

$$\text{Is } \|(z - A)^{-1}\| \text{ large?}$$

This pattern of thought leads naturally to our first definition of pseudo-spectra[2]:

Definition 2.1. *(First Definition of Pseudospectra)* Let $A \in \mathbb{C}^{N \times N}$ and $\epsilon \geq 0$ be arbitrary. The ϵ-**pseudospectrum** $\Lambda_\epsilon(A)$ of A is the set of $z \in \mathbb{C}$ such that

$$\|(z - A)^{-1}\| \geq \epsilon^{-1}. \tag{2.1}$$

The matrix $(z - A)^{-1}$ is known as the **resolvent** of A at z. In (2.1) and throughout these notes we employ the convention

$$\|(z - A)^{-1}\| = \infty \quad \text{for} \quad z \in \Lambda(A), \tag{2.2}$$

where $\Lambda(A)$ is the spectrum (set of eigenvalues) of A, so that in particular, the spectrum is contained in the ϵ-pseudospectrum for every $\epsilon \geq 0$. In words, *the ϵ-pseudospectrum is the subset of the complex plane bounded by the ϵ^{-1} level curve of the norm of the resolvent.* [3]

---

[2] Here and throughout these notes, we follow the usual pattern in the literature to date in defining pseudospectra via weak rather than strict inequalities (e.g. $\geq$ rather than $>$ in (2.1)). However, strict inequalities are sometimes simpler to deal with for problems involving operators rather than matrices, and it can be argued that definitions based on strict inequalities would be more consistent with mathematical convention. This route has been followed by Brian Davies in papers such as [12]. For applications, which choice one takes does not matter.

[3] Thus, unlike the spectrum, the pseudospectra depend on the norm. At first sight this lack of norm-invariance may seem a defect in the idea of pseudospectra, and it has certainly contributed to the fact that the development of a theory of non-normality has lagged far behind the development of standard spectral theory. (Pseudospectra are an idea of analysis; eigenvalues belong to algebra.) Yet what one really needs to know about in an applied problem is usually norm-dependent. For example, many non-normal operators can be made normal by a transformation to an exponentially weighted inner product and norm, but such a transformation may distort physical notions such as energy beyond recognition.

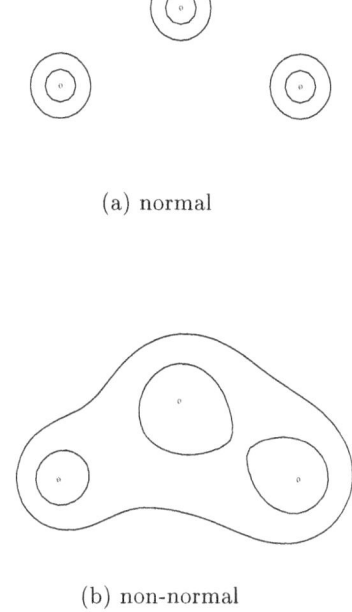

(a) normal

(b) non-normal

Fig. 2.1. The geometry of pseudospectra: schematic view.

It is perhaps not obvious at first whether the idea of pseudospectra serves much purpose. Is not $\|(z - A)^{-1}\|$ large precisely when z is close to an eigenvalue of A? For a normal matrix, when $\| \cdot \| = \| \cdot \|_2$, as we shall see in a moment, this intuition is correct (Fig. 2.1). The importance of pseudospectra arises for matrices that are far from normal, for which $\|(z - A)^{-1}\|$ may be large even when z is far from the spectrum, or more generally for matrices satisfying the conditions (1.7) or (1.8) discussed in the last section.

The norm of the resolvent may seem a rather technical notion to those unfamiliar with this way of thinking. Here is a more concrete definition:

Definition 2.2. *(Second Definition of Pseudospectra)* $\Lambda_\epsilon(A)$ *is the set of* $z \in \mathbb{C}$ *such that*

$$z \in \Lambda(A + E) \tag{2.3}$$

for some $E \in \mathbb{C}^{N \times N}$ *with* $\|E\| \leq \epsilon$.

In words, *the ϵ-pseudospectrum is the set of numbers that are eigenvalues of some perturbed matrix $A + E$ with $\|E\| \leq \epsilon$.*

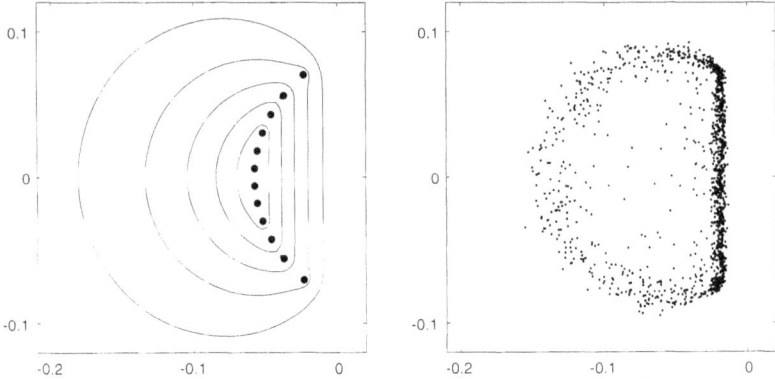

Fig. 2.2. Pseudospectra of a 12×12 Legendre spectral differentiation matrix. The left side shows the eigenvalues (solid dots) and the boundaries of the 2-norm ϵ-pseudospectra for $\epsilon = 10^{-3}, 10^{-4}, \ldots, 10^{-7}$ (from outer to inner). The right side shows 1200 of the 10^{-3}-pseudo-eigenvalues of A —specifically, a superposition of the eigenvalues of 100 randomly perturbed matrices $A + E$, where each E is a matrix with independent normally distributed complex entries of mean 0 scaled so that $\|E\| = 10^{-3}$. If all possible perturbations with $\|E\| = 10^{-3}$ were considered instead of just 100 of them, the dots on the right would exactly fill the outermost curve on the left.

From either of these definitions it follows that the pseudospectra associated with various ϵ are nested sets,

$$\Lambda_{\epsilon_1}(A) \subseteq \Lambda_{\epsilon_2}(A), \qquad 0 \leq \epsilon_1 \leq \epsilon_2, \qquad (2.4)$$

and that the 0-pseudospectrum is the same as the spectrum:

$$\Lambda_0(A) = \Lambda(A). \qquad (2.5)$$

Fig. 2.2 illustrates the equivalence between (2.1) and (2.3) for a highly non-normal 12×12 matrix arising in the field of spectral methods for partial differential equations [66],[72]. Contours of constant resolvent norm are plotted on the left, and eigenvalues of randomly perturbed matrices on the right. Evidently $(z - A)^{-1}$ has norm of order 10^5 or larger even when the distance from z to the spectrum is of order 1. Equivalently, the eigenvalues of A are highly sensitive to perturbations. This example also illustrates the fact that that there may be more geometric structure of a matrix or operator in the complex plane than is revealed by the spectrum alone. The conspicuous geometric feature is the group of almost exactly straight sections of pseudospectral boundaries near the origin of the complex plane. This feature reflects the construction of this matrix as an approximation to a certain differential operator (see §4).

Here is a third characterization of the ϵ-pseudospectrum:

Definition 2.3. *(Third Definition of Pseudospectra)* $\Lambda_\epsilon(A)$ *is the set of* $z \in \mathbb{C}$ *such that*

$$\|(z - A)v\| \leq \epsilon \tag{2.6}$$

for some $v \in \mathbb{C}^N$ *with* $\|v\| = 1$.

The number z in (2.6) (or equivalently in any of our definitions) is an ϵ-**pseudo-eigenvalue** of A, and v is a corresponding ϵ-**pseudo-eigenvector**. In words, *the ϵ-pseudospectrum is the set of ϵ-pseudo-eigenvalues.*

We now establish the equivalence of these three definitions.

Theorem 2.1 (Equivalent definitions of pseudospectra). *The* three *definitions above are equivalent.*

Proof. For $z \in \Lambda(A)$ the equivalence is trivial, so assume $z \notin \Lambda(A)$, implying the existence of $(z - A)^{-1}$. To prove $(2.3) \Rightarrow (2.6)$, suppose $(A + E)v = zv$ for some $E \in \mathbb{C}^{N \times N}$ with $\|E\| \leq \epsilon$ and some nonzero $v \in \mathbb{C}^N$, which we may take to be normalized by $\|v\| = 1$. Then $(z - A)v = \|Ev\| \leq \epsilon$, as required. To prove $(2.6) \Rightarrow (2.1)$, suppose $(z - A)v = \sigma u$ for some $v, u \in \mathbb{C}^N$ with $\|v\| = \|u\| = 1$ and $\sigma \leq \epsilon$. Then $(z - A)^{-1}u = \sigma^{-1}v$, so $\|(z - A)^{-1}\| \geq \sigma^{-1} \geq \epsilon^{-1}$. Finally, to prove $(2.1) \Rightarrow (2.3)$, suppose $\|(z - A)^{-1}\| \geq \epsilon^{-1}$. Then $(z - A)^{-1}u = \sigma^{-1}v$ and consequently $zv - Av = \sigma u$ for some $v, u \in \mathbb{C}^N$ with $\|v\| = \|u\| = 1$ and $\sigma \leq \epsilon$. To establish (2.3) it is enough to show that there exists a matrix $E \in \mathbb{C}^{N \times N}$ with $\|E\| = \sigma$ and $Ev = \sigma u$, for then v will be an eigenvector of $A + E$ with eigenvalue z. In fact E can be taken to be a rank-one matrix of the form $E = \sigma u w^*$ for some $w \in \mathbb{C}^N$ with $w^* v = 1$ (* denotes conjugate transpose). If $\| \cdot \|$ is the 2-norm, this is evident simply by taking $w = v$. In the case of an arbitrary norm $\| \cdot \|$, the existence of a vector w satisfying the required conditions can be interpreted as the existence of a linear functional L on $\mathbb{C}^N$ with $Lv = 1$ and $\|L\| = 1$, which is guaranteed by the Hahn–Banach theorem. A less highbrow version of the same proof can be carried out by the method of dual norms; see [36].[4]

One could proceed from here with a theory of pseudospectra of matrices defined with respect to an arbitrary norm $\| \cdot \|$; when we come to operators, this would amount to a theory in the setting of Banach spaces. However, we shall follow a simpler route that is general enough for most applications. From now on, we shall assume except where otherwise indicated that $\mathbb{C}^N$ is endowed with the standard inner product

$$(u, v) = u^* v \tag{2.7}$$

and that $\| \cdot \|$ is the corresponding 2-norm,

$$\|v\| = \|v\|_2 = \sqrt{v^* v}. \tag{2.8}$$

---

[4] This proof of $(2.1) \Rightarrow (2.3)$ for arbitrary $\| \cdot \|$ first appeared in [19].

Thus we shall restrict our attention to Hilbert spaces. With this choice of inner product and norm, the Hermitian conjugate (conjugate transpose) of a matrix is the same as its adjoint; we use the symbol A^*. Applications where more general weighted inner products and norms are desired can be handled within the framework of the 2-norm by introducing a similarity transformation $A \mapsto WAW^{-1}$, where W is nonsingular.

If $\| \cdot \| = \| \cdot \|_2$, the norm of a matrix is its largest singular value and the norm of the inverse is the inverse of the smallest singular value.[5] In particular.

$$\|(z - A)^{-1}\| = [\sigma_N(z - A)]^{-1}, \tag{2.9}$$

where $\sigma_N(z - A)$ denotes the smallest singular value of $z - A$, and this suggests a fourth definition of pseudospectra:

Definition 2.4. *(Fourth Definition of Pseudospectra) (assuming that the norm is $\| \cdot \|_2$). $\Lambda_\epsilon(A)$ is the set of $z \in \mathbb{C}$ such that*

$$\sigma_N(z - A) \leq \epsilon. \tag{2.10}$$

From (2.9) it is clear that (2.10) is equivalent to (2.1) and therefore also to our other characterizations of pseudospectra. In the proof of Theorem 2.1. the rank-1 matrix $E = \sigma uv^*$ can now be understood as follows: σ, u, and v are the smallest singular value and associated left and right singular vectors of $z - A$, respectively.

We come now to the matter of non-normality. First of all, note that if U is a unitary matrix (i.e., $U^* = U^{-1}$), then

$$(z - UAU^*)^{-1} = [U(z - A)U^*]^{-1} = U(z - A)^{-1}U^*, \tag{2.11}$$

and therefore

$$\|(z - UAU^*)^{-1}\| = \|(z - A)^{-1}\| \qquad \forall z \in \mathbb{C}.$$

Thus the resolvent norm is invariant with respect to unitary similarity transformations, which implies that the same is true of the pseudospectra:

$$\Lambda_\epsilon(A) = \Lambda_\epsilon(UAU^*) \qquad \forall \epsilon \geq 0. \tag{2.12}$$

A normal matrix is a matrix with the special property that there exists a unitary similarity transformation that makes it diagonal:

Definition 2.5. *A matrix $A \in \mathbb{C}^{N \times N}$ is **normal** if it has a complete set of orthogonal eigenvectors, that is, if it is unitarily diagonalizable:*

$$A = U\Lambda U^*. \tag{2.13}$$

(Here U is unitary and Λ is a diagonal matrix of eigenvalues—not to be confused with $\Lambda(A)$, the set of eigenvalues.)

---

[5] We assume the reader is familiar with the singular value decomposition (SVD); see [37] or [70].

Remark: An equivalent characterization is that A is normal if it commutes with its adjoint: $AA^* = A^*A$. This may seem a long way from (2.13), but the link is that both are equivalent to a third statement: A and A^* have the same eigenvectors, that is, they are **simultaneously diagonalizable**. For a long list of equivalent conditions for a matrix to be normal, see [30].

For a normal matrix, the ϵ-pseudospectrum is just the union of the closed ϵ-balls about the points of the spectrum, as suggested in Fig. 2.1. In other words, the eigenvalues all have condition number exactly 1 with respect to matrix perturbations; equivalently, the resolvent norm satisfies

$$\|(z - A)^{-1}\| = \frac{1}{\mathrm{dist}(z, \Lambda(A))}, \tag{2.14}$$

where $\mathrm{dist}(z, \Lambda(A))$ denotes the usual distance of a point to a set in the complex plane. The following theorem expresses these facts with the aid of the notation

$$\Delta_\epsilon = \{z \in \mathbb{C} : |z| \le \epsilon\}. \tag{2.15}$$

In this theorem a sum of sets has the usual meaning: $\Lambda(A) + \Delta_\epsilon = \{z : z = z_1 + z_2, \ z_1 \in \Lambda(A), \ z_2 \in \Delta_\epsilon\}$, which is equal to $\{z : \mathrm{dist}(z, \Lambda(A)) \le \epsilon\}$ since $\Lambda(A)$ and Δ_ϵ are closed.

Theorem 2.2 (Pseudospectra of a normal matrix). *For any* $A \in \mathbb{C}^{N \times N}$,

$$\Lambda_\epsilon(A) \supseteq \Lambda(A) + \Delta_\epsilon \qquad \forall \epsilon \ge 0, \tag{2.16}$$

and if A is normal, then

$$\Lambda_\epsilon(A) = \Lambda(A) + \Delta_\epsilon \qquad \forall \epsilon \ge 0. \tag{2.17}$$

Conversely, (2.17) implies that A is normal.

Proof. If z is an eigenvalue of A, then $z + \delta$ is an eigenvalue of $A + \delta$ for any $\delta \in \mathbb{C}$; since $\|\delta\| = |\delta|$, this establishes (2.16). For (2.17) we note that if A is normal. it can be assumed without loss of generality to be diagonal, with diagonal elements a_{jj} equal to the eigenvalues λ_j. In this case the resolvent is also diagonal, which implies that it satisfies (2.14), and as noted above, (2.1) implies that this is equivalent to (2.17). Finally, for the converse, here is a sketch of a proof that can be made precise. Eqn. (2.17) implies that each eigenvalue of A has condition number 1. By a standard formula for eigenvalue condition numbers, it follows that each right eigenvector of A is also a left eigenvector, i.e., that A and A^* are simultaneously diagonalizable. Therefore A is normal.

Now suppose A is diagonalizable but not necessarily normal, and let $V \in \mathbb{C}^{N \times N}$ be a **matrix of eigenvectors** of A as in (1.3) and (1.4). The condition number of this basis of eigenvectors, mentioned already in (1.8), is

$$\kappa(V) = \|V\| \, \|V^{-1}\| = \frac{\sigma_1}{\sigma_N}, \tag{2.18}$$

where σ_1 and σ_N are the largest and smallest singular values of V, respectively.[6] In general, $\kappa(V)$ may be any number in the range $1 \leq \kappa(V) < \infty$,[7] and the value $\kappa(V) = 1$ is possible if and only if A is normal.

The condition number provides an upper bound for the condition numbers of the individual eigenvalues of A. This fact is known as the Bauer–Fike theorem:

Theorem 2.3 (Bauer–Fike). *Let $A \in \mathbb{C}^{N \times N}$ be diagonalizable and let V be any matrix of eigenvectors of A. Then for each $\epsilon \geq 0$,*

$$\Lambda(A) + \Delta_\epsilon \ \subseteq \ \Lambda_\epsilon(A) \ \subseteq \ \Lambda(A) + \Delta_{\epsilon\kappa(V)}. \tag{2.19}$$

Proof. (cf. [1],[74],[65]) The first inclusion was established in (2.16). For the second we calculate

$$(z - A)^{-1} = (z - V\Lambda V^{-1})^{-1} = [V(z - \Lambda)V^{-1}]^{-1} = V(z - \Lambda)^{-1}V^{-1}.$$

which implies

$$\|(z - A)^{-1}\| \ \leq \ \kappa(V)\|(z - \Lambda)^{-1}\| = \frac{\kappa(V)}{\operatorname{dist}(z, \Lambda(A))},$$

and (2.1) completes the proof.

The following theorem collects some easy properties of the pseudospectrum, whose proofs we shall omit.

Theorem 2.4 (Properties of the pseudospectrum). *Let $A \in \mathbb{C}^{N \times N}$ and $\epsilon \geq 0$ be arbitrary.*

(i) $\Lambda_\epsilon(A)$ is nonempty, closed, and bounded, with at most N connected components.
(ii) $\Lambda_\epsilon(A^*) = (\Lambda_\epsilon(A))^*$.
(iii) For any $c \in \mathbb{C}$, $\Lambda_\epsilon(A + c) = c + \Lambda_\epsilon(A)$.
(iv) For any $c \in \mathbb{C}$, $\Lambda_{|c|\epsilon}(cA) = c\Lambda_\epsilon(A)$.
(v) $\Lambda_\epsilon(A_1 \oplus A_2) = \Lambda_\epsilon(A_1) \cup \Lambda_\epsilon(A_2)$.

---

[6] Since V is not unique, $\kappa(V)$ is not uniquely defined for a given A. If the eigenvalues of A are distinct, however, then $\kappa(V)$ becomes unique if the eigenvectors are normalized by $\|v_j\| = 1$, though this choice is not necessarily the one that minimizes $\kappa(V)$.

[7] We shall also write $\kappa(V) = \infty$ as a convenient shorthand in the case of nondiagonalizable A.

In part (v), $A_1 \oplus A_2$ denotes the direct sum of two matrices A_1 and A_2, whose dimensions need not be equal; in other words it is the block diagonal matrix

$$A_1 \oplus A_2 = \begin{pmatrix} A_1 & 0 \\ 0 & A_2 \end{pmatrix}.$$

Part (iv) is also worth a comment, for although elementary, it is surprising at first: the ϵ-pseudospectrum of $2A$ is not twice the ϵ-pseudospectrum of A, but twice the $\epsilon/2$-pseudospectrum of A.

Following the German Eigenwert and Eigenvektor, I have found it convenient in informal work to use the abbreviations ew and ev for eigenvalue and eigenvector, ψew and ψev for pseudo-eigenvalue and pseudo-eigenvector. (My abbreviations ψsm and ψsa for pseudospectrum and pseudospectra are less satisfactory.) Those who find it worthy of remark that the word "eigenvalue" is a blend of two languages may take pleasure in noting that "pseudoeigenvalue," for better or worse, is a combination of three.

3. A Matrix Example

Consider the tridiagonal Toeplitz matrix[8]

$$A = \begin{pmatrix} 0 & 1 & & & \\ \frac{1}{4} & 0 & 1 & & \\ & \frac{1}{4} & 0 & 1 & \\ & & \frac{1}{4} & 0 & 1 \\ & & & \frac{1}{4} & 0 \end{pmatrix} \in \mathbb{C}^{N \times N}. \tag{3.1}$$

This matrix is nonsymmetric, but it can be symmetrized by the diagonal similarity transformation

$$D^{-1}AD = S \tag{3.2}$$

with $D = \mathrm{diag}(1, 2^{-1}, \dots, 2^{-(N-1)})$ and

$$S = \begin{pmatrix} 0 & \frac{1}{2} & & & \\ \frac{1}{2} & 0 & \frac{1}{2} & & \\ & \frac{1}{2} & 0 & \frac{1}{2} & \\ & & \frac{1}{2} & 0 & \frac{1}{2} \\ & & & \frac{1}{2} & 0 \end{pmatrix} \in \mathbb{C}^{N \times N}. \tag{3.3}$$

It follows that the eigenvalues of A are the same as those of S, namely

$$\lambda_k(A) = \lambda_k(S) = \cos \frac{k\pi}{N+1}, \qquad 1 \le k \le N. \tag{3.4}$$

---

[8] Entries not shown are always zero.

Thus the spectrum of A consists of N distinct real numbers in the interval $(-1, 1)$.

The pseudospectra of A, however, lie far from the real axis. For $N = 64$, Fig. 3.1 plots the boundaries of $\Lambda_\epsilon(A)$ for $\epsilon = 10^{-2}, 10^{-3}, \ldots, 10^{-8}$, revealing wide oval-shaped regions in the complex plane. In fact, it can be shown that the ϵ-pseudospectrum of A is approximately equal to the ellipse that is the image of the circle $|z| = \epsilon^{1/N}$ under the mapping

$$f(z) = \tfrac{1}{4}z^{-1} + z, \qquad (3.5)$$

which is known as the **symbol** of A. Specifically, it is shown in [60] that for each z inside the ellipse $f(S)$, where $S = \{z \in \mathbb{C} : |z| = 1\}$, the resolvent norm $\|(z - A)^{-1}\|$ grows exponentially as $N \to \infty$, whereas for each z outside $f(S)$, $\|(z - A)^{-1}\|$ is bounded uniformly with respect to N.

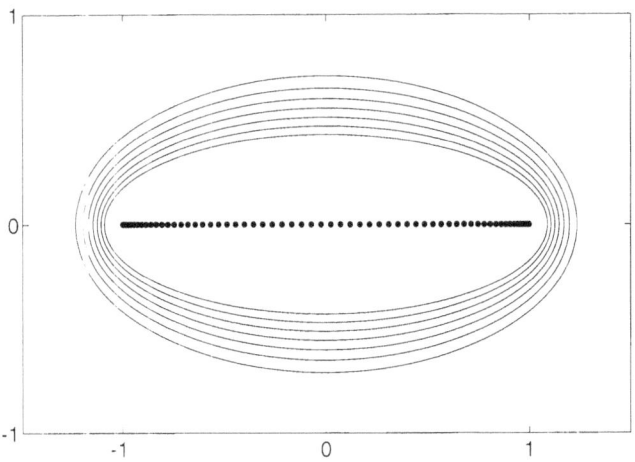

Fig. 3.1. Boundaries of pseudospectra Λ_ϵ, $\epsilon = 10^{-2}, 10^{-3}, \ldots, 10^{-8}$, for the matrix (3.1) of dimension $N = 64$. The eigenvalues are marked by solid dots.

Fig. 3.2 illustrates the pseudospectra of A in another way by presenting eigenvalues of randomly perturbed matrices. The figure shows the eigenvalues (3.4) in the complex plane as solid dots. Superimposed as smaller dots on the same plot are 6400 10^{-3}-pseudo-eigenvalues: the eigenvalues of 100 matrices $A + E$, where each E is a random matrix with $\|E\| = 10^{-3}$. The connection with ellipses is again obvious.

These pictures change quantitatively but not qualitatively if N is varied. To illustrate this, Fig. 3.3 shows nine sets of dots corresponding to the same experiment as in Fig. 3.2, but for $N = 16, 32, 64$ and $\|E\| = 10^{-2}, 10^{-3}, 10^{-4}$. Each tile of this figure depicts a superposition of eigenvalues of $640/N$ randomly perturbed matrices $A + E$—640 dots altogether. The sensitivity

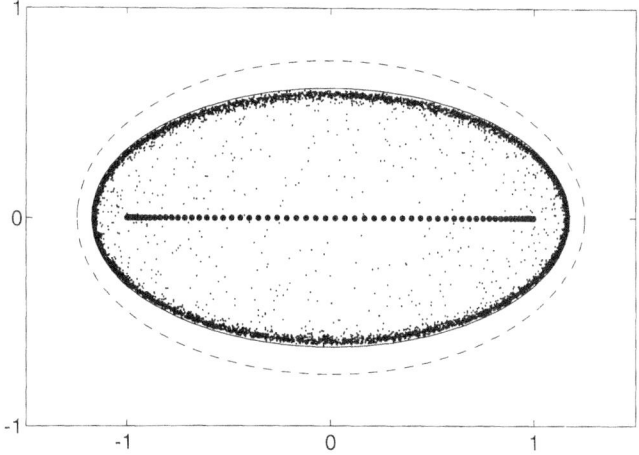

Fig. 3.2. Superposition of eigenvalues of 100 matrices $A + E$, where A is the tridiagonal Toeplitz matrix (3.1) of dimension $N = 64$ and each E is a random matrix with $\|E\| = 10^{-3}$. The eigenvalues of A are real (solid dots), but the perturbation introduced by E moves them far into the complex plane, close to the ellipse defined by (3.5) with $|z| = (.001)^{1/64}$ (solid curve). The dashed ellipse corresponds to $N \to \infty$ and $|z| = 1$.

of the eigenvalues obviously becomes more pronounced as N increases, but qualitatively, all nine pictures are much the same. For $N = 16, 32, 64$, the condition numbers of their eigenvector matrices are $\kappa(V) = 3.5 \times 10^4$, 2.3×10^9, 1.0×10^{19}; asymptotically as $N \to \infty$ we have $\kappa(V) \approx 2^N$.

The theme of these notes is that pseudospectra may reveal more than spectra about certain aspects of the behavior of matrices and operators. As an illustration of this principle for the present example, suppose we want to predict the norms of the powers A^n for various values of n. The standard approach to this problem is to consider eigenvalues. By (3.4), the spectral radii of A and S are equal,

$$\rho(A) = \rho(S) = \cos\left(\frac{\pi}{N+1}\right), \qquad (3.6)$$

and since this quantity is less than 1, both S and A must be power-bounded:

$$\|S^n\| \le C_S, \quad \|A^n\| \le C_A \qquad \forall n \ge 0, \qquad (3.7)$$

with $\|S^n\| \to 0$ and $\|A^n\| \to 0$ as $n \to \infty$. Fig. 3.4, however, shows that although these statements are true, they are only a part of the truth. In actuality, $\|A^n\|$ and $\|S^n\|$ bear little resemblance to one another. Whereas the powers S^n decrease smoothly, so that (3.7) holds with $C_S = 1$, the powers A^n grow exponentially for $n < N$ and achieve huge norms. Obviously, although

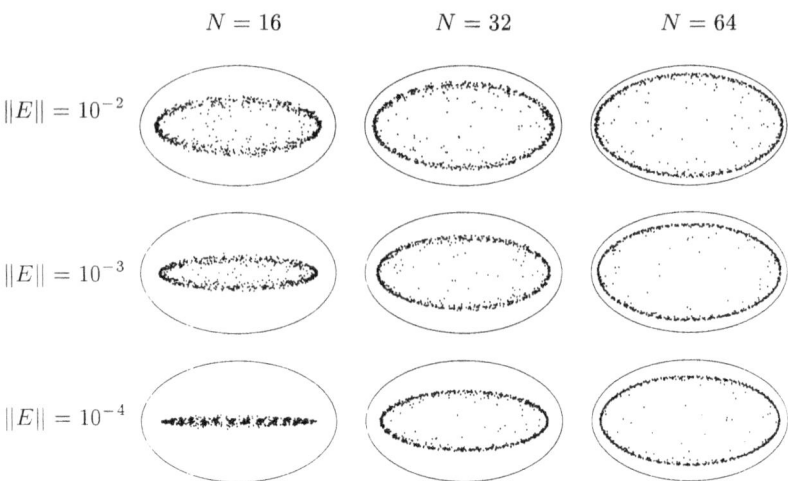

Fig. 3.3. Nine plots as in Fig. 3.2 corresponding to $N = 16, 32, 64$ and $\|E\| = 10^{-2}$. 10^{-3}, 10^{-4}. Each plot shows eigenvalues of $640/N$ matrices $A + E$, i.e., 640 dots. The ellipse is the image of the unit circle under the symbol $f(z) = \frac{1}{4}z^{-1} + z$.

the matrices A are power-bounded for each dimension N, they are not uniformly power-bounded. Even for fixed N, if $\|A^n\|$ becomes as great as 10^{15} for certain values of n it is doubtful whether the power-boundedness of A has much practical meaning.

On the other hand, Fig. 3.4 reveals that the rate of growth of $\|A^n\|$ for $n < N$ is very close to $(1.25)^n$. This number 1.25 is the largest absolute value of the points on the dashed ellipse in Fig. 3.2. Evidently in this example the slope of the $\|A^n\|$ curve for modest values of n can be accurately predicted by considering a **pseudospectral radius**.

4. An Operator Example

Figure 4.1 shows the pseudospectra of a highly non-normal differential operator, the Schrödinger operator for a harmonic potential. What is unusual about this Schrödinger operator is that the amplitude of the potential is complex rather than real:

$$\mathcal{A}u = -\frac{d^2 u}{dx^2} + cx^2 u, \qquad x \in \mathbb{R}, \qquad c = 1 + 3i. \tag{4.1}$$

The figure reveals that like the matrix of the last section, this operator deviates exponentially from normality, a discovery due to E. B. Davies [13],[14]. In fact, the deviation grows without bound as one moves further into the complex plane.

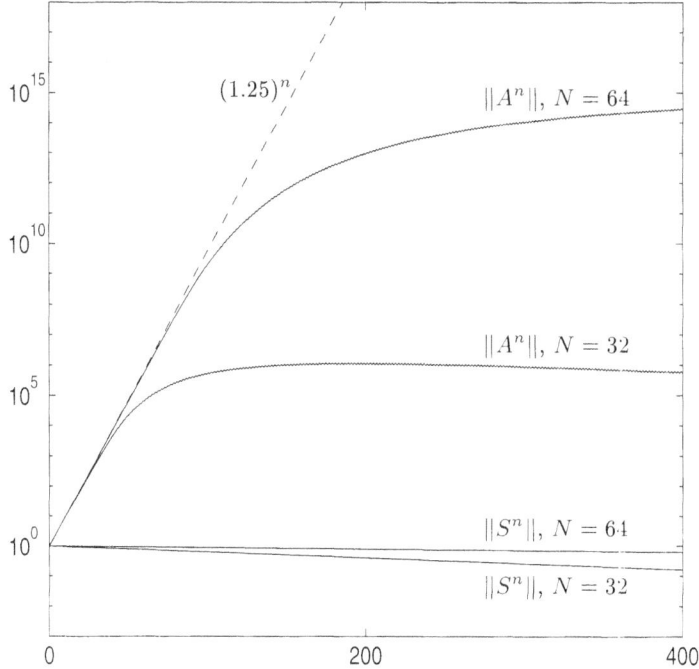

Fig. 3.4. Norms of powers $\|A^n\|$ and $\|S^n\|$ for the matrices A and S of (3.1)–(3.3), dimensions $N = 32$ and 64. Since A and S are similar for each N, the curves approach zero as $n \to \infty$ with equal asymptotic slopes determined by the spectral radius. For finite n, however, $\|A^n\|$ and $\|S^n\|$ are very different, and in particular, the powers $\|A^n\|$ are not bounded uniformly with respect to N. Note the logarithmic scale.

Rather than pursue this relatively complicated example here, in this section will shall study a simpler example that we can explain more fully.[9] For the space we take $L^2[0,d]$, and the operator will be the first derivative

$$\mathcal{A}u = u' = \frac{du}{dx} \qquad (4.2)$$

subject to the boundary condition

$$u(d) = 0. \qquad (4.3)$$

To be precise, $\mathcal{A}$ is the differentiation operator (4.2) with domain $D(\mathcal{A})$ equal to the set of continuous functions u on $[0,d]$ that satisfy (4.3) and possess a derivative in $L^2[0,d]$.

---

[9] For discussions of the same example from other points of view, see [33, p. 537], [41, p. 174], and [54, p. 44].

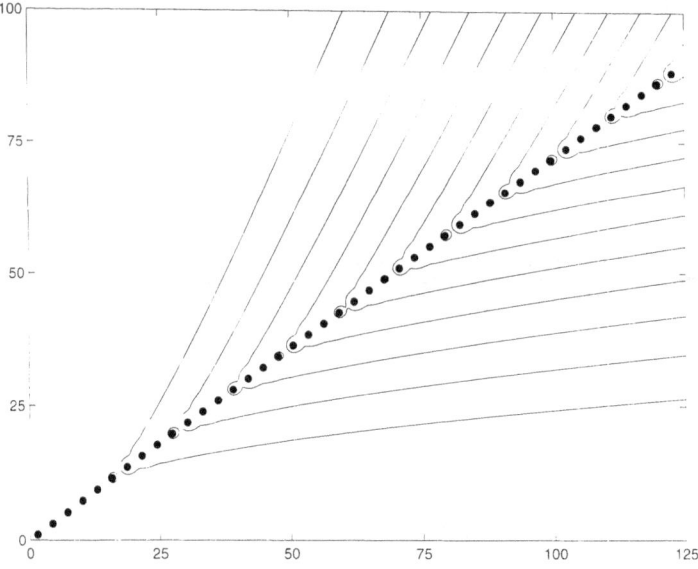

Fig. 4.1. ϵ-pseudospectra of the operator $\mathcal{A}$ of (4.1) corresponding to a harmonic oscillator with a complex potential. From outside in, the curves correspond to $\epsilon = 10^{-1}, 10^{-2}, \ldots, 10^{-11}$. The resolvent norm grows exponentially as $z \to \infty$ along rays in the complex plane satisfying $0 < \theta < \arg c$. Consequently, the higher eigenvalues of this operator would be of limited physical significance in applications. (This plot is based on a spectral discretization by a 200×200 matrix; the example comes from E. B. Davies.)

The spectrum of $\mathcal{A}$ is empty: $\Lambda(\mathcal{A}) = \emptyset$. Intuitively one sees this by noting that an eigenfunction would have to be of the form e^{zx} for some $z \in \mathbb{C}$, but since no such functions satisfy the boundary condition, there are no eigenfunctions. A proof can be obtained by showing that the resolvent $(z - \mathcal{A})^{-1}$ exists as a bounded operator for any $z \in \mathbb{C}$; it is given by

$$(z - \mathcal{A})^{-1}v(x) = \int_x^d e^{z(x-s)}v(s)\,ds. \tag{4.4}$$

This formula can be derived by the method of variation of parameters applied to the ordinary differential equation $zu - u' = v$. It can also be interpreted as the integral of $v(x)$ times the Green's function for the solution to $zu - u' = \delta(x)$, where $\delta(x)$ is the Dirac delta function.

The pseudospectra of $\mathcal{A}$, however, are another matter.[10] It follows from (4.4) that although the resolvent norm $\|(z - \mathcal{A})^{-1}\|$ is bounded for every z, it

---

[10] So far in these notes we have defined pseudospectra for matrices, not operators. The standard definition for operators is (2.1), which applies to any closed op-

is huge when z is in the left-half plane, growing exponentially as a function of $\exp(-d\Re z)$. It can also be seen from (4.4) that $\|(z - \mathcal{A})^{-1}\|$ depends only on $\Re z$, not $\Im z$. (Proof: for any $z \in \mathbb{C}$, $v(s)$, and $\alpha \in \mathbb{R}$, the pairs z, $v(s)$ and $z + i\alpha$, $e^{i\alpha s}v(s)$ lead to the same norm of the integral in (4.4).) Therefore for each ϵ, $\Lambda_\epsilon(\mathcal{A})$ is equal to the half-plane lying to the left of some line $\Re z = c_\epsilon$ in the complex plane. This situation is illustrated in Fig. 4.2, where the striking thing to note is the rapid decrease of ϵ as one moves into the left half-plane.

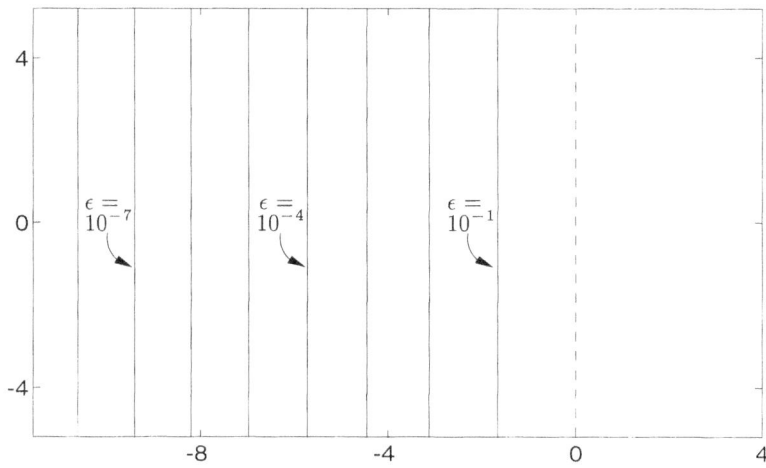

Fig. 4.2. Pseudospectra of the differentiation operator $\mathcal{A}$ of (4.2)–(4.3) for an interval of length $d = 2$. The solid lines are the right-hand boundaries of $\Lambda_\epsilon(\mathcal{A})$ for $\epsilon = 10^{-1}, 10^{-2}, \ldots, 10^{-8}$ (from right to left). The dashed line, the imaginary axis, is the right-hand boundary of the numerical range. If d were increased, the ϵ levels would decrease exponentially.

Why does $\mathcal{A}$ have a huge resolvent norm in the left half-plane? One explanation is suggested by Fig. 4.3. The function

$$u(x) = e^{zx}, \qquad z \in \mathbb{C}$$

does not satisfy (4.3), but if $\Re z \ll 0$, it almost does. Thus u is not an eigenfunction of $\mathcal{A}$, nor near to any eigenfunction, but it is "nearly an eigenfunction" in the sense that it is an eigenfunction of a slightly perturbed problem. It is tempting to call it a pseudo-eigenfunction, but we reserve this usage for functions that belong to the domain of the operator in question, which u does

---

erator in a Hilbert space. The alternative definitions (2.3) and (2.6) can also be applied for operators if they are modified by taking set closures.

not because it violates the boundary condition. However, $u(x)$ can be modified so as to become a true pseudo-eigenfunction by subtraction of a small term such as $e^{d\Re z+ix\Im z}$, and by this means a lower bound for $\|(z-\mathcal{A})^{-1}\|$ can be derived.

Instead of pursuing this idea, we shall obtain sharper estimates by working with (4.4) directly.[11]

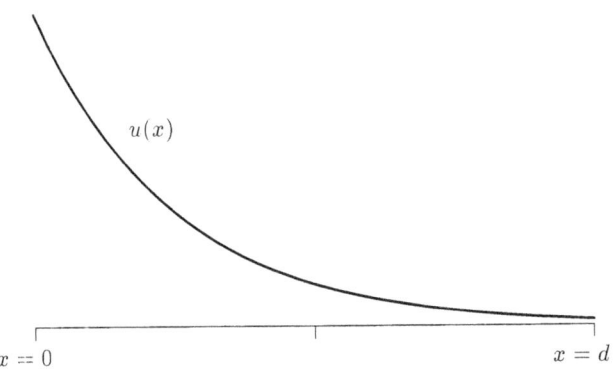

Fig. 4.3. For $d\Re z \ll 0$, the function $u(x) = e^{zx}$ is "nearly an eigenfunction" of $\mathcal{A}$, though it is not near any eigenfunction. Here $d = 2$, $z = -2$.

Theorem 4.1 (Pseudospectra of the differentiation operator). *The spectrum of the operator $\mathcal{A}$ is the empty set. The resolvent norm $\|(z-\mathcal{A})^{-1}\|$ depends on $\Re z$ but not $\Im z$, and satisfies*

$$\|(z-\mathcal{A})^{-1}\| \leq \frac{1}{\Re z} \tag{4.5}$$

for $\Re z > 0$ and

$$\|(z-\mathcal{A})^{-1}\| = \frac{e^{-d\Re z}}{-2\Re z} + O(\frac{1}{\Re z}) \tag{4.6}$$

for $\Re z < 0$, where the constant in the "O" is independent of z and d. The pseudospectra of $\mathcal{A}$ are half-planes of the form

$$\Lambda_\epsilon(\mathcal{A}) = \{z \in \mathbb{C} : \Re z \leq c_\epsilon\} \tag{4.7}$$

with

$$c_\epsilon \sim \begin{cases} (\log \epsilon)/d & \text{as } \epsilon \to 0, \\ \epsilon & \text{as } \epsilon \to \infty. \end{cases} \tag{4.8}$$

---

[11] It is also possible to determine $\|(z-\mathcal{A})^{-1}\|$ exactly by calculus of variations, though the result is not a closed formula. I am indebted to Satish Reddy for this observation.

Proof. If $u = (z - \mathcal{A})^{-1}v$ is given by (4.4), then $u(x)$ is the restriction to $[0, d]$ of the convolution $v * g$, where v and g are both regarded as functions in $L^2(-\infty, \infty)$ and $g(x) = e^{zx}$ for $x \in [-d, 0]$, 0 otherwise. Therefore by the Fourier transform, with $\| \cdot \|$ temporarily denoting the norm in $L^2(-\infty, \infty)$,

$$\|u\| \leq \|v * g\| = \|\widehat{v * g}\| = \|\widehat{v}\,\widehat{g}\| \leq \|\widehat{v}\| \sup_{\omega \in \mathbb{R}} |\widehat{g}(\omega)| = \|v\| \sup_{\omega \in \mathbb{R}} |\widehat{g}(\omega)|.$$

An elementary calculation gives $\widehat{g}(\omega) = (e^{d(i\omega - z)} - 1)/(i\omega - z)$, and this expression reaches a maximum at $\omega = \Im z$:

$$\sup_{\omega \in \mathbb{R}} |\widehat{g}(\omega)| = \frac{e^{-d\Re z} - 1}{|\Re z|}.$$

Thus we have

$$\|(z - \mathcal{A})^{-1}\| \leq \frac{1 - e^{-d\Re z}}{\Re z} \qquad (\Re z > 0), \tag{4.9}$$

which establishes (4.5).

On the other hand, assuming $\Re z < 0$, break (4.4) into two pieces

$$(z - \mathcal{A})^{-1}v(x) = R_1 v(x) - R_2 v(x) = \int_0^d e^{z(x-s)} v(s)\, ds - \int_0^x e^{z(x-s)} v(s)\, ds.$$

Then we have

$$\|R_1\| - \|R_2\| \leq \|(z - \mathcal{A})^{-1}\| \leq \|R_1\| + \|R_2\|, \tag{4.10}$$

and by an argument like the one just used for (4.9),

$$\|R_2\| \leq \frac{1}{-\Re z} \qquad (\Re z < 0). \tag{4.11}$$

The norm of R_1 can be evaluated exactly. Since $R_1 v(x) = e^{zx} \int_0^d e^{-zs} v(s)\, ds$, the dependence of $R_1 v(x)$ on x is independent of the choice of v. Thus if we find a function $v(x)$ that maximizes $|R_1 v(0)|/\|v\|$, this choice will also maximize $\|R_1 v\|/\|v\|$. By the Cauchy–Schwarz inequality, an appropriate choice is $v(s) = e^{-\bar{z}s}$, with which we calculate

$$\|R_1\| = \frac{\|R_1 v\|}{\|v\|} = \frac{|R_1 v(0)|}{|v(d)|} = \frac{\int_0^d e^{-2s\Re z}\, ds}{e^{-d\Re z}} = \frac{e^{-2d\Re z} - 1}{-2\Re z\, e^{-d\Re z}}. \tag{4.12}$$

Combining this result with (4.10) and (4.11) establishes (4.6).

A proof of (4.7) was sketched in the text. Finally, the upper half of (4.8) follows from (4.6), and the lower half follows from (4.5) (upper bound) and further estimates based on (4.4) (lower bound), which we omit.

In the last section we saw that the pseudospectra of the non-normal Toeplitz matrix A provided insight into the behavior of A as measured by the norms of powers $\|A^n\|$. For the present example, the analogous problem concerns the norms of the exponentials $\|e^{t\mathcal{A}}\|$. If $\mathcal{A}$ were a bounded operator, then $e^{t\mathcal{A}}$ could be defined by the usual power series. Since $\mathcal{A}$ is unbounded, $e^{t\mathcal{A}}$ must be defined in a more general manner as the solution operator for the continuous evolution problem $du/dt = \mathcal{A}u$, that is, the first-order partial differential equation $u_t = u_x$ on $[0, d]$ with boundary condition $u(d) = 0$. The solution to this problem is the leftward translation

$$e^{t\mathcal{A}}u(x) = \begin{cases} u(x + t) & \text{if } x + t < d, \\ 0 & \text{if } x + t \geq d. \end{cases} \tag{4.13}$$

Although $\mathcal{A}$ is unbounded, $e^{t\mathcal{A}}$ is a bounded operator on $L^2[0, d]$ for all $t \geq 0$. In the theory of semigroups, $\{e^{t\mathcal{A}}\}$ is a familiar example of a translation semigroup and $\mathcal{A}$ is known as its infinitesimal generator [33],[54].

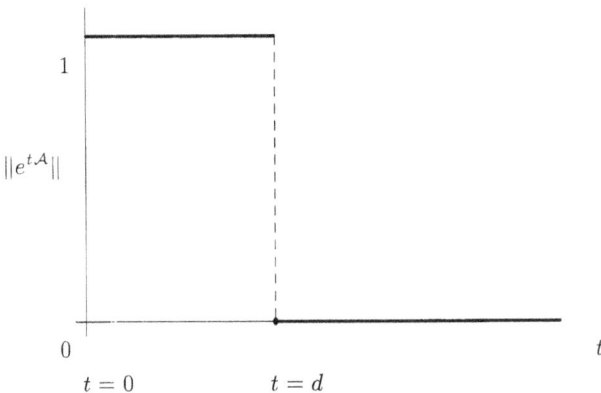

Fig. 4.4. Norm of the evolution operator $e^{t\mathcal{A}}$ as a function of t. The fact that $\|e^{t\mathcal{A}}\| \leq 1$ for all $t \geq 0$ ($\mathcal{A}$ is dissipative) can be inferred from (4.5). The fact that $e^{t\mathcal{A}} = 0$ for $t \geq d$ ($\mathcal{A}$ is nilpotent) can be inferred from (4.6).

We consider two aspects of the behavior of $\mathcal{A}$, both of which are illustrated in Fig. 4.4. First, from (4.13) we see that $\mathcal{A}$ is dissipative in the sense that the associated evolution process is a contraction: $\|e^{t\mathcal{A}}\| \leq 1$ for $t \geq 0$. This property can be inferred from the behavior of the ϵ-pseudospectra of $\mathcal{A}$ in the limit $\epsilon \to \infty$. Specifically, by the Hille-Yosida theorem of the theory of semigroups, any closed operator that satisfies (4.5) must be dissipative. Eqn. (4.5) is also equivalent to the statement that the numerical range of $\mathcal{A}$ is contained in the left half-plane.

Second and perhaps more interesting, $\mathcal{A}$ is nilpotent in the sense that for $t \geq d$, $e^{t\mathcal{A}} = 0$. This property can be inferred from the behavior of the pseudospectra of $\mathcal{A}$ in the limit $\epsilon \to 0$. In fact, by a Laplace transform argument analogous to the Paley-Wiener theorem, it follows that any closed operator with resolvent norm $O(e^{-d\Re z})$ as $\Re z \to -\infty$, as holds for $\mathcal{A}$ by (4.6), must satisfy $e^{t\mathcal{A}} = 0$ for $t \geq d$. See [21].

The operator $\mathcal{A}$ has appeared in these notes already: we saw a glimpse of it in Fig. 2.2. The matrix considered in that figure is a spectral discretization of $\mathcal{A}/144$, and this explains why the pseudospectra portrayed there contain nearly straight boundary segments near the origin. These segments represent high-accuracy approximations to the exactly straight lines of Fig. 4.2. Indeed, the data for Fig. 4.2 were calculated numerically via matrix approximations of just this kind.

5. History of Pseudospectra

The importance of non-normality has been recognized by certain people for many years. In this section I attempt to survey the narrower subject of the history of the sets known as pseudospectra. The following discussion includes all authors I am aware of who made use of pseudospectra before the years 1992 — at which point the idea took off, and it becomes difficult to be comprehensive. Very probably there are others of whom I am unaware.

In 1975 H. J. Landau of AT&T Bell Laboratories published a paper, "On Szegő's eigenvalue distribution theory and non-hermitian kernels," that introduced ϵ-pseudo-eigenvalues under the name ϵ-*approximate eigenvalues* [44]. The application was to the theory of Toeplitz matrices and associated integral operators. Two papers quickly followed on loss in unstable resonators [45] and mode selection in lasers [46]. Besides being early in the history of pseudospectra, these papers were notable in that, unlike much of the numerical analysis literature on pseudospectra that followed in the decade and a half afterwards, they recognized that the limitations of eigenvalues are not just a matter of rounding errors on computers. Landau wrote. "When we remember that, for ϵ sufficiently small, we cannot distinguish operationally between true and ϵ-approximate, the possibility arises that in certain non-Hermitian contexts it is the second notion that should replace the first at the center of the stage, even for purposes of theory."

Definition. λ is an ϵ-approximate eigenvalue of A, if there exists $\varphi \in \mathscr{S}(rQ)$, with $\|\varphi\| = 1$, such that $\|A\varphi - \lambda\varphi\| \leq \epsilon$. We call φ an ϵ-approximate eigenfunction corresponding to λ.

Fig. 5.1. First published definition of pseudospectra? From Landau, 1975 [44].

In 1979 J. M. Varah of the University of British Columbia published a paper "On the separation of two matrices," whose starting point was the study of the conditioning of the Sylvester equation $AX - XB = C$ [73]. This led to the question of under what circumstances the spectra of two matrices A and B could be said to be well separated. Varah defined the 2-norm ϵ-pseudospectrum in terms of the minimal singular value $\sigma_{\min}(A - \lambda I)$, giving it the name ϵ-$spectrum$ and the notation $S_\epsilon(A)$. He pointed out that there is an equivalent definition in terms of matrix perturbations, and emphasized that for a non-normal matrix, the pseudospectra may be very different from the spectrum.

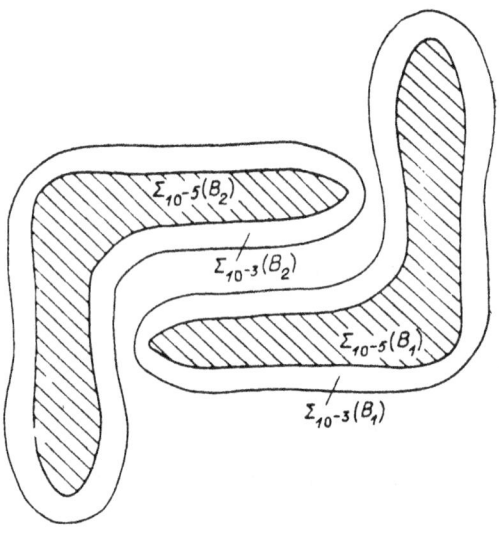

Fig. 5.2. First published sketch of pseudospectra? From Kostin and Razzakov. 1985 [43].

In Novosibirsk, there was work related to pseudospectra throughout the 1980s by S. K. Godunov and his colleagues, including A. G. Antonov, A. Y. Bulgakov, O. P. Kirilyuk, V. I. Kostin, A. N. Malyshev, and S. I. Razzakov. Godunov, together with Ryabenkii and others, had made significant contributions in the 1960s to the study of how non-normality affects numerical stability of discretizations of differential equations [26]. The subsequent work involving explicit discussion of pseudospectra was oriented instead to the aim of achieving "guaranteed accuracy" in computations of numerical linear algebra [27]. According to Malyshev and Kostin (personal communications, 1991), this work began around 1982. The Novosibirsk group defined the ϵ-$spectrum$ by relative rather than absolute perturbations:

$\Lambda_\epsilon(A) = \{z \in \mathbb{C} : \|(z - A)^{-1}\| \geq (\epsilon\|A\|)^{-1}\}$. [12] In [28] and [42], computed plots of pseudospectra were presented and called "spectral portraits of matrices" containing various "patches of spectrum," and computations based on both contour plotting and curve tracing were discussed. A hand-drawn sketch of a pseudospectrum had appeared earlier in a paper by Kostin and Razzakov in 1985 [43]. Some of the further work by this group appeared in a book in Russian by Godunov in 1997, with a color picture of pseudospectra on the cover [25].

One of the last papers of the eminent numerical analyst J. H. Wilkinson, "Sensitivity of eigenvalues II" (1986), defined the ϵ-pseudospectrum for an arbitrary norm $\| \cdot \|$ induced by a vector norm [75]. The set was denoted by $D(\eta)$ and given no name other than "the domain $D(\eta)$." Wilkinson described in his usual lucid fashion how $D(\eta)$ could be interpreted equivalently in terms of matrix perturbations or the norm of the resolvent. He discussed various applications and examples of small dimension, and mentioned at the end the extension to generalized eigenvalue problems. To my mind, the remarkable thing is that it took Wilkinson thirty years to come to the idea of pseudospectra, for given his lifelong dual interests in eigenvalue problems and backward error analysis, the idea would seem to have been hard to avoid. I can think of four partial explanations. First, computer graphics was not the effortless tool in Wilkinson's day that it later became. Second, the matrices he could handle were small, making the effects of pseudospectra less conspicuous. Third, perhaps Wilkinson was aware of the notion of pseudospectra for many years, but considered the idea a heuristic interpretation rather than something solid enough to be published. Fourth, a glance at any of his writings reveals that Wilkinson's habits of thought were resolutely algebraic, not visual. In the 662 pages of his magnum opus *The Algebraic Eigenvalue Problem* [74], there are only four figures, yet floating-point numbers seem to appear on every page. (In fact one of those figures, on p. 454, can be interpreted as a pseudospectrum—actually a structured pseudospectrum defined by real perturbations of a real matrix. Wilkinson introduces this as a "domain of indeterminacy" in his analysis of the method of bisection, but makes little use of the idea.)

Pseudospectra were investigated in several papers by J. W. Demmel in the mid-1980s, appearing with the labels $S(A, \epsilon)$ in [16] and $\Lambda_\epsilon(\epsilon, A)$ in [17]. The former paper contains the first published plot of computed pseudospectra that I know of. Demmel's starting point was the problem discussed in his 1983 thesis [15]: to devise an analogue of the Jordan canonical form that is robust enough to have meaning in the presence of rounding errors and other

---

[12] This definition is natural in those applications of backward error analysis where a numerical algorithm introduces perturbations on the scale of machine epsilon times A. For other applications, its suitability is less clear, and it has the disadvantage that according to this definition, the question of whether a point $z \in \mathbb{C}$ belongs to a particular ϵ-pseudospectrum $\Lambda_\epsilon(A)$ depends, via the norm, on the behavior of A at distant points of $\mathbb{C}$.

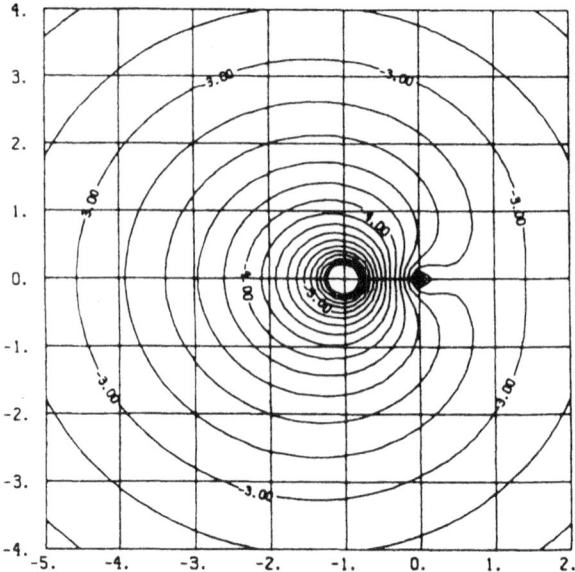

Fig. 5.3. First published computer plot of pseudospectra? From Demmel, 1987 [16]. This figure ©IEEE.

perturbations. For example, under what circumstances does it make sense to view several eigenvalues of a matrix as belonging to a cluster, which may itself perhaps be viewed as a perturbation of a more highly defective set of Jordan blocks? As in Varah's paper [73], the question arose here of the separation of the spectra of two (sub-)matrices. Demmel related pseudospectra to improvements of the Bauer–Fike theorem and to other results in matrix perturbation theory, and discussed applications in control theory, numerical computation of eigenvalues, and other areas.

Beginning in the mid-1980s, D. Hinrichsen and A. J. Pritchard wrote a number of papers on the "stability radius" of a matrix, i.e., the distance to the set of unstable matrices. In a paper in 1992 [34] they introduced the ϵ-pseudospectrum of a non-normal matrix with the name *spectral value set* and notation $\sigma(A, \rho)$. Only real perturbations were considered, making their spectral value set a structured pseudospectrum, but Hinrichsen and Pritchard were obviously well aware of the complex case. Their paper presented examples of spectral value sets of several matrices illustrated by plots of superpositions of eigenvalues of random perturbations.

Another body of early work related to pseudospectra was that of F. Chatelin, later Chaitin–Chatelin, and her colleagues in France. Beginning in the late 1980s, this group investigated questions of conditioning, stability, and floating-point arithmetic with the aid of random perturbations of problems [5],[6]. By perturbing a problem many times at random, they pointed out, one can acquire knowledge about the properties of the unperturbed problem. One of the many applications they considered was to matrix eigenvalue problems, where random perturbations may reveal, for example, if the original matrix has a nontrivial Jordan block. Though the pseudospectrum was apparently not explicitly defined in these writings, at least in the early years, the idea was implicit, and computed plots of eigenvalues of randomly perturbed matrices appeared in [5].

My own first publications that mentioned pseudospectra were [67] and [57] in 1990 (the first uses the term "ϵ-approximate eigenvalues"). These were an outgrowth of an earlier paper by myself and Trummer in 1987 [72], in which we found eigenvalues that were extraordinarily sensitive to perturbations but failed fully to appreciate their significance. Subsequent early papers by myself and my colleagues pertaining to pseudospectra included [49],[50],[58],[60]. A crucial collaborator in this work was my student S. C. Reddy, who began to work with pseudospectra in 1988 and made many contributions after that date. Reddy's early work on these topics was summarized in his 1991 thesis at MIT [56].

In 1992 my paper "Pseudospectra of matrices" appeared. which presented the idea of pseudospectra and exhibited thirteen examples [68]. It was after this point that the idea began to be widely known. A later companion paper "Pseudospectra of linear operators" presented ten examples involving operators on infinite-dimensional spaces [69].

The papers by Demmel and Wilkinson mentioned above cite each other, and they both cite the paper of Varah [73]. Aside from these cases, none of the papers discussed here published before 1990 cite any of the others. These data suggest that the notion of pseudospectra has been invented at least five times:

H. J. Landau	1975	*ϵ-approximate eigenvalues*
J. Varah	1979	*ϵ-spectrum*
S. K. Godunov et al.	1982	*spectral portrait*
L. N. Trefethen	1990	*ϵ-pseudospectrum*
D. Hinrichsen and A. J. Pritchard	1992	*spectral value set*

One should not trust this table too far, however, as even recent history is notoriously hard to pin down. It is entirely possible that Godunov or Wilkinson thought about pseudospectra in the 1960s, and indeed von Neumann may have thought about them in the 1930s. Nor were others such as Dunford and Schwartz, Gohberg, Halmos, Kato, Keldysch, or Kreiss far away.

Bibliography

1. F. L. Bauer and C. T. Fike, *Norms and exclusion theorems*, Numer. Math. **2** (1960), 137–141.

2. D. Borba, et al., *The pseudospectrum of the resistive magnetohydrodynamics operator: resolving the resistive Alfvén paradox*, Phys. Plasmas **1** (1994), 3151–3160.

3. A. Böttcher and B. Silbermann, *Introduction to Large Truncated Toeplitz Matrices*, Springer, Berlin, 1998.

4. N. Bourbaki, *Eléments d'Histoire des Mathématiques*, Masson, Paris, 1984.

5. F. Chatelin, *Resolution approchée d'équations sur ordinateur*, Lect. Notes, Lab. de Statistique Théorique et Appliquée, Univ. P. and M. Curie, Paris, 1989.

6. F. Chaitin-Chatelin and V. Fraysse, *Lectures on Finite Precision Computations*, SIAM, Philadelphia, 1996.

7. S. Chandrasekhar, *Hydrodynamic and Hydromagnetic Stability*, Clarendon Press, Oxford, 1961.

8. A. K. Chopra, *Dynamics of Structures: Theory and Applications to Earthquake Engineering*, Prentice Hall, Englewood Cliffs, NJ, 1995.

9. R. W. Clough and J. Penzien, *Dynamics of Structures*, McGraw-Hill, New York, 1975.

10. R. Courant and D. Hilbert, *Methods of Mathematical Physics, v. 1*, Wiley-Interscience, New York, 1953.

11. E. B. Davies, *Spectral Theory and Differential Operators*, Cambridge University Press, Cambridge, 1995.

12. E. B. Davies, *Pseudospectra, the harmonic oscillator and complex resonances*, to appear.

13. E. B. Davies, *Pseudospectra, the harmonic oscillator, and complex resonances*. Preprint KCL-MTH-98-03, Dept. of Maths., King's College London, February 1998.

14. E. B. Davies, *Semiclassical states for non-self-adjoint Schrödinger operators*, preprint, March 1998.

15. J. W. Demmel, *A Numerical Analyst's Jordan Canonical Form*, PhD diss., U. C. Berkeley, 1983.

16. J. W. Demmel, *A counterexample for two conjectures about stability*, IEEE Trans. Aut. Control **AC-32** (1987), 340–342.

17. J. W. Demmel, *Nearest defective matrices and the geometry of ill-conditioning*

18. J. Dieudonné. *History of Functional Analysis*, North-Holland, Amsterdam, 1981.

19. J. L. M. van Dorsselaer, J. F. B. M. Kraaijevanger, and M. N. Spijker, *Linear stability analysis in the numerical solution of initial value problems*, Acta Numerica 1993, 199–237.

20. P. G. Drazin and W. H. Reid, *Hydrodynamic Stability*, Cambridge University Press, Cambridge, England, 1981.

21. T. A. Driscoll and Lloyd N. Trefethen, *Pseudospectra for the wave equation with an absorbing boundary*, J. Comp. Appl. Math. **69** (1996), 125–142.

22. B. F. Farrell and P. J. Ioannou, *Stochastic dynamics of baroclinic waves*, J. Atmos. Sci. **50** (1993), 4044–4057.

23. N. H. Fletcher and T. D. Rossing, *The Physics of Musical Instruments*, Springer, New York, 1991.

24. B. Fornberg, *A Practical Guide to Pseudospectral Methods*, Cambridge University Press, Cambridge, 1996.

25. S. K. Godunov, *Modern Aspects of Linear Algebra* (Russian), Novosibirsk, 1997 (Russian).

26. S. K. Godunov and V. S. Ryabenkii, *Theory of Difference Schemes*, North-Holland, Amsterdam, 1964.

27. S. K. Godunov, A. G. Antonov, O. P. Kirilyuk, and V. I. Kostin, *Guaranteed Accuracy of the Solution to Systems of Linear Equations in Euclidean Spaces*, Nauka, Novosibirsk, 1988 (Russian).

28. S. K. Godunov, O. P. Kirilyuk, and V. I. Kostin, *Spectral portraits of Matrices*, Preprint 3, Inst. of Math., Sib. Branch of USSR Acad. Sci., 1990 (Russian).

29. D. O. Gough, J. W. Leibacher, P. H. Scherrer, and J. Toomre, *Perspectives in helioseismology*, Science **272** (1996), 1281–1283. (This is the lead article in a special issue of *Science* on helioseismology.)

30. R. Grone, et al., *Normal matrices*, Lin. Alg. Appl. **87** (1987), 213–225.

31. A. Greenbaum, *Iterative Methods for Solving Linear Systems*, SIAM, Philadelphia, 1997.

32. N. Hatano and D. R. Nelson, *Localization transitions in non-hermitian quantum mechanics*, Phys. Rev. Lett. **77** (1996), 570–573.

33. E. Hille and R. S. Phillips, *Functional Analysis and Semi-Groups*, American Mathematical Society, Providence, RI, 1957.

34. D. Hinrichsen and A. J. Pritchard, *On spectral variations under bounded real matrix perturbations*, Numer. Math. **60** (1992), 509–524.

35. D. Hinrichsen and A. J. Pritchard, *Stability of uncertain systems*, in Systems and Networks: Mathematical Theory and Applications, v. I, Akademie–Verlag, Berlin, 1994, pp. 159–182.

36. R. A. Horn and C. R. Johnson, *Matrix Analysis*, Cambridge University Press, New York, 1985.

37. R. A. Horn and C. R. Johnson, *Topics in Matrix Analysis*, Cambridge University Press, New York, 1991.

38. V. E. Howle and L. N. Trefethen, *Eigenvalues and musical instruments*, Rep. 97/20, Oxford U. Computing Lab., 1997.

39. G. F. Jónsson and L. N. Trefethen, *A numerical analyst looks at the "cutoff phenomenon" in card shuffling and other Markov chains*, in D. F. Griffiths, D. J. Higham, and G. A. Watson, eds., Numerical Analysis 1997, Longman, Harlow, Essex, UK, 1998.

40. T. Kailath, *Linear Systems*, Prentice Hall, Englewood Cliffs. NJ, 1980.

41. T. Kato, *Perturbation Theory for Linear Operators*, Springer, New York, 1976.

42. V. I. Kostin, *On definition of matrices' spectra*, in High Performance Computing II, North-Holland, 1991.

43. V. I. Kostin and S. I Razzakov, *On convergence of the power orthogonal method of spectrum computing*, Trans. Inst. Math. Sib. Branch Acad. Sci., v. 6, 55–84, translated in *Software Optimization*, New York, 1986.

44. H. J. Landau, *On Szegő's eigenvalue distribution theory and non-Hermitian Kernels*, J. d'Analyse Math. **28** (1975), 335–357.

45. H. J. Landau, *Loss in unstable resonators*, J. Opt. Soc. Amer. **66** (1976), 525–529.

46. H. J. Landau, *The notion of approximate eigenvalues applied to an integral equation of laser theory*, Quart. Appl. Math. **35** (1977), 165–172.

47. P. M. Morse and K. U. Ingard, *Theoretical Acoustics*, Princeton University Press, Princeton. 1968.

48. J. D. Murray, *Mathematical Biology*, Springer, Berlin, 1989.

49. N. M. Nachtigal, S. C. Reddy, and L. N. Trefethen, *How fast are nonsymmetric matrix iterations?*, SIAM J. Matrix Anal. Applics. **13** (1992), 778–795.

50. N. M. Nachtigal, L. Reichel, and L. N. Trefethen. *A hybrid GMRES algorithm for nonsymmetric linear systems*, SIAM J. Matrix Anal. Applics. **13** (1992), 796–825.

51. M. G. Neubert and H. Caswell, *Alternatives to resilience for measuring the responses of ecological systems to perturbations*, Ecology **78** (1997), 653–665.

52. J. R. Norris, *Markov Chains*, Cambridge University Press, Cambridge, 1997.
53. S. A. Orszag, *Accurate solution of the Orr–Sommerfeld stability equation*, J. Fluid Mech. **50** (1971), 689–703.
54. A. Pazy, *Semigroups of Linear Operators and Applications to Partial Differential Equations*, Springer-Verlag, New York, 1983.
55. Lord Rayleigh, *The Theory of Sound* (2 vols.), Dover, New York, 1945.
56. S.C. Reddy, *Pseudospectra of Operators and Discretization Matrices and an Application to Stability of the Method of Lines*, PhD thesis, MIT, 1991.
57. S. C. Reddy and L. N. Trefethen, *Lax-stability of fully discrete spectral methods via stability regions and pseudo-eigenvalues*, Comp. Meth. Appl. Mech. Engr. **80** (1990), 147–164.
58. S. C. Reddy and L. N. Trefethen, *Stability of the method of lines*, Numer. Math. **62** (1992), 235–267.
59. S. C. Reddy and Lloyd N. Trefethen, *Pseudospectra of the convection-diffusion operator*, SIAM J. Appl. Math. **54** (1994), 1634–1649.
60. L. Reichel and L. N. Trefethen, *Eigenvalues and pseudo-eigenvalues of Toeplitz matrices*, Lin. Alg. Applics. **162–164** (1992), 153–185.
61. R. D. Richtmyer and K. W. Morton, *Difference Methods for Initial-Value Problems*, Wiley-Interscience, New York, 1967.
62. A. E. Sigeman, *Nonorthogonal optical modes and resonators*, in R. Kossowky et al., eds., Optical Resonators—Science and Engineering, Kluwer, 1998, pp. 29–53.
63. F. H. Slaymaker and W. F. Meeker, *Measurements of the tonal characteristics of carillon bells*, J. Acoust. Soc. Am. **26** (1954), 515–522.
64. L. A. Steen, *Highlights in the history of spectral theory*, Amer. Math. Monthly **80** (1973), 359–381.
65. G. W. Stewart and J. Sun, *Matrix Perturbation Theory*, Academic Press, San Diego, 1990.
66. H. Tal-Ezer, *A pseudospectral Legendre method for hyperbolic equations with an improved stability condition*, J. Comp. Phys. **67** (1986), 145–172.
67. L. N. Trefethen, *Approximation theory and numerical linear algebra,* in J. C. Mason and M. G. Cox, eds., Algorithms for Approximation II, Chapman and Hall, London, 1990.
68. L. N. Trefethen, *Pseudospectra of matrices,* in D. F. Griffiths and G. A. Watson. Numerical Analysis 1991, Longman, 1992.
69. L. N. Trefethen, *Pseudospectra of linear operators*, SIAM Review **39** (1997). 383–406.
70. L. N. Trefethen and D. Bau, III, *Numerical Linear Algebra*, SIAM, Philadelphia, 1997.
71. L. N. Trefethen, A. E. Trefethen, S. C. Reddy, and T.A. Driscoll, *Hydrodynamic stability without eigenvalues*, Science **261** (1993), 578–584.
72. L. N. Trefethen and M. R. Trummer, *An instability phenomenon in spectral methods*, SIAM J. Numer. Anal. **24** (1987), 1008–1023.
73. J. M. Varah, *On the separation of two matrices*, SIAM J. Numer. Anal. **16** (1979), 216–222.
74. J. H. Wilkinson, *The Algebraic Eigenvalue Problem*, Clarendon Press, Oxford, 1965.
75. J. H. Wilkinson, *Sensitivity of Eigenvalues II*, Utilitas Math. **30** (1986), 243–286.

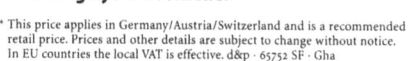

Lecture Notes
in Computational Science and Engineering

Vol. 1 D. Funaro, *Spectral Elements for Transport-Dominated Equations.* 1997. X, 211 pp. Softcover. ISBN 3-540-62649-2

Vol. 2 H. P. Langtangen, *Computational Partial Differential Equations.* Numerical Methods and Diffpack Programming. 1999. XXIII, 682 pp. Hardcover. ISBN 3-540-65274-4

Vol. 3 W. Hackbusch, G. Wittum (eds.), *Multigrid Methods V.* Proceedings of the Fifth European Multigrid Conference held in Stuttgart, Germany, October 1-4, 1996. 1998. VIII, 334 pp. Softcover. ISBN 3-540-63133-X

Vol. 4 P. Deuflhard, J. Hermans, B. Leimkuhler, A. E. Mark, S. Reich, R. D. Skeel (eds.), *Computational Molecular Dynamics: Challenges, Methods, Ideas.* Proceedings of the 2nd International Symposium on Algorithms for Macromolecular Modelling, Berlin, May 21-24, 1997. 1998. XI, 489 pp. Softcover. ISBN 3-540-63242-5

Vol. 5 D. Kröner, M. Ohlberger, C. Rohde (eds.), *An Introduction to Recent Developments in Theory and Numerics for Conservation Laws.* Proceedings of the International School on Theory and Numerics for Conservation Laws, Freiburg / Littenweiler, October 20-24, 1997. 1998. VII, 285 pp. Softcover. ISBN 3-540-65081-4

Vol. 6 S. Turek, *Efficient Solvers for Incompressible Flow Problems.* An Algorithmic and Computational Approach. 1999. XVII, 352 pp, with CD-ROM. Hardcover. ISBN 3-540-65433-X

Vol. 7 R. von Schwerin, *Multi Body System SIMulation.* Numerical Methods, Algorithms, and Software. 1999. XX, 338 pp. Softcover. ISBN 3-540-65662-6

Vol. 8 H.-J. Bungartz, F. Durst, C. Zenger (eds.), *High Performance Scientific and Engineering Computing.* Proceedings of the International FORTWIHR Conference on HPSEC, Munich, March 16-18, 1998. 1999. X, 471 pp. Softcover. 3-540-65730-4

Vol. 9 T. J. Barth, H. Deconinck (eds.), *High-Order Methods for Computational Physics.* 1999. VII, 582 pp. Hardcover. 3-540-65893-9

For further information on these books please have a look at our mathematics catalogue at the following URL: http://www.springer.de/math/index.html

Computer to Film: Saladruck, Berlin
Binding: H. Stürtz AG, Würzburg

GPSR Compliance

The European Union's (EU) General Product Safety Regulation (GPSR) is a set of rules that requires consumer products to be safe and our obligations to ensure this.

If you have any concerns about our products, you can contact us on ProductSafety@springernature.com

In case Publisher is established outside the EU, the EU authorized representative is:

Springer Nature Customer Service Center GmbH
Europaplatz 3
69115 Heidelberg, Germany

Batch number: 09625323

Printed by Printforce, the Netherlands